THE HOST-PLANT IN RELATION TO INSECT
BEHAVIOUR AND REPRODUCTION

THE HOST-PLANT IN RELATION TO INSECT BEHAVIOUR AND REPRODUCTION

Edited by

T. JERMY

Research Institute for Plant Protection
of the Hungarian Ministry of Agriculture and Food
Budapest

PLENUM PRESS · NEW YORK AND LONDON

Proceedings of the Symposium held at the
Biological Research Institute in Tihany
11 to 14 June, 1974

Symposia Biologica Hungarica 16

Technical editor
Á. SZENTESI, Ph. D.

Coedition
published by
Akadémiai Kiadó, Budapest, Hungary

and

Plenum Publishing Corporation
227 West 17th Street, New York N.Y. 10011, U.S.A.

ISBN 0-306-30909-2

Library of Congress Catalog Card Number 75-37209

Printed in Hungary

CONTENTS

5

6

LIST OF PARTICIPANTS AND CONTRIBUTORS

AGARWAL, R. A., Indian Agricultural Research Institute, New Delhi 110012, India

ALI, M., Laboratory of the Research Institute for Plant Protection, H–8360 Keszthely, Felszabadulás út 1/a, Hungary

AUCLAIR, J. L., Dépt. des Sciences Biologiques, Université de Montréal 101, Québec, Canada

BECK, S. D., Department of Entomology, University of Wisconsin, Madison, Wisconsin 53706, USA

BERNAYS, E. A., Centre for Overseas Pest Research, College House, Wrights Lane, London W8 5SJ, UK

BLANEY, W. M., Department of Zoology, Birbeck College, Malet Street, London W. C. 1., UK

CHAPMAN, R. F., Centre for Overseas Pest Research, College House, Wrights Lane, London W8 5SJ, UK

COAKER, T. H., Department of Applied Biology, Downing Street, Cambridge CB3 9LU, UK

COOK, A. G., Centre for Overseas Pest Research, College House, Wrights Lane, London W8 5SJ, UK

DĄBROWSKI, Z. T., Department of Applied Entomology, Agricultural University of Warsaw, 02—766 Warsaw–Ursynow, Poland

DESEŐ, K. V., Research Institute for Plant Protection, H–1525 Budapest, Pf. 102, Hungary

DETHIER, V. G., Department of Biology, Princeton University, Princeton, N. J. 08540, USA

DEZSŐ, G., Hungarian Academy of Sciences, H–1361 Budapest, Pf. 6, Hungary

EBERT, W., Institut für Pflanzenschutzforschung, Zweigstelle Eberswalde, 13, Eberswalde-Finow, Schicklerstrasse 5, DDR

FEKETE, É., Biological Centre of the Hungarian Academy of Sciences, H–6726 Szeged, Odessza krt. 62, Hungary

HANIOTAKIS, E. G., Department of Biology, "Democritos" Nuclear Research Center, Aghia Paraskevi, Attikis, Greece

HANSON, F. E., Department of Biological Sciences, University of Maryland, Baltimore County, Catonsville, Baltimore, Maryland 21228, USA

HARREWIJN, P., Institute for Phytopathological Research, Binnenhaven 12, Wageningen, The Netherlands

HAWKES, C., Department of Applied Biology, Downing Street, Cambridge CB3 9LU, UK

HERREBOUT, W. M., Department of Systematic Zoology, c. o. Rijksmuseum van Natuurlijke Historie, Raamsteg 2, Leiden, The Netherlands

HSIAO, T. H., Department of Biology, Utah State University, Logan, Utah 84322, USA

HUIGNARD, J., Laboratoire d' Ecologie Expérimentale, Université de Tours, Avenue Monge, Parc Grandmont, 37200 Tours, France

ISAEV, A. S., Institute of Forest and Wood, Acad. Sci. of the USSR, Siberian Branch, Krasnoyarsk 49, pr. Mira 53, USSR

JERMY, T., Research Institute for Plant Protection, H–1525 Budapest, Pf. 102, Hungary

KARASEV, V. S., Laboratoriya Zashtshity Drevesiny, UkrNIIMOD, ul. Bozhenko 84, Kiev 6, USSR

KENNEDY, J. S., Imperial College Field Station, Silwood Park, Ascot, Berks. SL5 7PY, UK

KISHIN, M., Research Institute for Plant Protection, H–1525 Budapest, Pf. 102, Hungary

KOZÁR, F., Research Institute for Plant Protection, H–1525 Budapest, Pf. 102, Hungary

KRISHNANANDA, N., Potato Research Institute, Ooteccamund, India

KRZYMAŃSKA, J., Institute for Plant Protection, 60318 Poznań, Miczurina 20, Poland

KUIJTEN, P. J., Department of Systematic Zoology and Evolutionary Biology, University, Leiden, The Netherlands

LABEYRIE, V., Laboratoire d'Ecologie Expérimentale, Université François Rabelais, Parc Grandmont, 37200 Tours, France

LASTER, M. L., Mississippi State University, Mississippi State, Mississippi 39762, USA

LAUNOIS-LUONG, H., Laboratoire d'Entomologie et d'Ecophysiologie Expérimentales, Faculté des Sciences, Université Paris–Sud, 91405 Orsay, France

LAUREMA, S., Agricultural Research Centre, Department of Pest Investigation, 01300 Tikkurila, Finland

LEBERRE, J. R., Laboratorie d'Entomologie et d'Ecophysiologie Expérimentale, Faculté des Sciences, Université Paris–Sud, 91405 Orsay, France

MA, WEI-CHUN, International Centre of Insect Physiology and Ecology, Nairobi, P. O. Box 30772, Kenya

MARKKULA, M., Agricultural Research Centre, Department of Pest Investigation, 01300 Tikkurila, Finland

MAXWELL, F. G., Department of Entomology, Drawer EM, Mississippi State University, Mississippi State, Mississippi 39762, USA

McCAFFERY, A. R., Centre for Overseas Pest Research, College House, Wrights Lane, London W8 5SJ, UK

MEREDITH, W. R., Delta Branch Experiment Station, Mississippi Agricultural and Forestry Experiment Station, USA

MITTLER, T. E., Division of Entomology and Parasitology, University of California, Berkeley, Calif. 94720, USA

MOREAU, J.-P., Station Centrale de Zoologie, C.N.R.A., Route de St Cyr, 78000 Versailles, France

MÜLLER, F. P., Forschungsgruppe Phyto-Entomologie, Sektion Biologie, Universität Rostock, Wismarsche Str. 8, 25 Rostock, DDR

NAGY, B., Research Institute for Plant Protection, H–1525 Budapest, Pf. 102, Hungary

NAGY, M., Research Institute for Plant Protection, H–1525 Budapest, Pf. 102, Hungary

NORRIS, D. M., Jr., 642 Russel Laboratory, University of Wisconsin, Madison, Wisconsin 53706, USA

PETTERSSON, J., Dept. of Plant Pathology and Entomology, Agr. College of Sweden, Uppsala 7, Sweden

PROKOPY, R. J., Route 1, Bailey's Harbor, Wisconsin, USA

RÁCZ, V., Research Institute for Plant Protection, H–1525 Budapest, Pf. 102, Hungary

RADKEVITCH, V. A., Pedagogical Institute, 6 Pushkin Street, 210026 Vitebsk, USSR

REESE, J. C., Department of Entomology, University of Wisconsin, Madison, Wisconsin 53706, USA

ROBERT, P., Ch., Station de Zoologie, I.N.R.A., B.P. 384, 68 Colmar, France

RÓZSA, K. S., Biological Research Institute of the Hungarian Academy of Sciences, H–8237 Tihany, Hungary

RUDINSKY, J. A., Entomology Department, Oregon State University, Corvallis, Oregon, USA

RUDNEV, D. F., Ukrainskiy Institut Zashtshity Rasteniy, ul. Vasilkovskaya 33, Kiev 22, USSR

RUSS, K., Bundesanstalt für Pflanzenschutz, Wien II., Trunnerstrasse 5, Austria

SÁRINGER, GY., Laboratory of the Research Institute for Plant Protection, H–8360 Keszthely, Felszabadulás út 1/a, Hungary

SAXENA, K. N., Department of Zoology, Delhi University, Delhi 7, India

SCHELTES, P., International Centre of Insect Physiology and Ecology, P.O. Box 30772, Nairobi, Kenya

SCHEURER, S., VVB Agrochemie und Zwischenprodukte, 7101 Cunnersdorf, DDR

SCHOONHOVEN, L. M., Department of Animal Physiology, Agricultural University, Haarweg 10, Wageningen, The Netherlands

SCHUSTER, M. F., Department of Entomology, USDA, ARS, USA

SHAPIRO, I. D., All-Union Research Institute for Plant Protection, Leningrad 1, ul. Gertzena 42, USSR

STÄDLER, E., Swiss Federal Research Station, CH–8820, Wädenswil, Switzerland

SZALAY-M., L., Research Institute for Plant Protection, H–1525 Budapest, Pf. 102, Hungary

SZENTESI, Á., Research Institute for Plant Protection, H–1525 Budapest, Pf. 102, Hungary

TIITTANEN, K., Agricultural Research Centre, Department of Pest Investigation, 01300 Tikkurila, Finland

TJALLINGII, W. F., Department of Animal Physiology, Agricultural University, Haarweg 10, Wageningen, The Netherlands

VARJAS, L., Research Institute for Plant Protection, H–1525 Budapest, Pf. 102, Hungary

VASETSCHKO, G. I., Ukrainskiy Institut Zashtshity Rasteniy, ul. Vasilkovskaya 33, Kiev 22, USSR

VILKOVA, N. A., All-Union Research Institute for Plant Protection, Leningrad 1, ul. Gertzena 42, USSR

VISSER, J. H., Department of Entomology, Agricultural University, Binnenhaven 7, Wageningen, The Netherlands

WĘGOREK, W., Institute for Plant Protection, 60318 Poznań, Miczurina 20, Poland

WIEBES, J. T., Department of Systematic Zoology, c. o. Rijksmuseum van Natuurlijke Historie, Raamsteg 2, Leiden, The Netherlands

WILDE, J., DE, Department of Entomology, Agricultural University, Binnenhaven 7, Wageningen, The Netherlands

WOOD, D. L., Department of Entomological Sciences, 137 Giannini Hall, University of California, Berkeley, California 94720, USA

ZWOLIŃSKA-ŚNIATAŁOWA, Z., Institute for Plant Protection, 60318 Poznań, Miczurina 20, Poland

Symp. Biol. Hung. 16, pp. 13–22 (1976)

PREFERENCE TO OVIPOSITION AND ANTIBIOSIS MECHANISM TO JASSIDS (*AMRASCA DEVASTANS* DIST.) IN COTTON (*GOSSYPIUM* sp.)

by

R. A. Agarwal[1] and N. Krishnananda[2]

[1] INDIAN AGRICULTURAL RESEARCH INSTITUTE, NEW DELHI, INDIA
[2] POTATO RESEARCH INSTITUTE, OOTECCAMUND, INDIA

Preference to oviposition and antibiosis mechanism to jassids (*Amrasca devastans* Dist.) in cotton (*Gossypium* sp.) were investigated. Six varieties of cotton: two of each highly and medium resistant, and two highly susceptible were studied during the peak period of jassid activity.

It was established that jassids preferred to oviposit on the susceptible varieties rather than on the medium resistant varieties. The latter, i. e. the medium resistant varieties, harboured more eggs than the highly resistant ones. Normally, the main leaf veins were preferred for oviposition. The different veins on the leaf were designated.

The studies on the antibiosis mechanism of resistance to jassids in cotton showed that the survival ratio of nymphs was almost one-third to half on resistant varieties, 75% on the medium resistant and 92–96% on the susceptible varieties. The nymphal period was longer by 3–4 days on resistant varieties compared to susceptible ones. There was no difference in the size of the same instar nymphs feeding on different varieties. But significant differences in the size were recorded between the nymphs of different instars, when reared on the same variety or on different varieties. Similar responses towards varieties were obtained with respect to the weight of nymphs of different instars.

Over 25 *Empoasca* species have been reported on cotton from different parts of the world. Amongst them *Amrasca* (*Empoasca*) *devastans* Dist. is one of the most serious pests of cotton in India. Besides cotton, it feeds on various other host-plants, the most common ones are okra (*Abelmoschus esculentus*) and castor (*Ricinus communis*). It has been observed that certain varieties of cotton were heavily damaged whereas others remained free of its attack, though grown under identical conditions.

REVIEW OF LITERATURE

The leafhoppers insert their eggs mainly into the veins on the lower surface of the leaf. Verma and Afzal (1940), and Yadav et al. (1967) found that *E. devastans* laid more eggs on the susceptible cotton varieties than on the resistant ones. Likewise, Cowland (1947) working on cotton and Jayaraj (1968) on castor observed that the fecundity of *E. lybica* and *E. flavescens* was considerably increased on susceptible than on the resistant varieties. Geering and Coaker (1960) working on cotton with *Dysdercus superstitiosus*, Holmes and Peterson (1961) on wheat with *Cephus cinctus*, Pond et al. (1962) on cabbage with *H. brassicae*, Curtis and Mc Coy (1964) on carrots with *Lygus lineolaris* and Pathak (1969) on paddy with *Chilo suppressalis* and *Tryporyza incentellus* suggested that the susceptible varieties of crops harboured more eggs than the resistant ones.

Poos and Smith (1931) experimenting with *Empoasca fabae* on cowpea, Reed and Adkisson (1961), and Bailey et al. (1967) with *Anthonomus grandis* on cotton and Jayaraj (1967) with *E. flavescens* on castor found that the survival rate, duration of different instars, size and weight of insects were considerably reduced when these insects were reared on resistant varieties compared to the susceptible ones.

MATERIAL AND METHODS

For our investigations six varieties of cotton were used, two highly resistant (B.1007 and Badnawar-1), two medium susceptible (H.14 and Reba-B.50) and two highly susceptible (Cocker-100 A and Lankart-57) during the peak period of jassid attack. Twenty leaves in each variety were dis-

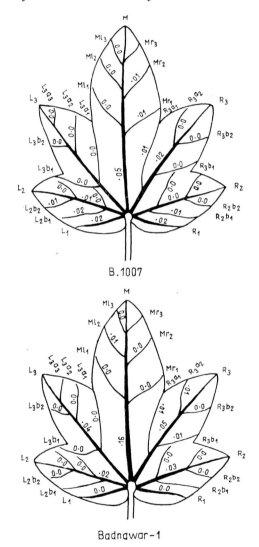

B.1007

Badnawar-1

Fig. 1. Number of egg punctures on different leaf veins of cotton varieties (resistant)

14

sected on five occasions at five-day intervals to recover the number of eggs deposited in the main and subveins. In most cases the eggs were not traceable. Therefore, the oviposition punctures evidenced by the rupture of the vein tissues and the subsequent reddening and thickening of the surrounding cells were counted under $\times 10$ magnification. Data are presented in Table 1.

The main, lateral and subveins were designated as follows. The central thick vein arising from the base of the leaf was named as Main vein (M). There are six other thick veins all originating from the apex of the petiole, i.e. three on either side of the main vein. The three right side veins were called as Right vein (R) and the three on the left side as Left vein (L).

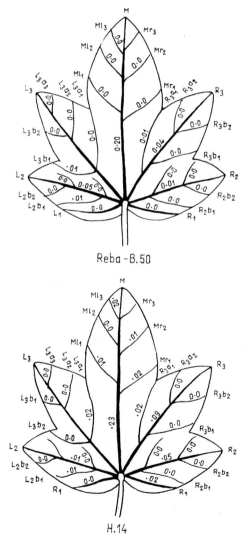

Fig. 2. Number of egg punctures on different leaf veins of cotton varieties
(medium resistant)

15

The first vein originating near the base of the leaf was named as R_1, the next one to it as R_2 and the one towards the apical side as R_3. Likewise, the vein on the left side originating near the base was named as L_1, the next one as L_2 and the third running towards the apical side as L_3.

The subveins arising from the Right (R) and Left (L) veins were further marked, i.e. those leading towards the apex as "a" and those leading towards the base as "b". The above system of nomenclature along with the average number of eggs found in each vein in the three different types of varieties are presented in Figs 1, 2, 3 and 4.

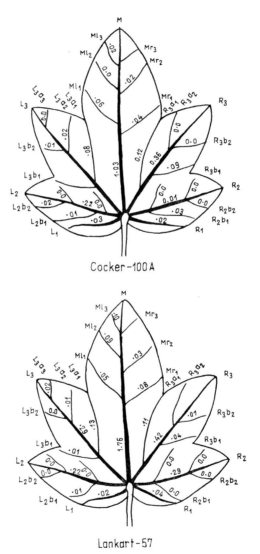

Fig. 3. Number of egg punctures on different leaf veins of cotton varieties (susceptible)

16

TABLE 1

Number of ovipositional punctures in leaf veins

Varieties	Main vein	Subvein	Total
B.1007	3.1	1.3	4.4
Badnawar-1	6.1	1.1	7.2
Reba-B.50	10.3	1.3	11.6
H.14	11.0	2.6	13.6
Cocker-100 A	43.0	10.3	53.2
Lankart-57	63.2	17.0	80.6
C.D. at 5%	1.7	1.2	1.6
C.D. at 1%	2.3	1.6	2.2

Average of 100 leaves, i.e. 20 in each experiment.

The experiments on antibiosis mechanism were repeated three times. Nine cages with fine muslin cloth were fixed on nine different plants of each variety. Two leaves were enclosed in a cage. The hatched nymphs obtained

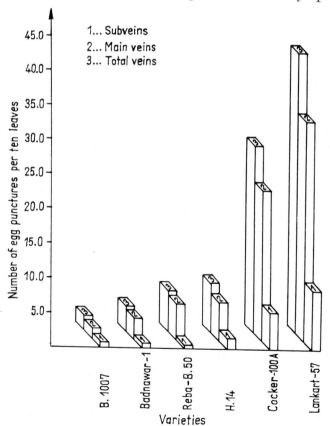

Fig. 4. Preference for oviposition on different leaf veins of cotton varieties by jassids

from culture, raised on Cocker-100 A, were immediately transferred onto the caged leaves. Daily records were made on the rate of development and the number of surviving nymphs, until they developed into adults (see Tables 2–4).

TABLE 2

Survival and duration of development on different cotton varieties

Varieties	Nymphs developed (%)	Nymphal period (days)	Generations in a year (approx. No.)
B.1007	37.6	11.6	31
Badnawar-1	52.5	10.6	34
Reba-B.50	78.5	7.5	48
H.14	75.6	8.6	42
Cocker-100 A	92.8	7.6	48
Lankart-57	96.2	7.1	51
C.D. at 5%	1.9	0.7	
C.D. at 1%	2.7	1.0	

Average of 81 nymphs, i.e. 9 in each replication in three sets of experiments.

TABLE 3

Size of nymphs and adults (mm) reared on different varieties of cotton

Varieties	Nymphal instar					Adults		C.D. at 5% level
	1st	2nd	3rd	4th	5th	female	male	
B.1007	0.61	1.03	1.24	1.56	2.11	2.26	2.18	0.38
Badnawar-1	0.63	0.99	1.23	1.54	2.14	2.23	2.16	0.36
Reba-B.50	0.61	1.03	1.23	1.55	2.22	2.30	2.24	0.33
H.14	0.63	1.00	1.26	1.53	2.26	2.31	2.30	0.32
Cocker-100 A	0.62	1.01	1.25	1.57	2.30	2.39	2.36	0.41
Lankart-57	0.61	0.99	1.23	1.54	2.32	2.48	2.42	0.43
Average	0.62	1.01	1.24	1.55	2.23	2.33	2.28	
C.D. at 5% level	N.S.	N.S.	N.S.	N.S.	N.S.	N.S.	N.S.	N.S.

TABLE 4

Weight* of nymphs and adults (mg) reared on different cotton varieties

Vareties	Nymphal instar					Adults		C.D. at 5% level
	1st	2nd	3rd	4th	5th	female	male	
B.1007	1.93	2.42	6.43	8.23	10.83	16.23	14.92	1.03
Badnawar-1	1.89	2.44	6.46	8.31	10.89	16.31	14.73	1.21
Reba-B.50	1.91	2.46	6.43	8.29	11.02	16.33	15.03	1.18
H.14	1.90	2.41	6.49	8.31	11.09	16.30	15.27	1.23
Cocker-100 A	1.93	2.45	6.41	8.33	11.91	17.34	16.17	1.27
Lankart-57	1.88	2.46	6.45	8.21	12.23	17.62	16.21	1.19
Average	1.91	2.44	6.44	8.28	11.33	16.71	15.40	
C.D. at 5% level	N.S.	N.S.	N.S.	N.S.	0.16	0.32	0.45	

* Average weight of 10 insects.

18

A. Preference to oviposition

The resistant varieties (B. 1007 and Badnawar-1) on an average gave 3.1 and 6.1 ovipositional punctures in the main veins, respectively, and 1.3 and 1.1 in the subveins, yielding a total of 4.4 and 7.2 punctures in 20 leaves, respectively. Correspondingly, the medium resistant varieties (Reba-B.50 and H.14) showed 10.3 and 11.0 punctures on the main veins, and 1.3 and 2.6 in the subveins. The two susceptible varieties (Cocker-100 A and Lankart-57) showed the highest number of ovipositional punctures, i.e. 43.0 and 63.2 on the main veins, and 10.3 and 17.0 in subveins. The differences in the number of ovipositional punctures between the resistant, medium resistant and susceptible varieties were highly significant (Table 1).

B. Antibiosis mechanism

1. *Survival ratio:* 37.6% of the nymphs completed development on the highly resistant variety (B.1007) whereas 52.5% developed into adults on the other resistant variety (Badnawar-1). The latter indicates that more than a half of the nymphs could survive and develop on Badnawar-1, and only one third of the total number of nymphs released could survive and develop on B.1007. On the two medium resistant varieties (Reba-B.50 and H-14) 78.5 and 75.6% of the nymphs developed into adults, whereas on the two susceptible varieties (Cocker-100 A and Lankart-57) 92.8 and 96.2% developed into adults.

The survival rate of jassids on the above six varieties was significantly different, i.e. the survival was low on resistant varieties and high on susceptible varieties. The higher survival rate of jassids on the susceptible varieties would consequently result in faster building up of the population of jassids than on the resistant varieties (Table 2).

2. *Duration of nymphal development:* To complete development from 1st instar nymphs to adults jassids took 7.1 days on the susceptible variety Lankart-57; on the other susceptible variety Cocker-100 A they took 7.6 days. But on medium resistant variety Reba-B.50 again 7.5 days were required while on H.14, another medium resistant variety, the jassid took 8.6 days, i.e. 1.2 to 2.0 days more than in the case of the susceptible variety (Lankart-57). On the other hand, jassids needed 10.6 and 11.6 days on the resistant varieties (Badnawar-1 and B.1007), i.e. 4 days more to complete development than on the two susceptible varieties.

The above data reveal that in one year jassids should complete 51 life cycles on the susceptible variety Lankart-57, 48 on Cocker-100 A, 42 on H-14, 48 on Reba-B.50, 34 on Badnawar-1 and 31 on B.1007 (Table 2).

3. *Size of different instars:* The size of the 1st instar nymphs ranged from 0.61 to 0.63 mm; the 2nd instar nymphs from 0.99 to 1.03 mm; the 3rd, 4th and 5th instar nymphs from 1.23 to 1.26, 1.53 to 1.57 and 2.11 to 2.32 mm, respectively. The length of females and males ranged from 2.23 to 2.48 and 2.16 to 2.42 mm, respectively. The differences in the size of nymphs within an instar, due to feeding on different varieties, were not significant. However, there were significant differences in the size of nymphs between the 1st and 2nd and between the 4th and 5th instar nymphs reared

on the same variety (Table 3). It was interesting to note that the average percentage increase in the size of nymphs was more pronounced (62.8%) between the 1st and the 2nd instar nymphs. This increase in size was only 22.7% between the 2nd and 3rd instar, 26.6% between the 3rd and 4th instar, and 43.5% between the 4th and 5th instar. There was proportionately greater increase in size from the 4th to 5th instar in susceptible varieties than in the resistant varieties. The increase in size from the 5th instar nymphs to adults was only 4.4% in the females and 2.3% in males.

4. *Weight of jassids reared on different varieties:* The weight of the 1st instar nymphs ranged from 1.88 to 1.93 mg, the 2nd instar from 2.41 to 2.46 mg, the 3rd instar from 6.41 to 6.49 mg, the 4th instar from 8.21 to 8.33 mg, and the 5th instar from 10.83 to 12.23 mg. It was observed that the increase in weight of the nymphs from 1st to 2nd instar nymphs was only 22.5 to 30.8%. There was a sudden increase in weight from 2nd to 3rd instar nymphs reared on different varieties, when 161.2 to 169.2% values were recorded. The percentage increase in weight from 3rd to 4th instar nymphs was 27.2 to 28.9%, from 4th to 5th instar nymphs it was 31.0 to 48.9%. The nymphs of the 5th instar reared on susceptible varieties (Cocker-100 A and Lankart-57) were significantly heavier than those reared on resistant varieties. The weights of the adults, both males and females, reared on susceptible varieties (Cocker-100 A and Lankart-57) were significantly higher than of those reared on resistant varieties (B.1007 and Badnawar-1). In other words the effect of feeding on resistant varieties was not reflected in the weight of nymphs until the 4th instar but became evident only from the 5th instar onwards.

The increase in weight of the nymph was not significant between the 1st and 2nd instar, irrespective of the variety on which they were reared. But there were significant differences in weight from 2nd to 3rd, 3rd to 4th, 4th to 5th instar and from 5th instar to adults. Evidently, the feeding of the nymphs on cotton was more intensive from 2nd instar onwards (Table 4).

DISCUSSION

The least number of ovipositional punctures were in B.1007. The second resistant variety (Badnawar-1) had significantly higher number of punctures than B.1007, but less than either Reba-B.50 or H.14. The two susceptible varieties (Cocker-100 A and Lankart-57) showed well marked differences in the number of ovipositional punctures compared to the resistant or medium resistant varieties.

Thus, it became obvious that the susceptible varieties (Cocker-100 A and Lankart-57) were highly preferred for oviposition by the jassids followed in decreasing order by H.14, Reba-B.50, Badnawar-1 and B.1007. This is the general pattern of preference in jassids, i.e. the former varieties were always more damaged than the latter under field conditions. The number of ovipositional punctures in the subveins of resistant and medium resistant varieties was not as high as in the susceptible varieties. Furthermore, the main veins harboured significantly higher number of ovipositional punctures than the corresponding subveins. It was established that while ovipositing the jassid preferred the main veins to the subveins.

The mechanical barrier created by the dense population of trichomes on B.1007 and Badnawar-1 may be a limiting factor to the feeding of early instar nymphs. This factor may be the cause of reduced development and rate of survival of jassids on resistant varieties compared to susceptible ones. Similar evidences were also obtained by Poos and Smith (1931) for *E. fabae* on hairy host plants and by Jayaraj (1967) for *E. flavescens* on castor.

The high survival ratio, development, size, weight and time required for the different nymphal instars on the susceptible varieties were the evidence that all the nutritional as well as environmental conditions conducive to the optimum growth, etc. were not only present but also available on these varieties. Insignificant differences in the weight and size of the nymphs reared on different varieties were indicative of the fact that although the nutrients needed for development were present in the resistant varieties these were not available to a large number of nymphs, resulting in their low survival and development rate. The fewer number of days necessary to complete the life cycle on susceptible varieties would mean quicker multiplication and a higher number of generations in a year than on resistant varieties. The latter would result in the feeding by a large number of nymphs and adults on susceptible varieties in the later generations compared to the resistant varieties. The high rate of feeding and cell sap depletion would lead to extreme curling, crinkling, yellowing, bronzing, browning and drying of leaves leading to defoliation more rapidly in the susceptible varieties than in the resistant varieties. This process would cause great setback of plants and losses in yield.

REFERENCES

BAILEY, J. C., MAXWELL, F. G. and JENKINS, J. N. (1967): Boll weevil antibiosis studies with selected cotton lines utilizing egg-plantation techniques. *J. econ. Ent.* **60** (5), 1275–1279.

COWLAND, J. W. (1947): The cotton jassid *Empoasca libyca* de Berg in the Anglo-Egyptian Sudan and experiments in its control. *Bull. ent. Res.* **38** (1), 99–115.

CURTIS, C. E. and MC COY, C. E. (1964): Some host plant preferences shown by *Lygus lineolaris* (Hemiptera–Miridae) in the laboratory. *Ann. ent. Soc. Am.* **57** (4), 511–513.

GERING, Q. A. and COAKER, T. H. (1960): The effects of different food plants on the fecundity, fertility and development of a cotton stainer, *Dysdercus superstitiosus* (F.). *Bull. ent. Res.* **51**, 61–76.

HOLMES, N. D. and PETERSON, L. K. (1961): Resistance of spring wheats to the stem sawfly, *Cephus cinctus* Nort. (Hymenoptera–Cephidae) I. Resistance to the egg. *Canad. Ent.* **93** (4), 250–260.

JAYARAJ, S. (1967): Antibiosis mechanism and resistance in castor varieties to the leafhopper, *Empoasca flavescens* F. (Homoptera–Jassidae). *Curr. Sci.* **25** (22), 572–573.

JAYARAJ, S. (1968): Preference of castor varieties for feeding and oviposition by the leafhopper, *Empoasca flavescens* (F.) (Homoptera–Jassidae) with particular reference to its honey dew excretion. *J. Bomb. Nat. Hist. Soc.* **65** (1), 64–74.

PATHAK, M. D. (1969): Stemborer and leaf hopper plant hopper resistance in rice varieties. *Proc. Inter. Symp. on Insect and Host Plant*, Wageningen, 789–800.

POND, D. D., DIONNE, L. A., WHITE, R. G. and MOORE, C. A. (1962): Note on egg laying response of two species of root maggots on turnips bred for resistance to the cabbage root maggots, *Hylemyia brassicae* (Bouché). *Canad. J. Pl. Sci.* **42** (3), 530–531.

Poos, F. W. and Smith, F. F. (1931): A comparison of oviposition and nymphal development of *Empoasca fabae* (Harris) on different host plants. *J. econ. Ent.* **24** (2), 361–371.

Reed, D. K. and Adkisson, P. L. (1961): Short day cotton stocks as possible sources of host plant resistance to the pink bollworm. *J. econ. Ent.* **54** (3), 484–486.

Verma, P. M. and Afzal, M. (1940): Studies on cotton jassid, *Empoasca devastans* Dist. in the Punjab. I. Varietal susceptibility and development of the pest on different varieties of cotton. *Ind. J. Agric. Sci.* **10,** 911–926.

Yadav, H. N., Mittal, R. K. and Singh, H. G. (1967): Correlation studies between leaf midrib structure and resistance to jassids (*E. devastans* Dist.) in cotton. *Ind. J. Agric. Sci.* **37,** 495–497.

Symp. Biol. Hung. 16, pp. 23–28 (1976)

STUDIES ON THE INDUCTION OF FOOD PREFERENCE IN ALFALFA LADYBIRD, *SUBCOCCINELLA 24-PUNCTATA* L. (COLEOPTERA: COCCINELLIDAE)

by

M. ALI

DEPARTMENT OF ECONOMIC ENTOMOLOGY, FACULTY OF AGRICULTURE, AL-AZHAR UNIVERSITY, NASR CITY, CAIRO, A.R. EGYPT

The conditioning of larvae of *Subcoccinella 24-punctata* L. to host plant selection could not be completely proven. Larvae persistently preferred alfalfa (*Medicago sativa* L.) to the other plants introduced. An exception was that *C. album*-reared larvae preferred *Chenopodium album* to *M. sativa*. On the other hand, the preference of adults for a given food plant could be induced by experience of larvae and young adults. It was possible to modify the host-plant selection of alfalfa ladybird, as a result of which *C. album* became more preferred than *M. sativa* which is the primary host-plant of this species. Studies on behaviour pointed out that the high preference for *C. album* by larvae or adults can be ascribed to the physiological condition of the plants tested.

INTRODUCTION

Although alfalfa ladybird, *Subcoccinella 24-punctata* L. attacks mainly alfalfa (*Medicago sativa* L.), it was recorded as a pest of more than 70 plant species belonging to the families: Leguminosae and Caryophyllaceae (Jang, 1964; Szelényi, 1944; Tanasijević, 1958).

The present work is an attempt to study the food selection and food preference of larva and adult to certain plant species. Within this discourse, the following questions are of importance:

1. Do larvae or adults reared on a certain food plant show a food preference for the same plant?
2. Is it possible to modify the host-plant preference of alfalfa ladybird?

MATERIAL AND METHODS

Food preference of larva and adult was tested for the plant species: *Chenopodium album* L., *Medicago sativa* L., *Saponaria officinalis* L., and *Trifolium pratense* L. To evaluate the food preference, the leaf discs test of Jermy (1961) was used. When beetles were given the opportunity to choose between leaf discs of two different plant species, six leaf discs of each plant were stuck into each glass dish alternately, forming a circle about 8 cm in diameter, while 4 leaf discs from each plant were used in case of testing the food preference among four different plant species. Because of the slow movement of larvae, leaf discs were cut in a square form and were stuck into the substratum adhering to each other.

The chief step in food choice experiments of larvae was to rear them from the first larval instar until the beginning of moulting into the fourth one on the leaves of the original plant. Larvae just moulted to the fourth instar were divided into two groups: the first was further reared on the leaves of the original plant while to the second group the leaves of the other plant were given for 2 days, then each group was given the opportunity to choose equally between the leaf discs of the original and tested plants together (Fig. 1).

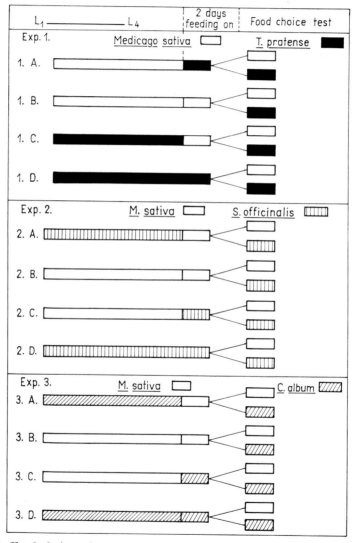

Fig. 1. Food choice scheme of experiments carried out with the 4th larval instar of *S. 24-punctata* L.

When the purpose was to test the food preference of adults, newly emerged adults (females and males) were reared for 7–10 days on the leaves of the original plant on which they fed as larvae, then they were treated in the same manner as described above with larvae. Food selection tests of larvae or adults were carried out to test the food preference between two different food plants or among four plant species.

To facilitate the explanation of the experiments symbols indicating the food plants are used as follows: C = *Chenopodium album* L., M = *Medicago sativa* L., S = *Saponaria officinalis* L., T = *Trifolium pratense* L. Food choice experiments were carried out at a constant temperature of 23 °C and at a photoperiod of LD 17/7 h.

RESULTS

Food choice by larvae

Larvae reared from L_1 to L_4 on *M. sativa* significantly preferred (P = = 0.05) it to *T. pratense* and *S. officinalis* (Fig. 2; 1 and 2). The interesting observations noticed in food choice tests with *T. pratense* were the high preference of larvae for *M. sativa* over *T. pratense* even in cases when larvae were reared from L_1 to L_4 on this plant (Fig. 2; 1D). Similar results were gained in food choice tests with *S. officinalis*.

When larvae were allowed to choose between *M. sativa* and *C. album*, they showed a preference for *M. sativa* only when they fed on it before starting the choice tests (Fig. 2; 3B). The most interesting results were the high preference of larvae for *C. album* to *M. sativa* either by larvae reared continually on *C. album* or by larvae fed on this plant for a short period of time which were previously reared from L_1 to L_4 on *M. sativa*. It was noticed that while a short feeding of larvae on *C. album* could induce a food

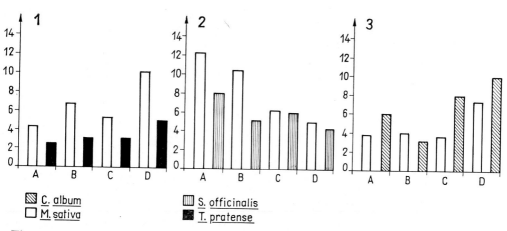

Fig. 2. Milligrams of dry weight eaten in food choice experiments at 23 °C by the 4th larval instar. See text for explanation

preference for this plant, similar feeding on *M. sativa* failed to induce such preference for this plant by larvae previously reared on *C. album* (Fig. 2; 3 A, C and D).

Further food choice experiments with larvae

Larvae reared continually on *M. sativa* and those moved to *T. pratense* for 2 days showed a food preference for *M. sativa* when they were given the opportunity to choose among the leaf discs of the four food plants together (Fig. 3; A, 1, 3).

Similar preference was noticed by *T. pratense*-reared larvae. On the other hand, it was interesting to find that larvae consumed the greatest amount of food from *C. album*, consequently, it was preferred to the other plants introduced in food choice tests.

S. officinalis-reared larvae preferred this plant only when they developed on it from L_1 to L_4 (Fig. 3; B, 4). One can find here also that the changing of original food by a new one for 2 days did not influence the preference of larvae for the original one (Fig. 3; B, 1, 2).

With regard to *C. album*-reared larvae, it was found that larvae always preferred it to the other introduced plants regardless of the manner by which the larvae were treated. An exception was recorded by *sativa*-reared larvae where *M. sativa* was more preferred than *C. album* (Fig. 3; B, 3).

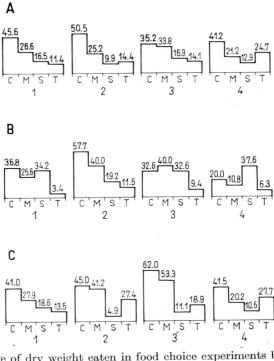

Fig. 3. Percentage of dry weight eaten in food choice experiments by the 4th larval instar at 23 °C. C: *Chenopodium album* L.; M: *Medicago sativa* L.; S: *Saponaria officinalis* L.; T: *Trifolium pratense* L.

Female beetles preferred the plant on which they were reared as larvae and young adults. *Sativa*-reared females showed a high preference as well as high consumption of *M. sativa* that exceeded the one consumed from other plants introduced in food choice tests (Fig. 4; A, 1). Females produced from larvae reared on *T. pratense* or on *S. officinalis* or on *C. album* did not behave differently. Results revealed that female beetles preferred as well as consumed the greatest amount of food from the original plant on which they developed as larvae and young adults. The moving of females to a new plant for a short period (2 days) did not influence the food preference for the original plant (Fig. 4; 2, 3 and 4).

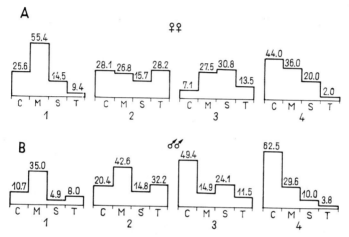

Fig. 4. Percentage dry weight eaten in food choice tests by male and female beetles. 1. Adults reared on *Medicago sativa* L.; 2. Adults reared on *Trifolium pratense* L.; 3. Adults reared on *Saponaria officinalis* L.; 4. Adults reared on *Chenopodium album* L.

The food selection of male beetles showed a slight variation as compared to that of females. Males reared on either *M. sativa* or on *C. album* always preferred the original plant on which they had been reared as larvae, while males reared on *S. officinalis* and those produced from larvae reared on *T. pratense* did not prefer the original food plant and preferred the other one (Fig. 4; B).

DISCUSSION AND CONCLUSION

According to the previously mentioned results on the food selection and food preference of alfalfa ladybird, it appeared that preference for food of larvae cannot be conditioned by short experience to a certain food plant. In other words, conditioning of larvae to either *S. officinalis* or to *T. pratense* by rearing them on those plants was unsuccessful, while it was possible for each of *M. sativa* and *C. album*. Similar observations were noticed in food choice tests of mature female and male beetles.

The preference of larva or adult for *M. sativa* can be expected because it is the main host plant of this species in the south-western region of Hungary. The preference of clover weevil (*Apion* sp.) for red clover to other plants (Deseő, 1967) supports the fact that insects may prefer the original host plant to others when introduced in food choice tests. The most interesting result was the possibility of modifying the food preference of alfalfa ladybird for *C. album* by rearing larvae and young adults on this plant even for a short period of time. On the other hand, in Colorado beetle, the preference for food of sexually mature beetles cannot be conditioned by experience on the part of larvae nor by any experience on the part of young adults (Bongers, 1965).

Food preference may be induced by temperature prevailing during the experiments or by the physiological condition of plants. Colorado beetle prefers potato at relatively lower temperatures and at higher temperatures bittersweet is relatively more eaten (Bongers, 1970). In our experiments, all food choice tests were carried out at a temperature of 23 °C, therefore we suggest that the preference of larva and adult for *C. album* may be ascribed to the variation in the physiological conditions of the introduced plants.

ACKNOWLEDGEMENTS

The author is much indebted to Dr. T. Jermy for his suggestions, guidance and kind encouragements. It is a great pleasure to thank Mrs. Enayate Salama, for her help and never desisting care of the insect material which was of dominant importance in the investigation.

REFERENCES

Bongers, W. (1965): External factors in the host plant selection of the Colorado beetle, *Leptinotarsa decemlineata* Say. *Meded. Landb. Hogesch. Wageningen,* **30** (3), 1516–1523.

Bongers, W. (1970): Aspects of host-plant relationship of the Colorado beetle. *Meded. Landb. Hogesch. Wageningen,* **70** (10), 1–77.

Deseő, V. K. (1967): Contribution to the biology of *Apion* species occurring on red clover in Hungary. *Acta Phytopath. Acad. Sci. Hung.* **2,** 141–152.

Jang, K. P. (1964): The effect of food plants on the development and viability of the 24-spotted lucerne ladybird. (In Bulgarian.) *Rastit. Zasht.* **12** (10), 16–23.

Jermy, T. (1961): On the nature of the oligophagy in *Leptinotarsa decemlineata* Say (Coleoptera: Chrysomelidae). *Acta Zool. Acad. Sci. Hung.* **7,** 119–132.

Szelényi, G. (1944): A lucernaböde. Der Luzernemarienkäfer. (*Subcoccinella viginti-quatuorpunctata* L.) *Növényegészségügyi Évkönyv,* **2–3,** 31–127.

Tanasijević, N. (1958): Zur Morphologie und Biologie des Luzernemarienkäfers *Subcoccinella vigintiquatuorpunctata* L. (Coleoptera: Coccinellidae). *Beitr. Ent.* **8,** 23–78.

Symp. Biol. Hung. 16, pp. 29–34 (1976)

FEEDING AND NUTRITION OF THE PEA APHID, *ACYRTHOSIPHON PISUM* (HARRIS), WITH SPECIAL REFERENCE TO AMINO ACIDS*

by

J. L. AUCLAIR

DÉPARTEMENT DES SCIENCES BIOLOGIQUES, UNIVERSITÉ DE MONTRÉAL, C.C. P. 6128, MONTRÉAL 101, QUÉBEC, CANADA

A brief review is given of the nutritional investigations made over the past twenty years on the pea aphid, *Acyrthosiphon pisum*, including the development of a chemically defined diet with a suitable amino acid mixture. The effects of total amino acid concentration on diet and amino acid uptake and growth of *A. pisum*, as well as the influence of individual amino compounds on phagostimulation are discussed. Correlations are drawn between results on diets and those obtained on pea plants. The concentrations of total soluble amino compounds (some 21 to 24 amino acids and amides), including those of 10 amino acids essential to pea aphid, and of two amides (asparagine and glutamine) are generally twice as high in water extracts of susceptible pea varieties than in those of resistant ones, and two to four times higher in terminal growth (susceptible) than in middle growth (resistant) samples of pea plants. Aphid sap uptake is also higher on susceptible than on resistant varieties. The conclusion remains that, under natural conditions, the establishment of aphid colonies of economic importance on cultivated plants is influenced markedly, among other things, by the nutritional superiority of the host, in which free amino compounds play an important part.

Research on the feeding and nutrition of aphids on chemically defined diets has been carried out well over the past twenty years. In preliminary tests made at the St. Jean Laboratory in Canada, a cage had been devised by Maltais (1952), with a feeding membrane made of natural rubber, which were satisfactory for rearing *Acyrthosiphon pisum* and *Myzus circumflexus*. We were able to show at the time that the stylets of aphids penetrated a rubber membrane and that their survival was increased by several days when given water or various chemical diets (Maltais et al., 1955). However, these diets contained glucose, dextrin or 1% sucrose as the carbohydrate source, and the amino acid mixture utilized was inappropriate. It was only in 1962–63 that a successful breakthrough was achieved when glucose was replaced by high concentrations of sucrose, together with a suitable amino acid mixture at an optimum pH. The dietary amino acid mixture that allowed rearing of the pea aphid was based primarily on the amino acid composition of the haemolymph and honeydew of that insect, composition published earlier (Auclair, 1960). The chemical diet finally prepared could maintain aphid life for two or three generations (Auclair and Cartier, 1963; Auclair, 1965).

The holidic diet originally devised for the pea aphid was modified later on (e.g. Srivastava and Auclair, 1971; Auclair and Srivastava, 1972, and unpublished results), but the principal refinements, on the composition of

---

* Because of space limitations, mention of many papers pertinent to this subject was deleted from the manuscript.

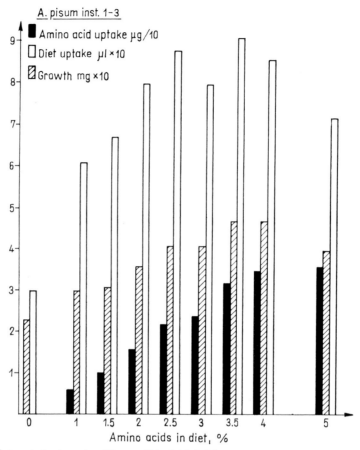

Fig. 1. Diet uptake in microliters×10/aphid (white columns), amino acid uptake in micrograms/10/aphid (black columns), and growth in mg×10/aphid (striped columns) of first to third-instar *A. pisum* over a four-day period, when served a complex diet containing different concentrations of a mixture of 20 amino acids

the various salts, were carried out by Akey and Beck (1971, 1972) who succeeded in maintaining the pea aphid continually for over 46 generations, or during some 20 months. Retnakaran and Beck (1968) had demonstrated earlier that the usual ten essential amino acids plus cysteine were dietary requirements for the pea aphid.

More recent work (Srivastava and Auclair, 1974) has dealt with the effect of total amino acid concentration on diet uptake and growth by *A. pisum*. Chemical diets including a mixture of 20 amino acids for a total concentration varying from 0 to 5%, in 1/2 to 1% increments, were used. In first to third-instar *A. pisum* (Fig. 1), diet uptake increased steadily with an increase in amino acid concentration from 0, where it was lowest, to 2.5%; it then more or less levelled off between 2.5 and 4%. This was followed by a decrease in uptake at 5%. The presence of amino acids up to a 2.5% concentration increased the acceptability of the diets which were ingested in larger amounts, uptake on the 2.5 and 3.5% amino acid diets being about 3 times higher than that on the free amino acid one. Thus the amino acid mixture acts as a phagostimulant in complex diets. Between 2.5 and 4%, there was a levelling off in uptake, indicating that phagostimulation was about the same in this range of concentration. At 5%, there was a deterrent effect on feeding, and uptake was reduced by almost 25%. This might occur directly through sensory stimulus, and/or by an internal negative feedback mechanism (e.g. Kennedy and Fosbrooke, 1973). The amino acid uptake in micrograms/aphid increased with increased concentration, but up to a point located at about 4%. At 5%, there was a trend toward levelling off, as if the aphid tended to keep amino acid intake steady at a dietary concentration above 4%. Growth was best at 3.5 to 4%, and, a further increase in amino acid concentration resulted in poorer growth, even though the amount of amino acids ingested at a 5% concentration was about equivalent to that ingested at 4%. Growth at 5% was comparable to that obtained between 2 and 3%. This suggests a toxic effect due to the high concentration of amino acids, but more probably a dietary imbalance between amino acids and other dietary constituents or water, resulting in metabolic disturbance.

The results obtained with fourth instar to adults were generally the same; diet uptake increased steadily with an increase in amino acid concentration from zero where it was lowest, to 2.5% where it was highest. There was then a steady decline in diet uptake with further increases in amino acid concentrations, as if here again, the aphid tended to keep amino acid intake fairly steady, in the 3 to 5% dietary range. Growth was best at the 3.5% concentration, but almost as good in the 2.5 to 5% range. It would appear that dietary uptake in older aphids is influenced more markedly by amino acid concentrations than in younger larvae, and that they tend to have better regulation of uptake.

By comparing results of pea aphid nutrition on holidic diets with those obtained on susceptible and resistant pea varieties, one can draw some interesting correlations. Determinations of the free and total amino acid contents of three susceptible pea varieties (Perfection, Daisy, and Lincoln) and of three resistant ones (Laurier, Champion of England, and Melting Sugar) were made by us many years ago (Auclair et al., 1957). In these studies, the parts of the plant most commonly infested, i.e. the terminal growth down to the fourth internode, were analysed at the stages of growth most susceptible to aphid attack in the field, i.e. first blossom to full bloom stages, from varieties grown in the greenhouse and in the field. The published results of our amino acid analyses have been re-examined in the light of new information on amino acid studies in aphids, and they are condensed in a different way in Table 1. They show that the concentration of total soluble amino compounds (some 21 to 24 amino acids and amides) from plant water extracts is generally almost twice as high in susceptible pea varieties than in resistant ones; the concentration of ten amino acids known to be dietary essential to the pea aphid is likewise almost twice as high in susceptible varieties; the concentration of the amides asparagine and glutamine is more than twice as high in susceptible varieties. This observation is interesting in view of the recent report by van Emden (1972) that amide concentration in Brussels sprout plants is positively correlated with the mean relative growth rate of two aphid species (*Myzus persicae* and *Bre-*

TABLE 1

Free amino acid and amide content (mg/100 ml of water extracts) of terminal and middle growth plant samples of three pea varieties susceptible (P, Li, D) and three resistant (La, C, M) to the pea aphid

Data condensed from Tables II, III, and IV from Auclair et al. (1957)

Amino compounds and stage of plant growth	Susceptible varieties			Resistant varieties		
	P	Li	D	La	M	C
21 to 24 amino compounds						
first blossom	151 (58)*	153 (65)	—	78 (31)	—	108 (32)
full bloom	142	—	116	72	64	—
*10 amino acids essential to pea aphid***						
first blossom	38 (14)	39 (19)	—	21 (7)	—	24 (6)
full bloom	43	—	35	27	24	—
2 amides asparagine and glutamine						
first blossom	40 (12)	42 (13)	—	11 (6)	—	25 (6)
full bloom	43	—	34	20	12	—
Gamma-aminobutyric acid						
first blossom	4.2 (3.7)	6.3 (4.5)	—	4.6 (2.9)	—	4.1 (2.3)
full bloom	2.1	—	8.8	5.1	3.7	—

* Data in parentheses are for middle growth samples.
** Arginine, cystine (and cysteine), leucine and/or isoleucine, lysine, methionine, phenylalanine, threonine, tryptophan, valine (less threonine and tryptophan for middle growth samples).

vicoryne brassicae). Maltais and Auclair (1962) had shown that these two amides (and homoserine) accelerated the growth of the pea aphid. The negative correlation reported with gamma-aminobutyric acid (GAB) for *M. persicae* (van Emden, 1972) was not observed in our analyses of pea plant samples (Table 1), the differences in the GAB content between varieties being insignificant. However, GAB represented only 3.8% of the soluble amino compounds in susceptible varieties as against 5.4% in resistant ones. Middle growth plant samples (including the fifth and sixth internodes of the pea stems) which represent a part of the plant not generally colonized by pea aphids, except in very heavy infestations, showed the concentrations of total soluble amino compounds, of the ten essential amino acids, and of amides, to be generally two to four times lower than the corresponding contents in more susceptible terminal growth samples of each variety, with their concentrations being again much lower in the resistant than in the susceptible varieties. GAB represented only 6.7% of the soluble amino compounds in susceptible varieties as against 8.3% in resistant ones.

Subsequent tests had shown that, on a resistant variety such as Melting Sugar, the rate of sap ingestion was about 68% of that recorded on a susceptible one (e.g. Perfection; Auclair, 1959), and nitrogen ingestion was also lower in about the same proportion. The higher nitrogen intake on a susceptible host was the result of a higher rate of sap ingestion, and of a slightly higher concentration of nitrogen (e.g. amino acids) in the sap (Auclair and Maltais, 1961). Calculations indicated that on a susceptible variety, pea aphid ingested a sap containing about 3.75% amino acids, whereas on a resistant variety, it ingested a sap containing about 2% amino acids. On the more resistant varieties, aphid growth and reproduction were reduced significantly in a way comparable to that of aphids reared on chemical diets having a similar reduction in amino acid level (Auclair, 1965; Srivastava and Auclair, 1974), or as for aphids starved for 10 hours daily (Auclair and Cartier, 1960).

Aphid performance on plants could be affected not only by the overall amino acid concentration, but also by the relative concentrations of different amino acids that may lead to possible *nutritional imbalance* and/or *reduced phagostimulation*. For instance, recent results obtained in our laboratory (Srivastava and Auclair, unpublished) indicate that many of the amino compounds present in pea plants, such as homoserine, leucine, phenylalanine, threonine, methionine, and glutamine, when added singly at certain concentrations (0.1 to 1%) to a 35% sucrose solution, are phagostimulatory at many of the concentrations tested, since the solutions are ingested in larger amounts than a 35% sucrose solution. A few other amino acids seem to have no influence on aphid ingestion, or may have a slight repellent effect at certain concentrations.

Other results on multiple choice tests (Auclair, 1965) in which pea aphids (biotype R1) were shown to select, feed on and colonize diets of certain amino acid levels (3.7–4.3%) in preference to lower or higher amino acid concentrations, bear importantly on a principle discussed by House (1967) which states that nutrient composition of a foodstuff can be responsible for food selection by an insect. Our multiple choice tests have shown rather conclusively that aphids will preferentially select diets that are nutrionally superior, on which their rate of growth and reproduction and their survival are highest.

In conclusion, there is little doubt that, under natural conditions, the act of selection of host-plants, but more probably the selection of specific feeding sites on a particular plant, and eventually the establishment of aphid colonies of economic importance, are influenced markedly, among other things, by the nutritional superiority of the host, in which amino acids may contribute significantly.

REFERENCES

Akey, D. H. and Beck, S. D. (1971): Continuous rearing of the pea aphid, *Acyrthosiphon pisum*, on a holidic diet. *Ann. ent. Soc. Am.* **64**, 353–356.

Akey, D. H. and Beck, S. D. (1972): Nutrition of the pea aphid, *Acyrthosiphon pisum:* Requirements for trace metals, sulphur, and cholesterol. *J. Insect Physiol.* **18**, 1901–1914.

Auclair, J. L. (1959): Feeding and excretion by the pea aphid, *Acyrthosiphon pisum* (Harr.) (Homoptera: Aphididae), reared on different varieties of peas. *Ent. exp. & appl.* **2**, 279–286.

Auclair, J. L. (1960): Teneur comparée en composés aminés libres de l'hémolymphe et du miellat du puceron du pois, *Acyrthosiphon pisum* (Harr.), à différents stades de développement. *Verh. XI. Int. Kongr. Ent., Vienna*, **3**, 134–140.

Auclair, J. L. (1965): Feeding and nutrition of the pea aphid, *Acyrthosiphon pisum* (Homoptera: Aphididae), on chemically defined diets of various pH and nutrient levels. *Ann. ent. Soc. Am.* **58**, 855–875.

Auclair, J. L. and Cartier, J. J. (1960): Effets comparés de jeûnes intermittents et de périodes équivalentes de subsistance sur des variétés résistantes ou sensibles de pois, *Pisum sativum* L., sur la croissance, la reproduction et l'excrétion du puceron du pois, *Acyrthosiphon pisum* (Harr.). *Ent. exp. & appl.* **3**, 315–326.

AUCLAIR, J. L. and CARTIER, J. J. (1963): Pea aphid: Rearing on a chemically defined diet. *Science,* **142,** 1068–1069.

AUCLAIR, J. L. and MALTAIS, J. B. (1961): The nitrogen economy of the pea aphid, *Acyrthosiphon pisum* (Harr.), on susceptible and resistant varieties of peas, *Pisum sativum* L. *Verh. XI. Int. Kongr. Ent., Vienna,* **1,** 740–743.

AUCLAIR, J. L., MALTAIS, J. B. and CARTIER, J. J. (1957): Factors in resistance of peas to the pea aphid, *Acyrthosiphon pisum* (Harr.). II. Amino acids. *Can. Ent.* **89,** 457–464.

AUCLAIR, J. L. and SRIVASTAVA, P. N. (1972): Some mineral requirements of the pea aphid, *Acyrthosiphon pisum* (Homoptera: Aphididae). *Can. Ent.* **104,** 927–936.

VAN EMDEN, H. F. (1972): Aphids as phytochemists. *In:* Harborne, J. B. (ed.): *Phytochemical Ecology.* Academic Press, N.Y.

HOUSE, H. L. (1967): The role of nutritional factors in food selection and preference as related to larval nutrition of an insect, *Pseudosarcophaga affinis* (Diptera, Sarcophagidae), on synthetic diets. *Can. Ent.* **99,** 1310–1321.

KENNEDY, J. S. and FOSBROOKE, I. H. M. (1973): The plant in the life of an aphid. *In:* van Emden, H. F. (ed.): *Insect/Plant Relationships.* Symp. Roy. Ent. Soc. London, **6,** 129–140.

MALTAIS, J. B. (1952): A simple apparatus for feeding aphids aseptically on chemically defined diets. *Can. Ent.* **84,** 291–294.

MALTAIS, J. B. and AUCLAIR, J. L. (1962): Free amino acids and amide composition of pea leaf juice, pea aphid haemolymph, and honeydew, following the rearing of aphids on single pea leaves treated with amino compounds. *J. Insect Physiol.* **8,** 391–399.

MALTAIS, J. B., AUCLAIR, J. L. and CARTIER, J. J. (1955): Artificial feeding of aphids. *In: Quinquennial Techn. Rep.* (1950–54) p. C40–C47, Science Service Laboratory, St. Jean, Qué., Canada.

RETNAKARAN, A. and BECK, S. D. (1968): Amino acid requirements and sulfur amino acid metabolism in the pea aphid, *Acyrthosiphon pisum* (Harr.). *Comp. Biochem. Physiol.* **24,** 611–619.

SRIVASTAVA, P. N. and AUCLAIR, J. L. (1971): An improved chemically defined diet for the pea aphid, *Acyrthosiphon pisum. Ann. ent. Soc. Am.* **64,** 474–478.

SRIVASTAVA, P. N. and AUCLAIR, J. L. (1974): Effect of amino acid concentration on diet uptake and performance by the pea aphid, *Acyrthosiphon pisum. Can. Ent.* **106,** 149–156.

Symp. Biol. Hung. 16, pp. 35–40 (1976)

THE ABILITY OF *LOCUSTA MIGRATORIA* L.
TO PERCEIVE PLANT SURFACE WAXES

by

E. A. BERNAYS,* W. M. BLANEY,** R. F. CHAPMAN,* and A. G. COOK*

* CENTRE FOR OVERSEAS PEST RESEARCH, COLLEGE HOUSE,
WRIGHTS LANE, LONDON W8 5SJ, UK
** DEPARTMENT OF ZOOLOGY, BIRKBECK COLLEGE, MALET STREET,
LONDON W. C. 1, UK

Extracts of the lipid-soluble material from the surface of leaves of *Poa annua* were made, taking care to avoid contamination with internal constituents. These extracts, when dried onto strips of filter paper, induced nymphs of *Locusta migratoria* L. to bite. Control strips of filter paper had no effect. It is concluded that the insects are able to perceive the surface waxes of leaves simply by contact with chemoreceptors on the tips of the maxillary palps. In the discussion it is argued that discrimination between different plant waxes may also occur.

INTRODUCTION

Locusts and grasshoppers which have not previously been deprived of food may reject some potential food-plants without biting them, simply as a result of the maxillary palps touching the plant surface. Other plants, however, which comprise the normal food of the insects, are bitten and eaten (Bernays and Chapman, 1970; Blaney and Chapman, 1970). This implies that the insects are responding to features of the leaf surface and this paper describes experiments which show that they can perceive and respond to the waxes in the leaf cuticle.

METHODS

Preparation of the leaf surface extract

Throughout the experiments *Poa annua* was used as food for the insects and surface extracts were made from this same grass. In making the surface extracts care was taken to avoid contamination with the internal constituents of the leaves and for this reason any damaged leaves were rejected. The grass was sorted into flat bundles of approximately 5 g and the cut ends of the stems were clamped between two 8 cm × 2.5 cm microscope slides held together by a bulldog clip. Only the undamaged parts of the leaves extending beyond the slides were exposed to the solvent. Each bundle was washed gently for 30 sec in each of the baths of distilled water. This removed obvious surface dirt. The bundle was then held over a clean, empty 1-liter beaker and sprayed with the solvent, drops of which ran off the leaf surface and were collected in the beaker. The precise nature of the spray was critical; a continuous stream was too vigorous and caused contamination with internal constituents of the leaves; too fine a spray resulted in more volatile solvents freezing on the leaf surfaces and not producing an effective extraction.

3*

A satisfactory spray was obtained with some chromatographic sprayers, but the suitability depended on the characteristics of the individual sprayer.

Each surface of the bundle was sprayed for 15 sec, the spray being directed so as to cover as evenly as possible all the exposed leaf-blade surfaces. This time was chosen because after longer periods of exposure to an acetone spray some chlorophyll was present in the extract, indicating that some extraction of the internal constituents of the leaf had occurred. After only 15 sec spraying no chlorophyll was revealed in the extract with an SP.800 spectrophotometer. This technique only removed a small proportion of the waxes of the leaf-surface which remained non-wettable, but it was considered preferable to remove only some of the wax rather than risk contamination with internal constituents.

For normal experimental work four 5 g bundles of grass were extracted and the run-off was evaporated to a standard volume of 4.5 ml at low temperatures so as to reduce the risk of decomposition of the dissolved materials. In later experiments the solution was evaporated to dryness by bubbling nitrogen through it. This had the advantage that known weights of solute could then be redissolved to make standard solutions.

Chemical analysis of extract

The organic chemical content of the extract was determined using standard test methods. The total amounts of lipids were determined using a Boehringer Mannheim Total Lipid test kit. Amino acids and sugars were tested after evaporating to dryness and redissolving in water. Amino acids were tested for using a modified ninhydrin analysis (Rosen, 1956) and sugars by an anthrone test (Roe, 1954). Theoretically no water soluble material should have been present in the extract, but some carry over of water droplets in emulsion in the solvent was possible.

Rearing and preparation of the insects

Throughout the experiments nymphs of *Locusta migratoria* L. were used. These were reared to the third instar in 12-litre cylindrical cages. The cages were kept at 30 °C, adjacent to a 100 watt electric bulb which provided a constant source of light and radiant heat. The insects were fed twice daily, once on Saturdays and Sundays, with *Poa annua* so that they had a constant supply of food.

All the tests were carried out with third instar nymphs which were carefully prepared for the experiments to ensure that they were ready to feed, but were not deprived of food for extended periods, since after prolonged periods of deprivation biting becomes indiscriminate (Blaney and Chapman, 1970). Individuals were taken from the stock cage and put separately into 8 cm × 2.5 cm specimen tubes arranged round a light so as to be at approximately 30 °C. Each insect was given a blade of *Poa* on which it fed sooner or later. The precise time at which feeding finished was recorded, the end of a meal being defined as a period of one minute without feeding following a previous period of at least two minutes feeding. Previous observations had shown that it was rare for an insect to begin feeding again for some time after such a pause.

36

At the end of the meal, remaining fragments of grass were removed and the insect kept without food for 30–60 minutes, a period approximating to the normal interfeed period (Blaney et al., 1973). At this stage the insect was ready to test, but it was essential, first, to establish that it was ready to feed if given suitable food. Hence after 30–60 min without food each insect was offered a fresh blade of *Poa*. If it started to feed the grass was immediately removed and the insect was tested with one of the extracts. Insects which did not feed were given a further period without food before testing again, but any which were not ready to feed within 60 min of the previous meal were discarded.

Method of bioassay

Extracts were presented to the insects on strips of Whatman No.1 filter paper 100 mm long and 4 mm wide. This size was chosen as approximating to the average size of a leaf-blade of *Poa*. The paper strip was soaked for 1 minute in the extract and then dried for 15 min in an airflow before being presented to the insect. The strip was inserted into a tube containing an insect ready to feed and the behaviour of the insect was recorded for 5 min noting whether it palpated on the strip and if, following palpation, it bit at the end of the strip. At the end of each test the insect was discarded and was not used again. Control strips were treated in exactly the same way, but were soaked in solvent only.

The earlier experiments were carried out in a lighted room, but much more consistent results, with less biting on control strips, were obtained when the insects were tested in a lighted arena and screened from the observer in a darkened room. This technique was employed in the experiments n which methylene dichloride was used as the solvent (see Table 2).

RESULTS

Bioassay of extracts

Nearly all insects which were ready to feed after 30 to 60 min of food deprivation bit and fed when offered a fresh leaf of *Poa*. Insects offered a dry untreated filter paper strip, however commonly rejected the strip after palpation without biting, irrespective of whether or not the strip had previously been soaked in water (Table 1). This experiment indicates that biting is induced by some feature of the leaf and not just by any object of appropriate dimensions.

TABLE 1

Percentage of insects biting on a Poa *leaf and on filter paper strips*

(Only insects which palpated on the test material are included)

Test material	No. of insects	% biting
Poa leaf	50	94
Untreated filter paper	25	52
Water dried filter paper	26	31

Extracts of the chemicals in the leaf surface were made with a number of solvents to determine whether or not they should stimulate biting. With water as the solvent there was no difference between extract and control, but with the lipid solvents chloroform, acetone and methylene dichloride the extracts produced a significant increase in the amount of biting on the filter paper (Table 2).

TABLE 2

*Percentage of insects biting on filter paper strips
treated with various extracts*
(Only insects which palpated on the test material are included)

Solvent		No. of insects	% biting	χ^2	P
Water	control	26	31	0.55	>0.05
	extract	27	44		
Acetone	control	52	33	24.45	<0.001
	extract	50	84		
$CHCl_3$	control	50	44	17.58	<0.001
	extract	50	86		
CH_2Cl_2	control	23	22	12.52	<0.001
	extract	23	78		

Lipid content and contamination of the extract

The total lipid concentration of the extracts was 10.3 ± 2.6 $\mu g/ml$ (mean of 12 determinations). In four out of five samples no trace of either amino acids or carbohydrates was observed. In the remaining sample the total concentration of amino acids was 0.02 $\mu g/ml$ and of carbohydrates 0.53 $\mu g/ml$.

DISCUSSION

The results show that materials extracted from the leaf surface of *Poa annua* promote biting activity by *Locusta* nymphs which are ready to feed but have not been deprived of food for long periods. Since care was taken in the extraction procedure to ensure that contamination of the extracts with internal constituents of the leaf did not occur it seems certain that components of the cuticle produced the response.

Although plant cuticle may contain carbohydrates and amino acids "leaking" from the plant cells (Martin and Juniper, 1970) the major component of the cuticle is wax and the method of extraction with a lipid solvent will tend to have excluded water-soluble materials. These may nevertheless have been extracted in water droplets emulsified within the solvents, but the chemical tests indicated either no carbohydrates or amino acids or very low concentration of the order of 0.000003 M, taking glucose and glycine as standards. Such a concentration is well below the threshold level

38

of sensitivity of insect contact chemoreceptors (Dethier, 1963; Ma, 1972) and Cook (in prep.) has shown with a number of carbohydrates and amino acids that concentrations below 0.001 M dispersed on elder pith discs are not effective phagostimulants. It is therefore concluded that the insects were responding to the lipid-soluble components extracted from the leaf surface waxes.

Previous experiments with extracts of *Poa annua* and *Bellis perennis* also show that *Locusta* can differentiate between the waxes of different plants (Blaney and Chapman, 1970) and this is supported by less precise experiments on *Chorthippus parallelus* (Zett.) and *Chortoicetes terminifera* (Walker) in which certain plants are rejected following contact with the leaf surface by the maxillary palps (Bernays and Chapman, 1970, 1973). In all these cases perception and discrimination is achieved by the terminal sensilla of the palps, but the experiments of Kendall (1971) indicate that in *Schistocerca gregaria* (Forskål) the tarsal sensilla have a similar capacity.

The ability to perceive and respond to plant waxes has also been recorded in *Acyrthosiphon pisum* (Harris) by Klingauf et al. (1971). They have shown that this insect responds to certain n-alkanes from the surface of *Vicia faba* by prolonged probing and have also found a differential response to varying chain lengths which could provide a basis for differentiation between plant species. It is also significant that a straight chain primary alcohol of the type occurring in leaf waxes provides a biting stimulant for the larva of *Bombyx mori* (Fraenkel et al., 1960).

Plant waxes are complex mixtures of long chain molecules including alkanes, alcohols and ketones (Martin and Juniper, 1970; Kolattukudy and Walton, 1972) and their perception and especially their differentiation by the insect implies a high degree of sensory sophistication. In the locust, at least, this appears to depend on the analysis of information from an array of sensilla rather than from the input of any one sensillum or sense cell (Blaney, in prep.).

REFERENCES

BERNAYS, E. A. and CHAPMAN, R. F. (1970): Experiments to determine the basis of food selection by *Chorthippus parallelus* (Zetterstedt) (Orthoptera: Acrididae) in the field. *J. Anim. Ecol.* **39**, 761–776.
BERNAYS, E. A. and CHAPMAN, R. F. (1973): The role of food plants in the survival and development of *Chortoicetes terminifera* (Walker) under drought conditions. *Aust. J. Zool.* **21**, 575–592.
BLANEY, W. M. and CHAPMAN, R. F. (1970): The function of the maxillary palps of Acrididae (Orthoptera). *Ent. exp. & appl.* **13**, 363–376.
BLANEY, W. M., CHAPMAN, R. F. and WILSON, A. (1973): The pattern of feeding of *Locusta migratoria* (L.) (Orthoptera, Acrididae). *Acrida*, **2**, 119–137.
DETHIER, V. G. (1963): *The Physiology of insect senses*. Methuen, London.
FRAENKEL, G. S., NAYAR, J., NALBANDOV, O. and YAMAMOTO, R. T. (1960): Further investigations into the chemical basis of the insect–hostplant relationship. *11th Int. Congr. Ent. Symp. Vienna*, **3**, 122–126.
KENDALL, M. D. (1971): *Studies on the tarsi of Schistocerca gregaria* Forskål. Unpublished thesis, University of London.
KLINGAUF, F., NÖCKER-WENZEL, K. and KLEIN, W. (1971): Einfluß einiger Wachskomponenten von *Vicia faba* L. auf das Wirtswahlverhalten von *Acyrthosiphon pisum* (Harris) (Homoptera: Aphididae). *Z. Pflkrankh. PflPath. PflSchutz.* **78**, 641–648.

KOLATTUKUDY, P. E. and WALTON, T. J. (1972): The biochemistry of plant cuticular lipids. In: Holman, R. T. (ed.): *Progress in the Chemistry of Fats and other Lipids*. Pergamon Press.

MA, W. C. (1972): Dynamics of feeding responses in *Pieris brassicae* L. as a function of chemosensory input: a behavioural, ultrastructural and electrophysiological study. *Meded. Landb. Hoogesch. Wageningen*, **72–11**, 162.

MARTIN, J. T. and JUNIPER, B. E. (1970): *The Cuticles of Plants*. Arnold, London.

ROE, J. H. (1954): The determination of sugar in blood and spinal fluid with anthrone reagents. *J. biol. Chem.* **208**, 335–343.

ROSEN, H. (1956): A modified ninhydrin colorimetric analysis for amino acids. *Arch. Biochem. Biophys.* **67**, 10–15.

Symp. Biol. Hung. 16, pp. 41–46 (1976)

ANTIFEEDANT PROPERTIES OF SEEDLING GRASSES

by

E. A. BERNAYS and R. F. CHAPMAN

CENTRE FOR OVERSEAS PEST RESEARCH, COLLEGE HOUSE,
WRIGHTS LANE, LONDON W8 5SJ, UK

Feeding by acridids is reduced on seedling grasses compared with that on mature grasses. Other groups of insects show preferences for the maturer stage, which has also been shown to contain lower concentrations of materials which inhibit feeding. Certain seedling chemicals have been examined, and some alkaloids found to be manufactured by the young plant from the time of germination. The importance of such plant resistance is discussed.

It is well known that many plants possess chemical constituents which render them distasteful to certain insects (Hsiao, 1969), but it is not always appreciated that the distasteful properties of the plant may vary with its stage of development. This paper deals with the distastefulness of seedling grasses to certain insect species and the possible nature and significance of this distastefulness.

INHIBITION OF FEEDING

Seedling grasses are relatively distasteful to acridids compared with mature leaves of the same species (Bernays et al., in press). In the case of nymphs of *Locusta migratoria*, meals eaten after five hours without food are up to nine times larger on mature leaves than on seedling leaves of the same species, over a range of 20 different grasses. It is not until four weeks or more after sowing that the young plant is eaten in the larger amounts normal for mature grasses (Fig. 1).

A variety of other insects which feed largely or wholly on Gramineae, also prefer mature grass leaves to seedling leaves of the same grass species. This has been casually reported by a number of authors concerning the attack of cereal crops by aphids and the chinch bug *Blissus* (Dahms, 1948a, 1948b; Viale, 1950; El-Ibrashy, El-Ziady and Riad, 1972), and further experiments have been carried out on aphids, Heteroptera and lepidopterous larvae.

Apterous adult maize aphids (*Rhopalosiphum maidis*) were collected from maize plants in the field and placed in jars of approximately one litre volume containing one maize leaf 9—13 days old and another similar sized apical leaf from a plant 9–11 weeks old. Each leaf was standing in water. Ten aphids were placed on each leaf, and there were ten separate experiments. Table 1 shows that the aphids tended to move off the younger leaves, and those that stayed produced fewer offspring than those on mature leaves. In a similar experiment with the cercopid *Poophilus* sp. the insects were

TABLE 1

Behaviour of Rhopalosiphum maidis *on leaves of mature and seedling* Zea mays
(Results are from ten separate experiments)

	Mature leaves	Seedling leaves	Wandering
Mean number on each initially	10	10	0
Mean number after 8 hours	11	5	4
Mean number after 2 days	12	3	3
Mean number of young per aphid surviving	10	5	0

TABLE 2

Behaviour of two Heteropterans on leaves of seedling and mature Lolium perenne
(Results are the means of three separate choice experiments, each using 50 insects)

	Leptoterna		*Calocoris*	
	seedling leaves	mature leaves	seedling leaves	mature leaves
Number after 10 minutes	14	14	6	6
Number after 2 hours	12	18	5	10
Number after 4 hours	6	28	7	17
Number after 6 hours	4	32	4	31

TABLE 3

Behaviour of Pararge aegeria *and* Lasciommata megera *caterpillars towards leaves of mature or seedling* Agropyron repens *and* Lolium perenne
(Numbers represent the means from three separate experiments with 27–49 individuals in each case)
Percentage of first instar larvae on leaves after 4 hours

	Pararge		*Lasciommata*	
	seedling leaves	mature leaves	seedling leaves	mature leaves
Agropyron	35	65	31	69
Lolium	14	86	29	71

TABLE 4

The amounts of feeding by larvae of Spodoptera exempta *and* S. littoralis *on leaf discs of mature and seedling* Zea mays
(Numbers refer to experiments in each category)

	Mature leaves eaten more than seedling leaves	No preference shown	Seedling leaves eaten more than mature leaves
S. exempta	11	11	9
S. littoralis	0	0	12

observed at intervals of 15–20 minutes and removed from the leaves. In 112 observations 80 showed insects feeding on the mature leaves, while six showed insects feeding on the seedling leaves.

Amongst the Heteroptera, *Leptoterna dolobrata* and *Calocoris norvegicus* were tested in plastic behaviour boxes in which were small vases containing mature and seedling leaves of *Lolium perenne* (Bernays et al., in press). Fifty insects were placed in each box midway between the two sets of leaves, and over a period of six hours the numbers on the mature leaves rose continuously, while few remained on the seedlings (Table 2).

The other major insect group, which includes graminivorous species, is the Lepidoptera. Two graminivorous satyrids were tested in behaviour boxes using seedling and mature leaves of *Lolium perenne* and *Agropyron repens* in a series of separate experiments. Newly hatched larvae of *Pararge aegeria* and *Lasciommata megera* were put onto the floor of the box, and after four hours, more were invariably present on the mature leaves (Table 3). Finally, *Spodoptera littoralis* and *Spodoptera exempta* larvae were collected in the field and given a choice of seedling and mature leaf discs. One larva was placed in the centre of a 9 cm petri dish with four symmetrically arranged discs of each age leaf. Examination of the discs 1–2 hours later showed however, that there was a preference for seedlings in the case of *S. littoralis* and no difference with *S. exempta* (Table 4).

BASIS OF FEEDING INHIBITION

A number of alkaloids have been found in grasses, especially in young leaves. Aasen et al. (1969) examined several species of *Lolium* and showed that of the various alkaloids present, one halostachine was present in high concentration in young seedlings of *L. perenne*, and that its concentration fell as the plant matured (see Fig. 1). Halostachine is an antifeedant for *Locusta* (Table 5) and the fall in its concentration appears to be matched by the rise in acceptability as the plant matures.

TABLE 5

Amounts of pith eaten by individual nymphs of Locusta *over 18 hours when sucrose or sucrose + halostachine have been added*

	Meal size (mg)	S.E.	Number of insects
Pith + 0.125 M sucrose	48	11	10
Pith + 0.125 M sucrose + 1% halostachine	12	8	10

The alkaloids gramine and hordenine, which occur in *Hordeum* spp., are also found in highest concentrations in the leaves of the seedling plants (Mann, Steinhart and Mudd, 1963). Both these chemicals have antifeedant properties for the acridid, *Melanoplus bivittatus* (Harley and Thorsteinson, 1967).

43

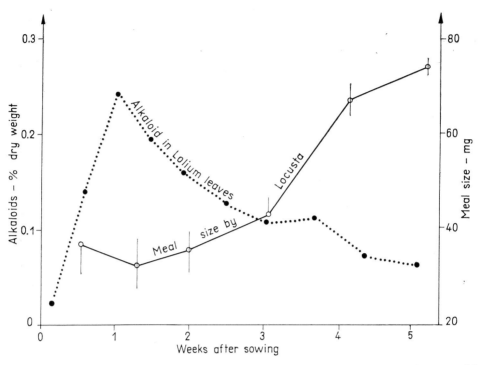

Fig. 1. The relationship of the meal size taken by *Locusta* nymphs and the alkaloid content of leaves of *Lolium perenne* (alkaloid levels after Aasen et al., 1969)

Other classes of material have also been found which decrease as the plant grows older. For example the glycosidic derivative 6-methoxybenzoxazolinone has been identified in *Zea mays* (Loomis, Beck and Stauffer, 1957; Klun and Robinson, 1969) as an antifeedant for *Pyrausta nubilalis* (Beck, 1960).

PRODUCTION OF ANTIFEEDANTS

It appears that grass seedling antifeedants are produced by the developing plant independently of the endosperm. In the case of 6-methoxybenzoxazolinone in maize, the embryonic plant at germination contains large amounts while the endosperm and scutellum contain only traces (Klun and Robinson, 1969). The plants continue to produce the chemicals but their concentration is reduced with time. Similarly in the case of *Lolium*, the alkaloid concentration is reduced with time, though later the plant can, under certain conditions, produce higher concentrations (Aasen et al., 1969). More certain evidence of the independent production of antifeedants by the embryo was obtained by growing seedlings of maize from isolated embryos on an agar culture medium, using the technique of Hartmann and Kester (1968). Embryos were grown in this way and healthy young seedlings were transferred from the culture medium to pots of soil and allowed to stabilize normally. The meal sizes of *Locusta* nymphs were similar on seedlings grown in this way or in the normal way from the whole seed (Table 6).

44

TABLE 6

Meal sizes taken by individual nymphs of Locusta
on Zea mays *seedlings grown under different conditions*

	Meal size (mg)	S.E.	Number of insects
Seedlings grown normally	20	6	15
Seedlings grown on agar from isolated embryos	17	6	15

THE SIGNIFICANCE OF ANTIFEEDANTS

Mature grasses may benefit from grazing (Culvenor, 1970) and it is possible for grazing to be utilized in tillering of wheat so that extra growing stems and subsequently more grain are produced. During the younger stages, however, there is a limit to the amount which can be eaten before regrowth ceases, and a small plant is very easily killed. Preliminary laboratory experiments with wheat indicate that during the period 3–4 weeks after sowing, the plant is most vulnerable, while younger plants will regrow unless they are eaten down to the seed. As pointed out by Culvenor (1970) cropping of seedlings robs the plant of its energy source when resources are low and also may lead to its being overshadowed by neighbouring plants. Undoubtedly the seedling is protected to some extent from being eaten at an important stage — a factor which may have contributed to the evolutionary success of the Gramineae. Plant species under heavy grazing pressure most commonly develop alkaloids (Mothes and Schutte, 1969), and the Gramineae provide the bulk of grazing for the world's herbivorous mammals as well as many insects. Thus at the one period of their life history when grasses can tolerate very little grazing they are protected by antifeedants which may be specifically developed for this purpose.

REFERENCES

AASEN, A. J., CULVENOR, C. C. J., FINNIE, E. P., KELLOCK, A. W. and SMITH, L. W. (1969): Alkaloids as a possible cause of ryegrass staggers in grazing livestock. *Aust. J. agric. Res.* **20**, 71–86.

BECK, S. D. (1960): The European corn borer, *Pyrausta nubilalis* (Hubn.), and its principal host plant. VII. Larval feeding behaviour and host plant resistance. *Ann. ent. Soc. Am.* **53**, 206–212.

BERNAYS, E. A., CHAPMAN, R. F., HORSEY, J. and LEATHER, E. M.: Inhibitory effects of seedling grasses on feeding and survival of acridids. *Bull. ent. Res.* **64**, 413–420.

CULVENOR, C. C. J. (1970): Toxic plants — a revaluation. *Search*, **1**, 103–110.

DAHMS, R. G. (1948a): Comparative tolerance of small grains to green bugs from Oklahoma and Mississippi. *J. econ. Ent.* **41**, 825–826.

DAHMS, R. G. (1948b): The effect of different varieties and ages of sorghum on the biology of the chinch bug. *J. agric. Res.* **76**, 277–288.

EL-IBRASHY, M. T., EL-ZIADY, S. and RIAD, A. A. (1972): Laboratory studies on the biology of the corn leaf aphid *Rhopalosiphum maidis* (Homoptera: Aphididae). *Ent. exp. & appl.* **15**, 166–174.

HARTMANN, H. T. and KESTER, D. E. (1968): *Plant propagation: principles and practices.* Englewood Cliffs, N.J., Prentice Hall.

HARLEY, K. L. S. and THORSTEINSON, A. J. (1967): The influence of plant chemicals on the feeding, behaviour, development, and survival of the two striped grasshopper, *Melanoplus bivittatus* (Say), Acrididae: Orthoptera. *Can. J. Zool.* **45**, 305–319.

HSIAO, T. H. (1969): Chemical basis of host selection and plant resistance in oligophagous insects. *Ent. exp. & appl.* **12**, 777–788.

KLUN, J. A. and ROBINSON, J. (1969): The concentration of two 1,4-benzoxazinones in dent corn at various stages of development of the plant and its relation to resistance of the host plant to the European corn borer. *J. econ. Ent.* **62**, 214–220.

LOOMIS, R. S., BECK, S. D. and STAUFFER, J. F. (1957): The European corn borer *Pyrausta nubilalis* (Hubn.), and its principal host plant. V. A chemical study of host plant resistance. *Plant Physiol.* **32**, 379–385.

MANN, J. D., STEINHART, C. E. and MUDD, S. H .(1963): Alkaloids and plant metabolism. V. The distribution and formation of tyramine methyltransferase during germination of barley. *J. biol. Chem.* **238**, 676–681.

MOTHES, K. and SCHUTTE, H. R. (1969): *Biosynthese der Alkaloide.* VEB Deutscher Verlag der Wissenschaften, Berlin.

VIALE, E. (1950): The biology of the corn leaf aphid *Aphis maidis* Fitch as affected by various strains of corn *Zea mays* and certain other environmental factors. *Ph. D. Thesis*, Kansas State College.

46

Symp. Biol. Hung. 16, pp. 47–54 (1976)

A CRITICAL REVIEW OF THE METHODOLOGY AND INTERPRETATION OF EXPERIMENTS DESIGNED TO ASSAY THE PHAGOSTIMULATORY ACTIVITY OF CHEMICALS TO PHYTOPHAGOUS INSECTS

by

A. G. Cook

CENTRE FOR OVERSEAS PEST RESEARCH, COLLEGE HOUSE,
WRIGHTS LANE, LONDON W8 5SJ, UK

An increasing number of studies on phagostimulatory chemicals for insects are being carried out, though experiments with *Locusta migratoria* demonstrated a need for care in the choice of experiments and the interpretation of results. This paper reviews some of the methods used and indicates some of the problems which may be encountered.

CHOICE OF SUBSTRATE

Several methods have been used for assaying the phagostimulatory activity of chemicals or plant extracts for phytophagous insects. The majority of these involve the presentation of the test material on an inert medium, the most widely used being elder pith, filter paper and agar/cellulose. Pith is usually used in the form of discs punched out of pith sections using a cork borer (Heron, 1965; Loschiavo, 1965; Norris and Baker, 1967; Ritter, 1967). Filter paper may be used in the same way or the test material may be applied to equally spaced areas inscribed round the circumference of a whole paper (Dadd, 1960; Goodhue, 1963; LaPidus et al., 1963; Mehrotra and Rao, 1972; Thorsteinson and Nayar, 1963; Wensler and Dudzinski, 1972; Yamamoto and Fraenkel, 1960). The agar/cellulose method involves the incorporation of a given amount of the test material into a mixture comprising 4% bacto-agar and 4% cellulose dissolved in water which, when cooled, sets in a thin layer. Discs may be punched from this (Hsiao and Fraenkel, 1968; Ma, 1972; Yamamoto and Fraenkel, 1960) or larger blocks of the medium used (Ma, 1972). Another material used as a substrate for phagostimulant studies is styropor, a foamed polystyrene. Thin lamellae are cut from styropor plates and discs (Meisner and Ascher, 1968) or rectangular laminae (Meisner et al., 1972) are prepared. Akeson et al. (1967) used root discs from the normal host plant of the insects under test as the substrate and Harris and Mohyuddin (1965) used vacuum infiltrated leaves.

Artificial substrates such as styropor and filter paper have the advantage for phagostimulatory studies of being botanically inert and uniform, but are possibly disadvantageous in having a texture less acceptable to the test insect than natural substrates. Also, lipid solvents cannot be used with styropor as a substrate. Elder pith, root discs and vacuum infiltrated leaves are likely to have a more acceptable texture, though leaves may lack uniformity and vary in thickness.

The use of a substrate has been avoided altogether by some workers, the test substance being applied directly in solution to the mouthparts (Barton Browne et al., 1974; Harley and Thorsteinson, 1967).

The manner of presentation of the substrate depends on whether the test insect feeds on the surface or at the edge of leaves. For surface feeders discs may be placed freely on the floor of the arena, but for edge feeders the discs should be raised off the surface of the arena supported on pins, thus making them accessible to the insect.

The form in which the substrate is presented may affect feeding behaviour. Ascher and Meisner (1973) found that the thickness and density of styropor affected the extent of feeding by *Spodoptera littoralis* larvae, though no differences in feeding levels on *Locusta migratoria* on sucrose impregnated pith discs of 1 mm and 5 mm thickness were observed.

Ascher and Meisner (1973) investigated the effect on feeding levels of having water available during the experimental period with *Spodoptera littoralis* larvae and found that the amounts of styropor lamellae consumed were doubled with both strongly and weakly phagostimulatory sugars. With *Locusta migratoria*, however, feeding on sucrose impregnated pith discs was not enhanced when water was available during the experimental period.

Some workers moistened the discs at the start of the experiment, but this may cause complications due to the progressive drying of the discs.

Attempts to keep the discs permanently moist in experiments with *Locusta migratoria* were unsuccessful. In these, fine capillary tubes were inserted through the base of the arena at the test disc locations into beakers of distilled water or sucrose solution. The supporting pin of each disc was inserted through the capillary tube so that the centre of each disc made contact with the tip of the capillary tube. The liquid in the beaker was drawn up the tube by capillarity and spread evenly over the surface of the disc. Initially distilled water was used with discs previously impregnated with 0.125 M sucrose, but autoradiography with ^{14}C labelled glucose showed that all the labelled glucose was concentrated at the periphery of the discs. Water spreading outwards from the centre of the discs, as it did throughout the experiment due to surface evaporation, eluted the sugar to the periphery, thus once the edges had been eaten the discs became effectively control discs and little more was eaten. When a 0.125 M sucrose solution was used instead of water, the sugar on the discs became progressively more concentrated throughout the experiment due to evaporation from the surface. Thus it was only satisfactory to present the discs dry.

APPLICATION OF THE TEST MATERIAL

Where pith discs and filter paper are used as the substrate the test material may be impregnated by soaking the discs in a solution of the test material for a constant period, or by pipetting a volume of the test solution on to individual discs. The latter method has the advantage that the amount of material applied to each disc is constant, whereas by soaking this may vary according to the thickness of the discs and the extent to which they are drained or blotted before being dried. In fact, the sucrose in discs soaked in a 0.125 M solution varied from 1.7 to 3.2 mg/disc. With either method, autoradiographic studies with ^{14}C labelled glucose showed the material to be evenly distributed over the discs.

Where agar/cellulose is used as the substrate, the test material is added to the agar/cellulose mixture before cooling and so is evenly distributed in the discs. Styropor lamellae are coated with the test substance in an ethanol/water solution.

PRETREATMENT OF THE TEST INSECTS

Some workers deprive their test insects of food prior to testing. This is necessary where the duration of the experiments is short to ensure a uniform nutritional condition but the possibility of a change in behaviour must be considered. Ascher and Meisner (1973) found no difference in feeding levels on sucrose impregnated styropor after being deprived of food for 6 hours for *Spodoptera littoralis*. The same was true of *Locusta migratoria* after 24 hours without food for both strongly and weakly phagostimulatory sugar impregnated pith discs.

Akeson et al. (1967) investigating feeding stimulants and deterrents for the sweet clover weevil, deprived the test insects of food for 48 hours, then allowed them to feed for 24 hours prior to testing on pith discs impregnated with a crude host plant extract, to condition the insects to the bioassay.

EXPERIMENTAL DESIGN

Most frequently, a test disc and a control disc are introduced into each arena, though some workers, finding that the test insects never feed on the control discs, use only test discs. With larger arenas more discs need to be used to ensure that the insect will come into contact with them. Here control and test discs are pinned alternately at the periphery of the arena. In multi-choice experiments, several discs with different treatments are placed in each arena.

Usually one insect is introduced into each arena, but some workers have used groups of insects especially where small insects are being tested, to obtain a more easily measurable feeding response.

The duration of the experiments is commonly a compromise between the time taken for the insect to eat an easily measurable amount of the substrate and one which is convenient to the observer, and has varied from 2 hours (Ritter, 1967) to 4 days (Wensler and Dudzinski, 1972).

Since the substrate is usually not a natural one for the insect, it is preferable to establish the duration of the tests on the basis of observed changes in behaviour over varying periods. With *Locusta migratoria*, time lapse films of five individual nymphs were taken over 18-hour periods, with one frame exposed every 30 seconds. Two insects fed mainly in the first 10 hours while three others continued to feed in a similar pattern over the whole period (Fig. 1). It was therefore concluded that though a 10-hour period would be adequate there was no disadvantage in continuing for 18 hours.

4

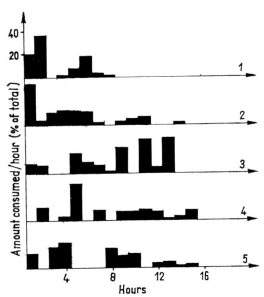

Fig. 1. The periodicity of feeding by *Locusta migratoria* on 0.125 M sucrose impregnated pith discs over an 18-hour period, as determined by time lapse cinephotography. Area loss of the discs was measured at hourly intervals for 5 insects and is expressed as a percentage of the total amount consumed

ASSESSMENTS OF PHAGOSTIMULATORY ACTIVITY
AND INTERPRETATION OF THE RESULTS

Methods adopted for assessing the activity of the test chemicals include weight loss of the substrate (Meisner et al., 1972) and area loss measured by graph paper (Akeson et al., 1967), planimeter (Heron, 1965) or photometer (Loschiavo, 1965; Soo Hoo, 1965; Wensler and Dudzinski, 1972). Area loss should be used only where the insects feed only at the edge as surface feeding will not be taken into account. Some workers have devised a subjective visual assessment (Hsiao and Fraenkel, 1968; Mehrotra and Rao, 1972; Ritter, 1967; Thorsteinson and Nayar, 1963). Faecal production has also been commonly used, either in terms of the number of pellets produced (Ascher and Meisner, 1973; Dadd, 1960) or the dry weight of the faeces (Ascher and Meisner, 1973; Dadd, 1960; Hsiao and Fraenkel, 1968; Ma, 1972). Faecal weight is a more accurate measure as Meisner et al. (1972) found that the faecal pellet size of *Spodoptera littoralis* larvae feeding on sugar impregnated styropor lamellae differed greatly for different sugars. LaPidus et al. (1963) assessed the attractiveness of test materials to the Mexican bean beetle by counting the number of feeding marks or ridges on impregnated filter paper. Attractiveness of test materials has also been determined by measuring the distribution of a group of insects on control and test discs at the end of the experiment or at intervals throughout the experiment (Beck and Hanec, 1958; Ito, 1961; Loschiavo, 1965) though this method assesses the responses to orientation stimuli rather than feeding stimuli. In experiments involving the direct application of materials to

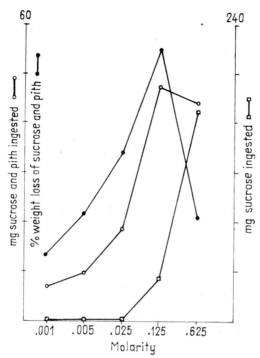

Fig. 2. Different methods of expressing the feeding responses of *Locusta migratoria* to sucrose impregnated pith discs over a logarithmic range of molarities

the mouthparts, attractiveness was measured by the number of drops of the liquid consumed or the increase in body weight (Barton Browne et al., 1974) or by a behavioural response such as drinking (Harley and Thorsteinson, 1967). In many cases, two or more of these criteria were measured during a series of experiments.

Several problems in interpretation of the results of these types of experiments have arisen during investigations into the phagostimulatory activity of sugars and amino acids for *Locusta migratoria*. Pith discs of 25 mm diameter and 1 mm thick were used as the substrate. They were soaked in the test solution for one hour, dried, and four test and four control discs were arranged alternately, supported by pins, at the periphery of a 30 cm diameter arena. Mid-fifth instar male *Locusta migratoria* nymphs which had not been deprived of food were used for the experiments, which ran for 18 hours at $28-30$ °C. Sugars and amino acids were tested over a logarithmic range of molarities and the amount eaten measured by weight loss of the pith discs.

Taking sucrose as an example, the amount of feeding assessed by weight loss of the pith disc plus sucrose increases with increasing molarity and reaches a peak at 0.125 M, then decreased at 0.625 M. However, if the amount of sucrose/mg pith is estimated at each concentration, it becomes clear that more sucrose is consumed at 0.625 M than at 0.125 M (Fig. 2). Thus is it the amount of the test chemical eaten in this type of experiment that is im-

4*

51

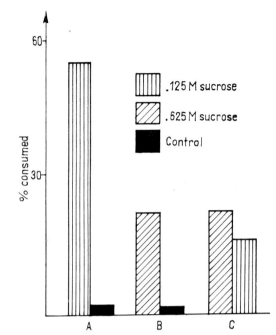

Fig. 3. Differences in feeding levels of *Locusta migratoria* on 0.625 M and 0.125 M sucrose impregnated pith discs in different experimental situations. In experiments A and B, the insect has a choice between one concentration of sucrose discs and control discs and in C it has a choice between the two concentrations of sucrose

portant, or how much the test chemical promotes feeding on the substrate? If phagostimulants are important in promoting ingestion of nutritionally important but non-stimulatory materials, the amount of pith that is eaten is important. Presumably different factors are causing the end of a feed on the discs of different concentrations. From the time-lapse films of *Locusta migratoria* feeding on 0.125 M sucrose pith discs it was estimated that up to 20 mg of pith was consumed in one hour, an intake sufficient to cause crop expansion (unpubl. results) and so meal size may be regulated in this way (Bernays and Chapman, 1973). On 0.625 M sucrose discs, only 20% of the discs are eaten and it is likely that osmotic or chemosensory factors control feeding.

Although 0.625 M sucrose discs appear to be less favourable than 0.125 M discs in terms of percentage weight loss when the insect has a straight choice between one concentration and the control, in a choice situation, where four 0.125 M and four 0.625 M discs were placed alternately in the arenas, the 0.625 M discs were eaten preferentially (Fig. 3). The amount of feeding on the 0.625 M discs was the same as in a 0.625 M/control experiment, whereas the amount of feeding on the 0.125 M discs was reduced to 29% of that recorded in a 0.125 M/control experiment. Probably sensory adaptation after feeding on the higher concentration discs influences the subsequent feeding response to the lower concentration discs.

The results from these experiments demonstrate that more care is needed than is commonly given to experimental design and interpretation if the results from different experiments and for different insects are to be comparable.

REFERENCES

AKESON, W. R., MANGLITZ, G. R., GORZ, H. J. and HASKINS, F. A. (1967): A bioassay for detecting compounds which stimulate or deter feeding by the sweet clover weevil. *J. econ. Ent.* **60**, 1082–1084.

ASCHER, K. R. S. and MEISNER, J. (1973): Evaluation of a method for assay of phagostimulants with *Spodoptera littoralis* larvae under various conditions. *Ent. exp. & appl.* **16**, 101–114.

BARTON BROWNE, L., MOORHOUSE, J. E. and VAN GERWEN, A. C. M. (1974): Sensory adaptation and the regulation of meal size in the Australian plague locust *Chortoicetes terminifera*. (In press.)

BECK, S. D. and HANEC, W. (1958): Effect of amino acids on feeding behaviour of the European corn borer, *Pyrausta nubilalis* (Hubn). *J. Insect Physiol.* **2**, 85–96.

BERNAYS, E. A. and CHAPMAN, R. F. (1973): The regulation of feeding in *Locusta migratoria:* Internal inhibitory mechanisms. *Ent. exp. & appl.* **16**, 329—342.

DADD, R. H. (1960): Observations on the palatability and utilization of food by locusts, with particular reference to the interpretation of performance in growth trials using synthetic diets. *Ent. exp. & appl.* **3**, 283–304.

GOODHUE, D. (1963): Feeding stimulants required by a polyphagous insect, *Schistocerca gregaria. Nature, Lond.* **197**, 405–406.

HARLEY, K. L. S. and THORSTEINSON, A. J. (1967): The influence of plant chemicals on the feeding behaviour, development, and survival of the two-striped grasshopper, *Melanoplus bivittatus* (Say), Acrididae: Orthoptera. *Can. J. Zool.* **45**, 305–19.

HARRIS, P. and MOHYUDDIN, A. I. (1965): The bioassay of insect feeding tokens. *Can. Ent.* **97**, 830–833.

HERON, R. J. (1965): The role of chemotactic stimuli in the feeding behaviour of spruce budworm larvae on white spruce. *Can. J. Zool.* **43**, 247–269.

HSIAO, T. H. and FRAENKEL, G. (1968): The influence of nutrient chemicals on the feeding behaviour of the Colorado potato beetle, *Leptinotarsa decemlineata* (Coleoptera: Chrysomelidae). *Ann. ent. Soc. Am.* **61**, 44–54.

ITO, T. (1961): Effect of dietary ascorbic acid on the silkworm *Bombyx mori. Nature, Lond.* **192**, 951–952.

LAPIDUS, J.B., DAVIDSON, R. H., FISK, F. W. and AUGUSTINE, M. G. (1963): Chemical factors influencing host selection by the Mexican bean beetle *Epilachna varivestis* Muls. *J. agric. Food Chem.* **11**, 462–463.

LOSCHIAVO, S. R. (1965): Methods for studying aggregation and feeding behaviour of the confused flour beetle, *Tribolium confusum* (Coleoptera: Tenebrionidae). *Ann. ent. Soc. Am.* **58**, 383–388.

MA, W. C. (1972): Dynamics of feeding responses in *Pieris brassicae* Linn. as a function of chemosensory input: a behavioural, ultrastructural and electrophysiological study. *Meded. LandbHoogesch. Wageningen*, **72–11**, 162.

MEHROTRA, K. N. and RAO, P. J. (1972): Phagostimulants for locusts: studies with edible oils. *Ent. exp. & appl.* **15**, 208–212.

MEISNER, J. and ASCHER, K. R. S. (1968): A method to assay the phagostimulatory effect towards insects of plant extracts applied to styropor discs. *Riv. Parassit.* **29**, 74–77.

MEISNER, J., ASCHER, K. R. S. and FLOWERS, H. M. (1972): The feeding response of the larvae of the Egyptian cotton leafworm, *Spodoptera littoralis* Boisd., to sugars and related compounds. 1. Phagostimulatory and deterrent effects. *Comp. Biochem. Physiol.* **42**, 899–914.

NORRIS, D. M. and BAKER, J. E. (1967): Feeding responses of the beetle *Scolytus* to chemical stimuli in the bark of *Ulmus. J. Insect Physiol.* **13**, 955–962.

RITTER, F. J. (1967): Feeding stimulants for the Colorado beetle. *Meded. Fac. Landb. Wet. Rijksuniv. Gent.* **32**, 291–327.

Soo Hoo, C. F. (1965): A light sensitive apparatus for the rapid measurement of experimental feeding by insects. *Bull. ent. Res.* **56**, 227–229.

THORSTEINSON, A. J. and NAYAR, J. K. (1963): Plant phospholipids as feeding stimulants for grasshoppers. *Can. J. Zool.* **41**, 931–935.

WENSLER, R. J. and DUDZINSKI, A. E. (1972): Gustation of sugars, amino acids and lipids by larvae of the Scarabeid, *Sericesthis geminata* (Coleoptera). *Ent. exp. & appl.* **15**, 155–165.

YAMAMOTO, R. T. and FRAENKEL, G. (1960): Assay of the principal gustatory stimulants for the tobacco hornworm, *Protoparce sexta* from Solanaceous plants. *Ann. ent. Soc. Am.* **53**, 499–503.

Symp. Biol. Hung. 16, pp. 55–60 (1976)

SOME NEW ASPECTS OF HOST-PLANT RELATION TO BEHAVIOUR AND REPRODUCTION OF SPIDER MITES (ACARINA: TETRANYCHIDAE)

by

Z. T. DĄBROWSKI

DEPARTMENT OF APPLIED ENTOMOLOGY, AGRICULTURAL UNIVERSITY OF WARSAW, 02-766 WARSZAWA–URSYNOW, POLAND

Various plant species are not equal in their acceptance by, and suitability for *Tetranychus urticae* Koch and *Panonychus ulmi* (Koch). The behaviour of females of these mites was studied using leaf disks 15 mm in diameter. One female was placed onto the centre of each disk and its manner of movement was observed and drawn on paper. The times of initial feeding and of permanent feeding were measured and recorded on the drawing indicating the points of feeding. Other group of test involved using ^{32}P incorporated to water solution of various plant chemicals.

The differences observed in mite development and especially in fecundity probably depend on the nutritional value of the species studied for the spider mites. Some plant species have been gustatorily accepted by the mites, however, the development of *P. ulmi* larvae and nymphs was restrained.

The hypothetical pattern of plant finding and acceptance by *T. urticae* females could be described on the basis of the author's own results as well as data found in the literature on the role of some ecological factors influencing the spider mite behaviour in natural conditions.

INTRODUCTION

The fundamental basis of host-plant–spider mite relationships, including host specificity and resistance, is a new field of study in acarology. Our present knowledge of these relationships, with few exceptions, has fallen behind the information available for insects. Most of the data published in acarological literature deals with the influence of the mineral fertilization of plants upon the tetranychid fecundity.

The main objective of the author's present investigation was to determine the mechanism of food plant acceptance by *Tetranychus urticae* Koch and by *Panonychus ulmi* (Koch) and the effect of some plant species or varieties upon the biology and behaviour of these two mite species.

HOST-PLANT CHOOSING AND ACCEPTANCE

The observation of the behaviour of hungry *T. urticae* females both on leaf disks of certain plant species and the comparison of feeding rates of females reared on some plant extracts and homogenates as well as on water solutions of various chemicals enabled to conclude that the feeding behaviour of *T. urticae* females involves a sequence of behavioural components, including: (a) initial piercing, (b) initiation of feeding, and (c) maintenance of feeding (Dąbrowski, 1973a). On acceptable host-plants the period between transferring the hungry mites onto a leaf and the start of initial feeding and later permanent feeding was short, as for example on *Phaseolus vulgaris* L. foliage. On non-preferred plant species the phases were much longer and permanent feeding began only after several initial piercings.

Unpublished observations of Dąbrowski and Bielak on *Panonychus ulmi* behaviour confirmed those observations for other species of tetranychids. Leaves of *Malus domestica* Borkh., *Prunus domestica* subsp. *syriaca* (Borkh.) Jan., *Prunus avium* L. and *Phaseolus vulgaris* L. were the species well accepted gustatorily by the females. Some of them, however, did not provide suitable food for the growth and development of the preimaginal instars of *Panonychus ulmi* (*Phaseolus vulgaris* L. and *Rubus idaeus* L.). Some of other plants possess both the attributes that supply all the required sorts of feeding stimuli and nutrients. They were quickly gustatorily accepted by *P. ulmi* females and the rate of development was high (Table 1). Most of the plant species which were rejected by *P. ulmi* females provided also poor nutritional conditions for mite development. The abnormal long movement of mites on leaf disks cut from some plants could suggest that the plants possess some compounds that inhibited the feeding of *P. ulmi*.

Such inhibitors or even feeding suppressants have been disclosed by Dąbrowski and Rodriguez (1972) and by Dąbrowski (1973b, c) for *Tetranychus urticae*. L-alanine, used in 1% concentration and kaempferol, D-catechin, p-hydroxyphenylacetic acid, 3,4,5-trihydroxybenzoic acid, p-hydroxybenzoic acid in 10^{-3} M concentration acted as feeding suppressants. L-glycine, L-aspartic acid, L-glutamic acid, L-serine, L-threonine, L-cystine and L-proline in 0.1% solution and 10^{-4}—10^{-6} M concentrations

TABLE 1

Plant species tested offering the best conditions for development, fecundity and survival of Panonychus ulmi (Koch)

Plant species	Population that reached adulthood (%)	Plant species	Fecundity (avg. no. of eggs/female/week)	Plant species	Female survival after a week (%)
Prunus domestica L.	46	*Prunus armeniaca* L.	8.0	*Rosa polyantha* cv. Joseph Guy	20
Rosa polyantha cv. Cocorico	48	*Prunus persica* Batsch.	8.0	*Crataegus monogyna* Jacq.	25
cv. Lilli Marlen	50	*Rosa polyantha* cv. Joseph Guy	8.7	*Rosa polyantha* cv. Moulin Rouge	25
cv. Concerto	52	*Sorbus aucuparia* L.	9.0	cv. Fashion	40
Prunus armeniaca L.	58	*Prunus domestica* subsp. *syriaca* (Borkh.) Jan.	9.3	*Prunus domestica* L.	40
Rosa polyantha cv. Elysium	66	*Rosa polyantha* cv. Fashion	14.9	*Prunus domestica* subsp. *syriaca*	45
Prunus domestica subsp. *syriaca*	70	*Prunus domestica* L.	16.7	*Prunus avium* L.	45
Rosa polyantha cv. Irene of Danemarc, cv. Fashion	72	*Prunus avium* L.	18.2	*Rosa polyantha* cv. Irene of Danemarc	40
Prunus padus L.	80	*Prunus padus* L.	21.4	*Prunus padus* L.	85

of most phenolic compounds tested acted as feeding deterrents of *T. urticae* females.

On the other hand, low concentrations of the amino acids tested stimulated the feeding rate of *T. urticae*. The response threshold, however, depends upon the kind of amino acid and upon changes in concentration below 1% and to 0.1 or even below 0.01%. L-alanine and L-arginine used in 0.1% solution, L-serine, L-threonine, L-glutamic acid, and L-histidine tested in 0.01% and L-leucine, L-isoleucine, L-lysine, L-phenylalanine and L-tyrosine used in both concentrations elicited stimulation of feeding. None of the amino acids tested evoked as high a feeding rate as was observed in the case of a homogenate or water extract of bean leaves (Dąbrowski, 1973a). Therefore the author started to test the gustatory response of *T. urticae* females to some carbohydrates. It was found that most of the sugars tested had a stimulatory effect upon the feeding rate and in some cases the degree of stimulation was much higher than that described for amino acids. More than a quadruple increase in food ingestion as measured by female radioactivity (79.8–47.1 counts/female/min) was observed after 24 hours of feeding on the following treatments: 1% sucrose, 2% sucrose, 3% sucrose, 2% fructose, and 1% fructose as compared to the feeding rate on the distilled water used herein as the control. Of the nine sugars tested, only one, cellobiose, at 3 and 2% concentrations had a significant deterrent effect upon feeding. Some stimulation was also elicited by certain vitamins soluble in water (Dąbrowski, unpublished).

Because most of the compounds acting as feeding stimulants generally occur in tissues of many plant species, they have not been described as specific phagostimulants governing plant acceptance or rejection by *T. urticae* mites. The statement has been also made that the permanent feeding of the mites occurs rather on the leaves without compounds which acted as feeding deterrents than on leaves possessing one specific phagostimulus. The phenomenon is known for food acceptance by some insect species (Jermy, 1968).

It was also demonstrated that various volatile substances as olfactory stimuli given off by foliage could play some role in the acceptance or rejection mechanism of host-plant by *T. urticae* females (Dąbrowski and Rodriguez, 1971). The response, however, after Dąbrowski's (1974) more extensive observation, has been classified as a kinesis and it was weakly expressed in the mite's behaviour. Such factors as light, humidity, surface gradient significantly modified the female response toward the source of odour.

In the case of *P. ulmi* it was also stated that the long distance between the point where the females were placed onto the leaf disk and the point of the initial piercing could suggest that the explanation of the mechanism of food acceptance by the females could not be limited only to gustatory stimuli emanated from leaves.

The food value of a plant species or variety for the spider mites was measured by the survival rate of preimaginal and adult stages and by the reproductive capacity. Even such polyphagous species as *Tetranychus urticae* increased its fecundity only on some plants (*Rosa dilecta* Rehd., *Heliopsis scabra* Dunal, *Phaseolus vulgaris* L., *Malus domestica* Borkh.). More distinct differences in development or fecundity were found for *Panonychus ulmi*, even between species belonging to the *Rosaceae* family. The most suitable

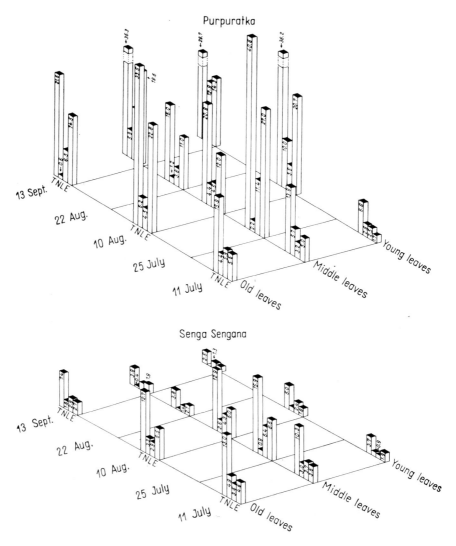

Fig. 1. The effect of leaf age and season upon the fecundity of *T. urticae* females after 7 days on leaf disks of two strawberry cultivars. Average number of: E — eggs, L — larvae, N — nymphs/female, T — total number of preimaginal instars/female

feeding conditions for *P. ulmi* were provided by leaves of plant species listed in Table 1. The differences could be only affected by the nutritional values of plants.

Our knowledge on effects of chemical composition of the host upon the spider mites is still fragmentary. Dąbrowski et al. (1975) compared the composition of free amino acids and sugars in apple and bean foliage. Arginine and ornithine were found only in the apple foliage, and tryptophane only in the bean leaves. It must be added that both leaves were well gustatory and accepted by *P. ulmi* females, but the development of most of the mites was restricted on the bean foliage.

The differences in the fecundity of *P. ulmi* on various apple varieties also could suggest the existence of nutritional divergences between them. Dąbrowski and Rejman (unpublished) described the leaf colonization and damages caused by *P. ulmi* on six apple varieties grafted on two different rootstocks under field conditions in Poland. They found in the average 16.0 eggs and mites per leaf on Jonathan cultivar grafted on 106 rootstock; on Spartan = 32.8; on McIntosh = 37.3; on Bankroft = 48.3; on Fantazja = 48.8 and on Red Delicious = 70.4. The Fantazja variety has been selected some years ago by Rejman from seedling of McIntosh × Linda cross progeny. The density of mites was similar for trees grafted on two rootstocks used, with one exception of Bankroft variety.

Similar great differences in the varietal suitability of crop plants for spider mites were found for *Tetranychus urticae*. Our unpublished data showed that variety, age of leaf and season could significantly affect mite fecundity. Generally, lower fecundity was observed mostly on young leaves of all cultivars tested (Fig. 1) than on old or moderately old leaves. It might suggest that the leaves did not supply all the nutrients essential for mite development and for egg formation. A high content of dissolved chemicals in young leaves could also deter the mites from feeding on them.

The differences in the fecundity of *T. urticae* females on the leaves of strawberry varieties tested were significant. The lowest fecundity rate was found during all periods for females fed on foliage of Macherauch's Frühernte, and the highest on Purpuratka (American old cultivar, unknown origin). The range of suitability of Senga Sengana, Talizman and Regina for *T. urticae* lies between the two cultivars mentioned above. The differences could be partly affected by various levels of nitrogen content in the foliage of cultivars tested. The 2.03% of total nitrogen and the 1.98% of protein nitrogen was identified in Purpuratka foliage and only 1.56% and 1.49%, respectively in Senga Sengana leaves.

Our present work is designed to find an explanation for the differences observed in the acceptability of plants with special emphasis on the chemical content (amino acids, carbohydrates and phenolic components) and anatomy of leaves of certain strawberry, apple and rose cultivars.

REFERENCES

DĄBROWSKI, Z. T. (1973a): Studies on the relationships of *Tetranychus urticae* Koch and host plants. II. Gustatory effect of some plant species. *Pol. Pismo Ent.* **43**, 127–138.

DĄBROWSKI, Z. T. (1973b): Studies on the relationships of *Tetranychus urticae* Koch and host plants. III. Gustatory effect of some amino acids. *Pol. Pismo Ent.* **43**, 309–330.

DĄBROWSKI, Z. T. (1973c): Studies on the relationships of *Tetranychus urticae* Koch and host plants. IV. Gustatory effect of some carbohydrates. *Pol. Pismo Ent.* **43**, 521–533.

DĄBROWSKI, Z. T. (1974): Basis of host plant choosing and acceptance by the two-spotted spider mite *Tetranychus urticae* Koch (Acarina: Tetranychidae). *Zesz. Nauk. Akad. Roln. Warsz.* **37**, 125.

DĄBROWSKI, Z. T. and RODRIGUEZ, J. G. (1971): Studies on resistance of strawberries to mites. 2. Preference and nonpreference responses of *Tetranychus urticae* and *T. turkestani* to essential oils of foliage. *J. econ. Ent.* **64**, 387–391.

DĄBROWSKI, Z. T. and RODRIGUEZ, J. G. (1972): Gustatory responses of *Tetranychus urticae* Koch to phenolic compounds of strawberry foliage. *Zesz. Probl. Post. Nauk. Roln.* **129**, 69–78.

DĄBROWSKI, Z. T., NIRAZ, S. and STANKIEWICZ, Cz. (1975): The composition of free amino acids and sugars of apple and bean foliage and their effect upon the feeding behaviour of *Tetranychus urticae* Koch. In: Agricultural Acarology in Poland, *Zesz. Probl. Post. Nauk Roln.* (in press)

JERMY, T. (1968): On some behavioural aspects of host specificity in phytophagous insects. *Proc. XIIIth Int. Congr. Ent.* **1**, 330–331.

Symp. Biol. Hung. 16, pp. 61–65 (1976)

THE OVIPOSITION OF THE INDIAN MEAL MOTH (*PLODIA INTERPUNCTELLA* HBN., LEP., PHYTICIDAE) INFLUENCED BY OLFACTORY STIMULI AND ANTENNECTOMY

by

K. V. Deseő

RESEARCH INSTITUTE FOR PLANT PROTECTION, H–1525 BUDAPEST, PF. 102, HUNGARY

Fecundity is increased in the Indian meal moth by the odour of certain nutrients-the effect of which is independent of nutritional experiences of the parental genera tions. The eggs are laid in batches near the source of the odour.

Following 50% antennectomy, the olfactory stimuli lose their effect and fecundity becomes the same as if there were no stimulation. Thus, increase of fecundity seems to be due to the response of receptors on the distal part of antennae. After more than 50% antennectomy fecundity is further reduced.

The change in the scent-orientation occurs only in case of total antennectomy. Thus, for the choice of oviposition-site the olfactory receptors on the proximal part of the antennae seem to be responsible, the role of receptors on tarsi is not excluded either. Thus, scent acts both on fecundity, and choice of egglaying place, however, changes in olfactory input differently affect the two processes.

In case of total antennectomy mating activity is also reduced. This and the amputation of one antenna to 75% which reduces fecundity significantly suggest that all the changes in oviposition are due to inhibition elicited by antennectomy to more than 50%. Whether this inhibition has a traumatic character must be investigated further.

INTRODUCTION

Plodia interpunctella Hbn. is a cosmopolitan insect, harmful in various stored products. Richards and Thomson (1932) listed 83 different kinds of food in which larvae fed. Females lay their eggs on the nutrients where larvae develop. Pupation may take place in these diets but mostly in the surrounding. Diapause in prepupae was observed only recently in an African population (Prevett, 1971) but it could be induced by short illumination (Bell and Walker, 1973). The effect of photoperiod on fecundity during preadult development was shown by Lum and Flaherty (1970) and the influence of food quality, temperature, and relative humidity was observed by Abdel-Rahman (1971).

Some of our earlier publications dealt with the oviposition of the Indian meal moth (Deseő, 1970a, 1970b). It was established that one female lays about 80 eggs but there always remain about 168–480 ripe eggs in the abdomen. The odour of certain nutrients stimulated egglaying, so fecundity increased. The scent of the adequate nutrients influenced also the site of egglaying; the eggs were found mostly in batches and around the source of the odour, whereas in the jars without nutrients these were scattered singly all over the jar.

The starting point was to answer the question whether the increase in fecundity by the odour is due to the preference to the nutrient in which the adults as larvae had developed, and whether the same receptors are involved in stimulating oviposition and choosing the egglaying place.

MATERIALS AND METHODS

Indian meal moth eggs were placed on wheat flour, chocolate, and pea-nuts. These populations were continuously reared half year long in these diets. Diapause did not occur. After 5–6 generations three pairs of the newly emerged adults in 25 repetitions were placed in jars supplied with the three types of diet. The insects could not touch their diets but perceived the odour as described earlier (Deseő, 1970a; Fig. 1). The eggs laid as well as the places of the batches were counted, and noted, respectively.

Antennectomy to different degrees was performed on chilled females. The exact performance of amputation to certain degrees was impossible, because the antennal segments are very numerous (98) and very small. Therefore, antennectomy was performed only to about 50% and about 75% of the segments. In case of 100% antennectomy the basal joint still remained. To state whether the change in behaviour was caused by the amputation or was due to the lack of receptors, antennectomy was carried out only on one antenna. Three males with three antennectomized females were kept in jars supplied with the odour of peanut. Dead females were dissected and matings were verified on the presence of spermatophores.

RESULTS AND DISCUSSION

I. The mean fecundity of females developed through many parental generations in chocolate is shown in Table 1. The data indicate that the fecundity of females kept in jars supplied with peanut odour were signif-icantly higher than that of those kept with the odour of chocolate or flour. Thus, this observation confirms our earlier statement, according to which fecundity is stimulated by the odour of special nutrients, so the increased fecundity is independent of the experiences by the larvae of parental gen-erations. Furthermore, it has been found that the sweet smell of fig and corn-meal had the most efficient stimulating effect on oviposition (Deseő, 1970a).

TABLE 1

*The effect of odour of different nutrients on the fecundity
of the Indian meal moth*

Nutrient	No. of eggs laid/1 female	t-test
Peanut	69.07 ± 2.35	$p < 0.05$
Chocolate	59.65 ± 3.77	$p < 0.001$
Meal	54.94 ± 1.29	$p < 0.1$

II. Figure 1 shows the effect of the different degrees of antennectomy on fecundity. The dissection of females indicated that mating was influenced only by whole antennectomy (including basal joint); in this case sperma-tophores could not be found in about 50% of females.

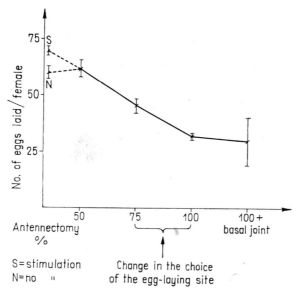

Fig. 1. The effect of different degrees of antennectomy on the fecundity
of the Indian meal moth

These data show that when about 50% of the antennal segments were
cut off, the stimulation by the odour ceased, although the difference between the values of fecundity is not significant, presumably due to the
difficulty mentioned above in the exact performance of 50% antennectomy.
But anyway, the fecundity in case of females antennectomized to 50%
remains "normal", there is no difference between this and the number of
eggs laid by females without stimuli (see Table 1, chocolate). However, we
can conclude that the receptors on the distal half of antennae are necessary
in the response to stimulate oviposition.

In case of 50% antennectomy performed only on one antenna, the fecundity was 74.5 ± 5.6 eggs per female. This means that the receptors on
the distal half of the antennae are responsible for the perception of olfactory
stimuli. Thus, not the number of the receptors, but the distal receptors
themselves are important. In case of more than 50% antennectomy (75%,
"100%") fecundity became already significantly reduced in comparison
with normal fecundity ($p < 0.1$, $p < 0.01$).

III. In Table 2 the effect of antennectomy on the choice of egglaying site
and on the pattern of oviposition is shown.

The data indicate the places of the eggs laid: whether they are near to the
source of the odour or are scattered singly all around in the jar. Even in the
case of 75% antennectomy there cannot be seen any change in the choice
of the egglaying site: more than half of the egg output is near the holes
made on the paper cups covering the diet.

In our earlier experiment (Deseő, 1970a) none of the eggs were laid near
the holes in jars without any nutrients. This fact indicates that tactile
stimuli, represented by the holes, did not affect the choice of the oviposition place. So the odour of nutrients determine the choice of oviposition

63

TABLE 2

*The effect of different degrees of antennectomy
on the choice of egglaying place of the Indian
meal moth*

Antennectomy	Near the source in batches (%)	Scattered singly all over the jar (%)
50%	57.36	42.64
75%	57.00	43.00
100%	24.30	75.70
100% + basal joint	7.88	92.12

site even in case when fecundity becomes already reduced. Perhaps in the choice of adequate ovipositing place olfactory receptors on other organs, e.g., on the tarsi may play a role, but changes in olfactory input differently affect by all means the two processes.

IV. In case of whole antennectomy the females lost their ability for finding the adequate egglaying place, furthermore the normal pattern of egglaying changed and even the receptivity for mating was blocked in 50% of the females. The question arises that perhaps antennectomy itself has an inhibiting effect on the oviposition as well.

The control experiment in which only one antenna was cut off to 75% indicated that this amputation resulted already in decrease of fecundity. The value of fecundity (42.00 ± 4.7 eggs per female) was significantly lower than "normal" fecundity (59.65 ± 3.77).

However, in other insect species, the effect of antennectomy on the oviposition differed from that of *P. interpunctella*. In *Acanthoscelides obtectus* Say (Pouzat, 1969) 100% antennectomy elicited egglaying. So, antennae seemed to exert an inhibitory influence on oviposition in that bean weevil population. In the Hungarian strain of the same species Szentesi (1975, in this volume) found that antennectomy reduced fecundity. In *Ceutorrhynchus maculaalba* Hbst. 100% antennectomy entirely inhibited oviposition, for egglaying at least one segment was necessary (Sáringer, 1970).

In *Grapholitha funebrana* Tr. (Lep., Tortricidae) antennectomy had somewhat similar effect on the oviposition as in the Indian meal moth. For the beginning of oviposition this species requires olfactory stimuli, but for normal oviposition also contact chemical stimuli are necessary (Deseő, 1967). When antennectomy was performed to 50%, 61% of the females did not begin oviposition (in the control only 14%) and fecundity was also reduced by 75% (Deseő, unpublished data). This latter observation suggests that the amputation inhibits the influence of both the olfactory stimuli necessary for the beginning of oviposition and the contact chemical stimuli important for normal egg output.

The reason of the restrained oviposition is not yet cleared up, it can be due to the lack of the antenna or, perhaps, to the trauma elicited by the amputation.

REFERENCES

ABDEL-RAHMAN, A. H. (1971): Some factors influencing the abundance of the Indian meal moth, *Plodia interpunctella* Hbn. on stored shelled corn. *Bull. Soc. Ent. Egypte* **55**, 321–330.

BELL, C. H. and WALKER, D. J. (1973): Diapause induction in *Ephestia elutella* Hbn. and *Plodia interpunctella* Hbn. (Lep., Pyralidae) with a dawn-dusk lighting system. *J. stored Prod. Res.* **9** (3), 149–158.

DESEŐ, K. V. (1967): The role of olfactorial stimuli in the egglaying behaviour of the plum moth (*Grapholitha funebrana* Tr.). *Acta Phytopath. Acad. Sci. Hung.* **2**, 243–250.

DESEŐ, K. V. (1970a): Az illathatás szerepe az aszalványmoly (*Plodia interpunctella* Hbn., Phyticidae) tojásprodukciójában (The role of olfactory stimuli in the egg-production of the Indian meal moth [*Plodia interpunctella* Hbn.]). *Növényvédelem* **6** (10), 461–464.

DESEŐ, K. V. (1970b): The effect of olfactory stimuli on the oviposition behaviour and egg production of some microlepidopterous species. In: *L'influence des stimuli externes sur la gamétogenèse des insectes.* Colloques International du CNRS No. **189**, 162–174.

LUM, P. T. M. and FLAHERTY, B. R. (1970): Regulating oviposition by *Plodia interpunctella* in the laboratory by light and dark conditions (Lep., Phyticidae). *J. econ. Ent.* **63**, 236–239.

POUZAT, J. (1970): Rôle des organes sensoriels céphaliques dans l'ovogenèse et l'émission chez la bruche du haricot *Acanthoscelides obtectus* Say. In: *L'influence des stimuli externes sur la gamétogenèse des insectes.* Colloques International du CNRS No. **189**, 381–400.

PREVETT, P. F. (1971): Some laboratory observations on the development of two African strains of *Plodia interpunctella* Hbn. (Lep., Phyticidae) with particular reference to the incidence of diapause. *J. stored Prod. Res.* **7**, 253–260.

RICHARDS, O. W. and THOMSON, W. S. (1932): Contribution to the study of the genera *Ephestia* and *Plodia. Roy. Ent. Soc. London. Trans.* **53**, 169—246.

SÁRINGER, GY. (1970): Role played by the contact receptors of antennae in the egg-laying process of *Ceuthorrhynchus macula-alba* Herbst (Col., Curculionidae). *Acta Agron. Acad. Sci. Hung.* **19**, 393–394.

SZENTESI, Á. (1975): The effect of amputation of head appendages on the oviposition of the bean weevil, *Acanthoscelides obtectus* Say (Coleoptera: Bruchidae). *Symp. Biol. Hung.* **16**, 275–281.

Symp. Biol. Hung. 16, pp. 67–70 (1976)

THE IMPORTANCE OF STIMULUS PATTERNS FOR HOST-PLANT RECOGNITION AND ACCEPTANCE

by

V. G. Dethier

DEPARTMENT OF BIOLOGY, PRINCETON UNIVERSITY,
PRINCETON, N.J. 08540, USA

Studies of lepidopterous larvae have indicated that at least for these plant feeding insects host-plant recognition and acceptance are based upon complex mixed olfactory and gustatory sensory information. Although a particular compound or category of compounds may dominate the chemical composition of a plant and may contribute the major sensory cue, it is the total chemical complex that forms the basis for perception. The receptor cells that are responsive to the vapors and solutions of which the plant consists are not narrowly specific in their sensitivities. Accordingly they transmit to the central nervous system a vast amount of information that forms the basis for central integration. Thus caterpillars are able to appreciate "flavor" in an analogous way that vertebrates sense it. Electrophysiological studies of the olfactory and gustatory receptors of several species of caterpillars, especially *Danaus plexippus*, species of *Papilio*, and *Malacosoma americana* indicate that these receptors are sensitive to a very wide variety of plants both within and without the normal host range. This paper is concerned primarily with electrophysiological responses of olfactory receptors to natural plant odors.

The process of finding a plant upon which to feed or oviposit, of biting it, of ingesting and swallowing it, is governed at each step by various arrays of sense organs of which the chemoreceptors play a most important role. Stimulation of the olfactory and gustatory receptors is accomplished by one or more volatile or non-volatile chemicals elaborated by the plant. Stimulation by some of the compounds (deterrents) prevents oviposition, biting, or ingestion. Stimulation by others (attractants, arrestants, feeding stimulants, ovipositional stimulants) facilitates orientation to the plant, arrests locomotion, initiates or drives oviposition, biting, or ingestion. Some of these are of nutritional value (e.g., sucrose); others may not be (e.g., sinigrin). The non-nutritious compounds, commonly referred to as token stimuli, secondary plant substances, allochemics, have been conceived of as acting in the ethological sense of sign stimuli. That is, it is generally held that a certain specific compound among all those present in the plant may be sufficient to initiate one or more of the steps in the ovipositional or feeding sequence or, similarly, that a single compound acting as a deterrent might block such behaviour. It has also been held that the two categories of compounds may act in concert. Compounds in the first category have been thought to be specific and more or less unique to those plants that are acceptable to a given insect. Deterrents are less specific, and many compounds may fill the role.

The classical example of a sign stimulus has been sinigrin. It serves as a model for other insect/plant relationships, and the evidence supporting the concept of which it is a model is impressive. Sinigrin occurs in most of the plants eaten by those oligophagous insects that select cruciferae; when

5*

applied to a neutral substrate or a normally non-preferred plant which does not contain overwhelmingly effective deterrents, it initiates or enhances feeding; a receptor has been described which is highly specific to sinigrin.

Another striking example is hypericin, a compound that is present in *Hypericum hirsutum*, the food plant of the beetle *Chrysolina brunsvicensis*. The beetle possesses a receptor which is especially sensitive to hypericin (Rees, 1969).

On the other hand, when electrophysiological analyses are made of the response characteristics of a number of gustatory and olfactory receptors of lepidopterous larvae to individual compounds and to freshly expressed (10–15 seconds old) sap of leaves or to the volatile emanations of intact as well as recently bitten leaves, the results are difficult to reconcile with the model exemplified by sinigrin and hypericin. For ease of presentation gustation and olfaction will be discussed separately.

Many of the results pertaining to gustation have recently been published (Dethier, 1973), so they will be reviewed but briefly. The responses of seven species of larvae to a variety of chemicals and to plant saps were analyzed. Although the gustatory receptors were not characterized as elegantly as Schoonhoven (1967) had done with the sinigrin receptor and Ishikawa (1963) with the receptors of *Bombyx mori*, the evidence suggests that the receptors are not rigidly specific. Each responds to a variety of compounds in a manner that is not constrained by chemical relationships. There is, for example, in some species a receptor that responds equally well to sinigrin and sodium chloride. These are not species for which sinigrin is a gustatory stimulant, and the possibility cannot be excluded that the receptor in question is a deterrent receptor responding to many compounds. Nonetheless there seemed to be no receptor exhibiting rigid specificity. The original paper should be consulted for details.

More compelling was the finding that stimulation with the saps of various plants, acceptable and unacceptable alike, caused all taste receptors to respond (as might be expected from the fact that the saps of leaves are complex chemical mixtures) but that the character of the overall pattern of response bore no orderly relationship to the acceptability or non-acceptability of the plants. Differences among plants in the acceptable series were as great as between acceptable and unacceptable plants.

Recently an extensive electrophysiological study (which will be reported elsewhere in detail) has been made of the olfactory responses of the eastern tent caterpillar (*Malacosoma americana*) to a variety of compounds in the vapor phase and to volatile constituents of living leaves at ambient concentrations. One example selected at random may serve as a model of the results obtained. In this case a tungsten electrode was inserted near a group of four olfactory receptors associated with the medial sensillum basiconicum of the antenna. The electrode was in an extracellular position and picked up responses from four cells. The amplitude of the action potential of each cell was unique so that activity from each could be identified and monitored over a recording period lasting three and a half hours during which time the electrode remained in the same position relative to the cells. In the absence of deliberate olfactory stimulation there was a low level of "spontaneous" activity which may or may not have truly represented the activity state of the cells in the resting state. In any case it served as a measure of the stability of the preparation because at the end of three and a half

68

hours of recording, during which the cells were stimulated forty-five times with fifteen different stimuli presented in random order, the unstimulated level of activity of each cell did not differ from the activity at the beginning of the experiment.

The results indicated that the cells were "generalists" (i.e., were sensitive to a wide variety of compounds and volatile emanations from plants), that no two were similar with respect to their spectra of response, and that for any given cell the responses were reproducible.

Considered together the results of analyses of gustatory and olfactory receptors strongly suggest that the receptors of the species studied are preferentially specific rather than rigidly specific and that the chemosensory system delivers to the central nervous system complex patterns of activity. It has been proposed that these patterns constitute the sensory picture that caterpillars receive of the plant and that acceptance, rejection, and preference are made on the basis of assessment of these patterns by the central nervous system (Dethier, 1973).

At the moment this hypothesis has several obvious weaknesses. First of all, the electrophysiological responses of chemoreceptors are highly variable. To what extent this variation is intrinsic or artefactual is not known (cf. Dethier, 1974, and Schoonhoven, this symposium). If the variation is of biological origin, patterns can be meaningful only if their identity and resolution are not obscured by noise. This point bears further investigation. Second, the fact remains that single compounds (e.g., sinigrin, hypericin, sucrose, etc.) can by themselves exert a powerful effect on behaviour. Third, knowledge of the character of sensory input cannot define the manner in which the plant affects the feeding and ovipositional behaviour of the insect partly because we know nothing of how the central nervous system "reads" and interprets the inflowing information, partly because we know little or nothing of the chemical moieties of the plant that causes the sense cells to respond as they do. Nonetheless, any concept of how the plant and the insect interact cannot ignore the nature of sensory input.

Consider again the case of sinigrin, the compelling example of a single compound affecting feeding. If there is a receptor that is specific for sinigrin and if there is a group of plants for which the common denominator is sinigrin, then it would seem that one compound is all that is required (insofar as chemistry alone is concerned) for discrimination and identification of plants. If some of the acceptable plants are preferred to others and if all contain the same concentration of sinigrin, then it is reasonable to assume that some other compound or compounds contribute to the behaviour. For example, there could be a balance between sinigrin and a deterrent in any proportion. This would be the simplest case. The balance could involve sinigrin, sucrose, and a deterrent in any proportion. The simplest case now begins to approximate a pattern especially if, in behavioural terms, a high concentration of sucrose could compensate for a low concentration of sinigrin in one case and vice versa in another.

The hypothesis that sign stimuli play a highly significant role in the recognition, discrimination, and acceptance of plants and the hypothesis that patterns are important are not mutually exclusive. Reality may lie somewhere between the two, and one mechanism may be controlling in some species and the other mechanism in others.

REFERENCES

DETHIER, V. G. (1973): Electrophysiological studies of gustation in lepidopterous larvae. II. Taste spectra in relation to food-plant discrimination. *J. Comp. Physiol.* **82**, 103–134.

DETHIER, V. G. (1974): Sensory input and the inconstant fly. In: Browne, L. B. (ed.): *Experimental Analysis of Insect Behaviour.* Springer-Verlag, Berlin, 21–31.

ISHIKAWA, S. (1963): Responses of maxillary chemoreceptors in the larva of the silkworm, *Bombyx mori,* to stimulation by carbohydrates. *J. Cell. Comp. Physiol.* **61**, 99–107.

REES, C. J. C. (1969): Chemoreceptor specificity associated with choice of feeding site by the beetle, *Chrysolina brunsvicensis* on its foodplant, *Hypericum hirsutum. Ent. exp. & appl.* **12**, 565–583.

SCHOONHOVEN, L. M. (1967): Chemoreception of mustard oil glucosides in larvae of *Pieris brassicae. Proc. K. Ned. Akad. Wet. (Ser. C)* **70**, 556–568.

Symp. Biol. Hung. 16, pp. 71–77 (1976)

COMPARATIVE STUDIES ON INDUCTION OF FOOD CHOICE PREFERENCES IN LEPIDOPTEROUS LARVAE

by

F. E. HANSON

DEPARTMENT OF BIOLOGICAL SCIENCES, UNIVERSITY OF MARYLAND, BALTIMORE COUNTY, CATONSVILLE, MARYLAND 21228, USA

Several species of lepidopterous larvae were studied to determine the extent of induced food preference modifications. Raising animals on certain host-plants enhanced feeding preference for that plant in a two- or three-choice situation. Two species of moths were studied (*Callosamia promethea* and *Antheraea polyphemus*) as well as three butterfly species and a hybrid (*Polygonia interrogationis, Limenitis archippus, Limenitis astyanax,* and *Limenitis* hyb. *rubidus*). Responses ranged from a very strong induction of preference (in *C. promethea*) to no detectable effect (in *L. archippus*). Preliminary results show that the feeding preferences of the hybrid are intermediate between its parental species.

INTRODUCTION

Modifications of insect feeding preferences have been occasionally reported in the literature over the past several decades (see references in Jermy et al., 1968). Several investigators have noted that larvae collected from various host-plants preferred plants on which they had been feeding. But because these animals were collected in the field, these experiments could not be controlled for genetic differences, nor could the investigators be certain of their feeding history. More recently, laboratory experiments have been done which confirm the basic findings of these field observations: feeding behavior of lepidopterous larvae can be modified in a manner that is dependent upon feeding experience (Jermy et al., 1968; Hanson and Dethier, 1973). Although highly significant and repeatable preference modifications were seen in these experiments, nevertheless in the species studied the absolute differences were not large. Therefore the question arises whether these are the only species in which it could be positively demonstrated and whether this small a preference change could modify the animal's behavior in a natural feeding situation. If not, perhaps the problem is purely academic. Therefore, the following study was undertaken to investigate the generality of this phenomenon among lepidopterans, to ascertain if different species differ in the ability to show modifications of preference, and whether the effect might be great enough in some species to be unquestionably of ecological significance.

MATERIAL AND METHODS

Eggs from a single ovipositing female were divided equally onto leaves of the various food plants. The animals were then raised in the laboratory in transparent plastic boxes under 20 hours of light. Shortly after a larva reached fifth instar, it was placed in a preference test adapted from the procedures of Jermy et al. (1968). Four leaf disks of each plant species (18 mm in diameter) were placed in the ABAB fashion around the circumference of circular testing containers (10 cm in diameter). The floor of the containers had a layer of paraffin wax to hold the pins that kept the leaf disks in place (5 mm from the bottom). One animal was placed into the center of each container for the test. When the animal hade aten 50% of the total area of one of the two plant species, the test was stopped and the final reading was recorded by visual estimations.

The Student's t-test or analysis of variance was used to determine the significance of differences in the food choice made by the larvae raised on the different leaves.

RESULTS

I. *Callosamia promethea* (Saturniidae). Raising the larvae of the promethea moth on one of its host-plants induces a very strong preference for that plant, as is shown by Fig. 1. The first four columns show that those raised on wild (black) cherry (*Prunus serotina*) prefer cherry almost exclusively, whereas those raised on poplar (*Liriodendron tulipifera*, the tulip poplar) prefer poplar. Of the 61 animals raised on each plant, only one from each group showed a preference for the opposite plant.

The other groups of tests show a similar pattern. As expected, there are slight differences due to different plants. For example, sassafras (*Sassafras albidum*) appears to be rejected to a lesser degree than tulip poplar by cherry-raised animals. Similarly, animals raised on spicebush (*Lindera Benzoin*) perhaps show less distaste for cherry than do larvae raised on other plants. Nevertheless, it is quite clear from these data that the preferences of the *promethea* larvae are completely determined by the plant on which they were raised (P < .001).

II. *Polygonia interrogationis* (Nymphalidae). Larvae of the question mark butterfly were raised and tested on American elm (*Ulmus americana*) and hackberry (*Celtis* sp.). As shown in Fig. 2, feeding on one or the other of its host plants induced a preference for that plant that is highly significant (P < .001). The degree to which induction is shown is not, however, as great as that of *C. promethea* but is more than that of *A. polyphemus* and the *Limenitis* complex (below) and probably more than that of the tobacco hornworm on which most of the previous laboratory studies on this phenomenon have been done.

III. *Antheraea polyphemus* (Saturniidae). The polyphemus moth feeds on a variety of trees and shrubs, among them oak, maple, and elm. Although we were not successful with raising them on elm, the larvae grew equally well on both oak (*Quercus alba*) and maple (*Acer platanoides*). The preference test scores depicted in Fig. 3 show that the animal can easily distinguish between them and that oak is preferred. The effect of prior feeding experiences is also seen in this animal: the preference for maple is enhanced (and oak slightly depressed) in maple-raised animals compared with oak-raised animals. The t-test shows these differences between the groups of animals raised on different foods to be highly significant (P < .001).

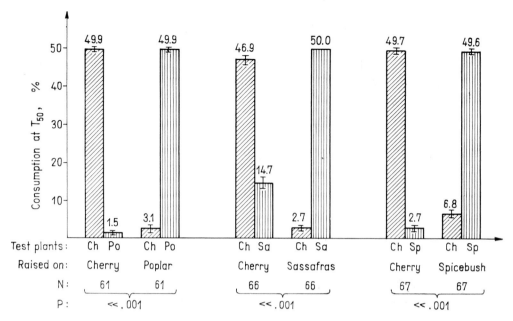

Fig. 1. Food preference of *Callosamia promethea*. Six groups of larvae raised on three different foodplants were placed in a 2-choice food selection preference test. The first group was raised on wild cherry (*Prunus serotina*) and tested for preference between cherry and poplar (*Liriodendron tulipifera*); the second group was raised on poplar tested on cherry vs. poplar, etc. Height of columns represents the percentage of each test plant eaten when the test was terminated at T_{50} (see text for further explanation). Numbers above each column represent the calculated value of the mean. Bars represent $\pm$ standard errors of the mean. N = number of animals in each group. P = probability that such a difference between groups could have occurred by chance (*t*-test)

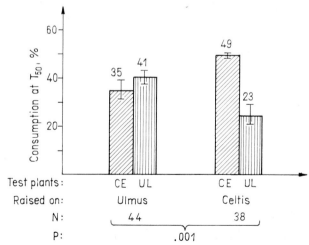

Fig. 2. Food preference of *Polygonia interrogationis*. One group of animals was raised on American elm (*Ulmus americana*) and the other on hackberry (*Celtis* sp.). Preference tests were done with both plants. Details as in Fig. 1

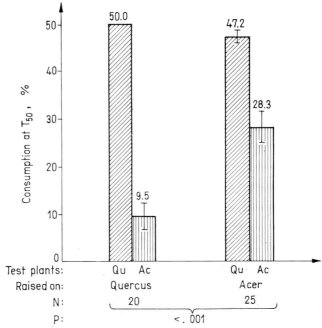

Fig. 3. Food preference of *Antheraea polyphemus*. One group of animals was raised on white oak (*Quercus alba*) and the other on Norway maple (*Acer platanoides*). Preference tests were done with both plants. Details as in Fig. 1.

IV. *Limenitis* (Nymphalidae). Two species of *Limenitis* and an interspecies hybrid were used in this study: *L. archippus* Cramer, the viceroy butterfly; *L. astyanax* Fabricus, the red-spotted purple butterfly; and the hybrid between them, *L.* hyb. *rubidus*. These animals were raised and tested on weeping willow (*Salix babylonica*), wild cherry (*Prunus serotina*), and quaking aspen (*Populus tremuloides*). The only exception was that *L. archippus* could not be raised on cherry despite repeated efforts.

The results in Fig. 4A clearly depict the rejection of cherry by *L. archippus* and suggest why it was not possible to raise these animals on this plant. On the other hand, *L. astyanax* feeds very well on cherry (Fig. 4B). A few results have been obtained from the hybrid *L.* hyb. *rubidus* (one brood of *L. archippus* ♂ × *L. astyanax* ♀) which show that food choice preferences are intermediate between those of its parental species. Figure 4C shows that, unlike its parent *L. archippus*, the hybrid will choose *Prunus* to some degree, but to a lesser extent than its other parent, *L. astyanax*.

Other differences among these species are noted in the data analysed to detect an induction of preference. *L. archippus* does not appear to induce (*t*-test, P = .9), whereas our preliminary data show that *L. astyanax* probably does (analysis of variance, P < .01). The hybrid *L. rubidus* also shows induction of preference (analysis of variance, P < .01) and as illustrated in Fig. 4, appears to be intermediate in this respect as well. However, these are tentative conclusions which must be taken with caution until more repetitions of preference tests can be done.

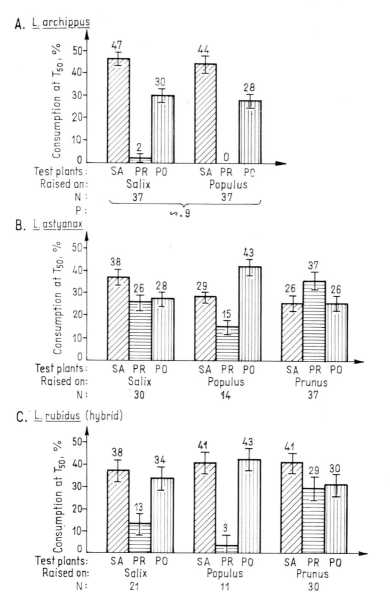

Fig. 4. Food preference of members of the genus *Limenitis*. Details as in Fig. 1. Larvae were raised and tested on willow (*Salix babylonica*), quaking aspen (*Populus tremuloides*), and wild cherry (*Prunus serotina*)

DISCUSSION

The foregoing presents conclusive evidence that larval food preferences can be strongly dependent on feeding experience. This is particularly true in *C. promethea*, where the preference is modified to such an extent by 5th instar that almost no feeding occurs on the "wrong" plant in the preference tests. If these tests are allowed to run well past the time when 50% of the first plant is eaten (T_{50}), most animals will eat 100% of the "right" plant and refuse the "wrong" plant for 24 hours or more. If animals raised on one plant are later switched to another, the resulting mortality can be very high and may approach 100% in some broods for some combinations of plants. Thus, the induction of preference as indicated here seems clearly strong enough to regulate food choice in natural feeding situations.

One of the main unanswered questions about this phenomenon is its physiological mechanism. Some behavioral experiments (Hanson and Dethier, 1973) link it closely with the sensory system, and electrophysiological evidence suggests that modification of firing frequency by chemoreceptors could account for the observed behavioral change in the tobacco hornworm (Schoonhoven, 1969; Städler, in this symposium). Because of the pronounced induced preference shown by *C. promethea*, it would appear to be a prime candidate for exhibiting such differences in sensory responses. However, preliminary data from electrophysiological recordings show that the differences, if any, are slight (Städler and Hanson, unpublished). Our tentative conclusion pending further experimentation and analysis is that the physiological basis of the induction of preference in *C. promethea* is due to either a modification of the central nervous system or a very subtle change in the sensory system (or both).

The occurrence of the phenomenon of induced preference appears to have some universality among lepidopterans and quite possibly among other insect orders as well. In addition to the foregoing are the two moths studied by Jermy et al. (1968), *Manduca sexta* and *Heliothis zea*; also a nymphalid butterfly *Chlosyne lacinia* has been shown to induce quite strongly under controlled laboratory conditions (Ting, 1970; Hanson and Ting, unpublished). Field reports indicate that several other insect species may also respond in this manner (see references in Jermy et al., 1968). It seems likely that a phenomenon this pervasive would have considerable importance in regulating insect feeding behavior.

One noteworthy aspect of the data presented is that the degree of the induction of preference is highly variable. Note in this regard the differences between these two silk moths: *C. promethea* shows a very strong preference modification, *A. polyphemus* a weak (but statistically significant) one. The two butterflies *P. interrogationis* and *L. astyanax* also show weak (but significant) effects, whereas *L. archippus* apparently has none at all.

The causes for this wide latitude in the degree of manifestation of the induction of preference are not at all apparent. In some cases there may be a correlation with increased polyphagy. For example, *L. archippus*, which is the least polyphagous of those mentioned, also induces least well, if at all (Fig. 4A). Slightly more polyphagous is *L. astyanax*, which also shows some induction of preference. The other animals reported here are more polyphagous and induce more strongly. In addition, the two moths studied by

Jermy et al. (1968) also follow this pattern: the more polyphagous (*Heliothis zea*) showed a greater induced preference than the less polyphagous (*Manduca sexta*). However, an important exception to this trend is represented by *C. promethea* which shows by far the greatest degree of induction of preference, yet is probably less, or at least not proportionately more, polyphagous than *A. polyphemus* or *H. zea*. Until more data are available, we must tentatively regard any correlation this type that may exist as spurious rather than causal, and to ascribe the basis of this variability to ecological or physiological constraints which are presently still obscure.

Finally, the studies of the *Limenitis* complex are interesting because of the inferences that may be drawn concerning the genetic basis of host feeding preferences. These preliminary data showing that the hybrid is intermediate between the two parents suggest that host-plant discrimination and possibly the capability of induction of preference itself is under multigenic control.

Thus we conclude that the manifestation of host preference in insects is not entirely a rigid, genetically programmed behavior, but that considerable plasticity remains. It is clear from the data presented here and in other studies (e.g. Yamamoto, 1974) that past experiences play a role in influencing future decisions concerning food selection behavior.

The extent to which this occurs in various species, its genetic and physiological basis, and its ecological implications remain to be explored. Should it prove amenable to manipulation by chemical agents, its potential for economic uses may be quite interesting.

ACKNOWLEDGEMENTS

I wish to acknowledge the help of the following people in these time consuming experiments: Patricia Jaris Filip, Phyllis Martin, Darlene and Robert Keane, Steve Kloetzel, Robert Laprade, and Austin Platt. This work was supported in part by U.S.P.H.S. Grant No. NS-10760.

REFERENCES

HANSON, F. E. and DETHIER, V. G. (1973): Role of gustation and olfaction in food plant discrimination in the tobacco hornworm, *Manduca sexta*. *J. Insect Physiol.* **19**, 1019–1034.

JERMY, T., HANSON, F. E. and DETHIER, V. G. (1968): Induction of specific food preference in lepidopterous larvae. *Ent. exp. & appl.* **11**, 211–230.

SCHOONHOVEN, L. M. (1969): Sensitivity changes in some insect chemoreceptors and their effect on food selection behavior. *Proc. K. Ned. Akad. Wet. (Sec. C)*, **72**, 491–498.

TING, A. Y. (1970): *The induction of feeding preference in the butterfly* Chlosyne lacinia. Masters Degree Thesis, University of Texas, Austin, Texas.

YAMAMOTO, R. T. (1974): Induction of hostplant specificity in the tobacco hornworm, *Manduca sexta*. *J. Insect Physiol.* **20**, 641–650.

Symp. Biol. Hung. 16, pp. 79–83 (1976)

HOST-PLANT FACTORS REGULATING WING PRODUCTION IN *MYZUS PERSICAE*

by

P. Harrewijn

INSTITUTE FOR PHYTOPATHOLOGICAL RESEARCH (I.P.O.), BINNENHAVEN 12, WAGENINGEN, THE NETHERLANDS

Suboptimal nutrition is often thought to stimulate wing formation in aphids by influencing the "milieu intérieur" in a way that stimulates the process of wing determination.

A few authors, however, report that imbalanced nutrition may cause a diversion from the alate to the apterous morph even when the aphids are reared in high densities.

By using a continuous flow artificial feeding apparatus for aphids it is demonstrated that poor nutrition may stimulate the production of alatae by the aphid *Myzus persicae*, but only in case enhanced restlessness results from reduced diet acceptance. "Low energy" diets which are accepted do not appear to comply with physiological conditions necessary for wing development.

On the other hand the apterous morph may be induced on a suitable host in a situation where production of apterae will be profitable. In this situation a factor not influencing growth and reproduction seems to "set" the endocrine system by acting on a functional antagonism between tyrosine and tryptophan metabolism, resulting in an inhibition of wing bud development.

Host-plant regulation of wing dimorphism can be simulated by the uptake of tyrosine and tryptophan derivatives by the aphids. The theoretical and practical implications of these results are discussed.

INTRODUCTION

It is generally known that the quality of the food source influences besides growth and reproduction also wing dimorphism in aphids (Johnson, 1966; Sutherland, 1969; Dadd, 1968; Mittler and Kleinjan, 1970). Although several authors report even a dominant effect of the nutritional status of a natural host or artificial diet on wing production, there appears to be a remarkable contradiction in their conclusions whether optimal nutrition induces the alate or the apterous morph.

To give an example, Sutherland (1969) postulates that optimal nutrition enhances apterae production in *Acyrthosiphon pisum*, while Schaefers and Judge (1971) found that *Chaetosiphon fragaefolii* produced a higher percentage of alatae when aphids were larger and had a higher reproductive rate on parts of *Fragaria vesca* L. than those which were small and produced fewer offspring.

Mittler and Kleinjan (1970) working with artificial diets observed a high proportion of apterous forms of *Myzus persicae* on nutritionally deficient diets, whereas Raccah et al. (1971) found diets of presumably nutritional imbalance to favour the alate course of development in the same aphid species.

The aim of this paper is to give a contribution to the knowledge of the complex relationships between nutrition and the possible role of specific factors regulating wing formation in aphids.

NUTRITIONAL AND CHEMICAL FACTORS INFLUENCING
WING DIMORPHISM

Using the system of Steiner (1961) to influence the nutritional status of the potato plant via mineral nutrition, we found food quality of mature leaves to be positively correlated with wing production of *M. persicae* when the crowding-stimulus was kept as constant as possible (Harrewijn, 1972). On the other hand, severe malnutrition can result in an insufficient amount of energy to produce alatae (Mittler, 1972; Harrewijn, 1972).

By using a continuous flow artificial feeding apparatus for aphids (Harrewijn, 1973) it could be shown that both imbalanced and well-balanced nutrition may modify the course of wing dimorphism in *M. persicae*. With this apparatus it could be demonstrated that short-term food deprivation may favour the production of alatae by stimulation of the crowding response, although the nutritional imbalance reduces alatae production. Obviously the crowding response needs a minimum amount of energy to express itself. It may be stated here that the occurrence of a high proportion of apterae or alatae on a nutritionally deficient diet or host-plant does not necessarily imply a specific adaptational response to the situation.

A host-plant or artificial diet offering excellent conditions for growth does not guarantee the production of mainly apterae or alatae either. The experiments of Mittler and Sutherland (1969), Mittler and Kleinjan (1970), Harrewijn (1972) indicate that optimal nutrition enables the aphids to respond maximally to a crowding stimulus. Especially the amount of amides and the amino nitrogen balance can have a dominating effect on wing production (Harrewijn, 1973).

In all these cases, one does not necessarily have to assume some specific alatae- or apterae-inducing factor. There are, however, two main "exceptions" to be discussed.

One is, that on young foliage, especially on seedlings, which provide good nutrition, wing dimorphism is directed towards the apterous course, even when a strong crowding stimulus is presented (Schaefers and Judge, 1971; Harrewijn, 1972, 1973). In this situation, Kunkel and Mittler (1971) assume that a crowding stimulus is counteracted by a yet unknown apterous promoting principle. This factor seems to be of chemical nature and is probably ingested (Mittler, 1972). We were unable to prove that differences in the free amino acid content of radish seedlings are responsible for this "seedling effect" (Harrewijn, 1972).

The other "exception" is, that indeed positive effects exist of substrates lacking certain nutrients on wing formation of aphids. A nutritionally slightly deficient diet may for instance be more acceptable for aphids than a complete one (Mittler and Kleinjan, 1970), but in this case a negative effect on wing production may be compensated by enhanced food uptake, making this "exception" explainable in terms of energy (Harrewijn and Noordink, 1971).

Omission of the amino acid tryptophan from the diets, however, resulted in a reduced production of apterae, whereas food uptake differed less than 10% from that on a complete diet (Harrewijn, 1972). This tryptophan effect is also found in *Aphis fabae* (Leckstein and Llewellyn, 1973).

The question arises whether a host plant signal exists, perceived by the

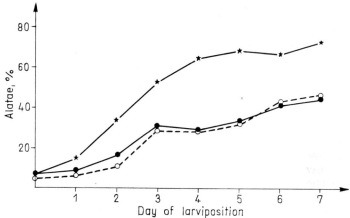

Fig. 1. Effect of the amount of tryptophan in artificial diets on wing production in *M. persicae*. Mothers reared on radish seedlings; ●——● larvae on basal diet (7, Harrewijn, 1973); ○——○ larvae on diet with 1.5×10^{-2} M tryptophan; ★——★ larvae on diet without tryptophan

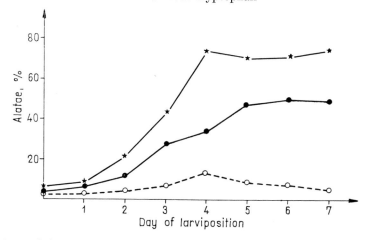

Fig. 2. Effect of the amount of 5-hydroxytryptamine and phenylalanine in artificial diets on wing production in *M. persicae*. Mothers reared on radish seedlings; ●——● larvae on diet with 10^{-2} M tryptophan; ○——○ larvae on diet with 10^{-2} M 5-HTP; ★——★ larvae on diet with 1.2×10^{-2} M phenylalanine

aphids and resulting in a stimulus to produce apterous offspring. In a previous paper (Harrewijn, 1973) it was shown that the seedling effect, counteracting crowding, can be simulated by influencing the tryptophan metabolism of new-born larvae of *M. persicae*. A derivative of tryptophan, 5-hydroxytryptamine (5-HT, serotonin) causes a rapid increase of the proportion of apterae, as in the case of seedlings, whereas inhibition of the formation of this compound in the aphid by drugs results in a high proportion of alatae.

This does not mean that tryptophan is in itself a morph-active amino acid. Figure 1 shows that although omission of tryptophan causes a rapid increase in the production of alatae, a high amount of tryptophan in the diet has no apterizing effect. 5-Hydroxytryptamine is highly effective in this respect (Harrewijn, 1973). It appears from Fig. 2 that 5-hydroxytryptophan (5-HTP), though less active than 5-HT, also initiates the apterous course of development. It is important to notice that 5-HTP, provided in an equal molarity as tryptophan, has no influence on growth (mean weights of sixth-day larvae after one week on diet with 10^{-2} M TP 382 μg; after one week on diet with 10^{-2} M 5-HTP 396 μg).

We demonstrated already that derivatives of tyrosine like 2-methyl-3-(3,4-dihydroxyphenylalanine) can counteract an apterous course of development (Harrewijn, 1973). The same is true for the amino acid phenylalanine. We now report that large amounts of phenylalanine in the diet are able to keep the switch mechanism of wing dimorphism set at the alatae course of development, as shown in Fig. 2. This effect occurs in spite of a reduction in growth caused by the large amount of phenylalanine used in this experiment (mean weights of second-day larvae on basal diet 546 μg; on the phenylalanine diet 256 μg). Not only do tryptophan and phenylalanine act in an antagonistic way, their presence or absence in the diet also has an opposite effect. Absence of phenylalanine has little specific effect on wing formation (Harrewijn, 1972), whereas large amounts of phenylalanine do have.

Microinjections of monoamines into the haemolymph of mature aphids and new-born larvae of *M. persicae* (Harrewijn, in prep.) enabled us to demonstrate that the balance between tyrosine and tryptophan derivatives can regulate wing bud development. At the present time, it is not yet possible to distinguish whether the endocrine system by reacting at a yet unknown signal influences monoamine metabolism or is regulated by a factor primary influencing the tyrosine/tryptophan equilibrium.

DISCUSSION

Although it is evident that the alate morph does not simply occur because escape from nutritionally inadequate host plants is urgent, it should be clear that well-balanced nutrition in itself does not stimulate wing production either.

Wing initiation probably takes place already before birth (Johnson and Birks, 1960). It is suggested that a yet unknown signal activates serotoninergic compounds which probably inhibit the development of wing buds already present in new-born larvae (Harrewijn, 1973). In *M. persicae* there seems to be an antagonistic action between derivatives of tyrosine and tryptophan. It is interesting to note that Sang (1972) found gene expression in *Drosophila* to be enhanced by derivatives of phenylalanine and to be suppressed by derivatives of tryptophan.

Activity of biogenic amines is found throughout the Animal Kingdom. Time has not yet come that we were able to compare the function of biogenic amines in gene expression of an insect and their proved function in human psychology (Wise et al., 1972) but it is the aim of the present paper to demonstrate that cyclic amino acids and their derivatives deserve special attention with respect to differentiation in insects.

REFERENCES

DADD, R. H. (1968): Dietary amino acids and wing determination in the aphid *Myzus persicae*. *Ann. ent. Soc. Am.* **61**, 1201–1210.

HARREWIJN, P. (1972): Wing production by the aphid *Myzus persicae* related to nutritional factors in potato plants and artificial diets. *Proc. Int. Conf. Insect and Mite Nutrition*, 575–588.

HARREWIJN, P. (1973): Functional significance of indole alkylamines linked to nutritional factors in wing development of the aphid *Myzus persicae*. *Ent. exp. & appl.* **16**, 499–513.

HARREWIJN, P. and NOORDINK, J. P. W. (1971): Taste perception of *Myzus persicae* in relation to food uptake and developmental processes. *Ent. exp. & appl.* **14**, 413–419.

JOHNSON, B. (1966): Wing polymorphism in aphids. III. The influence of the host plant. *Ent. exp. & appl.* **9**, 212—223.

JOHNSON, B. and BIRKS, P. R. (1960): Studies on wing polymorphism in aphids. I. The developmental process involved in the production of the different forms. *Ent. exp. & appl.* **3**, 327–339.

KUNKEL, H. and MITTLER, T. E. (1971): Einfluß der Ernährung bei Junglarven von *Myzus persicae* (Sulz.) (Aphididae) auf ihre Entwicklung zu Geflügelten oder Ungeflügelten. *Oecologia* (Berl.) **8**, 110—134.

LECKSTEIN, P. M. and LLEWELLYN, M. (1973): Effect of dietary amino acids on the size and alary polymorphism of *Aphis fabae*. *J. Insect Physiol.* **19**, 973–980.

MITTLER, T. E. (1972): Aphid polymorphism as affected by diet. *Perspectives in Aphid Biology*, **2**, 65–75.

MITTLER, T. E. and KLEINJAN, J. E. (1970): Effect of artificial diet composition on wing production by the aphid *Myzus persicae*. *J. Insect Physiol.* **16**, 833–850.

MITTLER, T. E. and SUTHERLAND, O. R. W. (1969): Dietary influences on aphid polymorphism. *Ent. exp. & appl.* **12**, 703–713.

RACCAH, B., TAHORI, A. S. and APPLEBAUM, S. W. (1971): Effect of nutritional factors in synthetic diet on increase of alate forms in *Myzus persicae*. *J. Insect Physiol.* **17**, 1385–1390.

SANG, J. H. (1972): The use of mutants in nutritional research. *Proc. Int. Conf. Insect and Mite Nutrition*, 9–17.

SCHAEFERS, G. A. and JUDGE, F. D. (1971): Effects of temperature, photoperiod and host plant on alary polymorphism in the aphid, *Chaetosiphon fragaefolii*. *J. Insect Physiol.* **17**, 365–379.

STEINER, A. A. (1961): A universal method for preparing nutrient solutions of a certain desired composition. *Pl. and Soil*, **15**, 134–154.

SUTHERLAND, O. R. W. (1969): The role of the host plant in the production of winged forms by two strains of the pea aphid *Acyrthosiphon pisum*. *J. Insect Physiol.* **15**, 2179–2201.

WISE, C., BERGER, B. D. and STEIN, L. (1972): Benzodiazepines: Anxiety-reducing activity by reduction of serotonin turnover in the brain. *Science* **177**, 180—183.

Symp. Biol. Hung. 16, pp. 85–89 (1976)

BEHAVIOURAL RESPONSES TO HOST-PLANT ODOURS IN ADULT CABBAGE ROOT FLY [*ERIOISCHIA BRASSICAE* (BOUCHÉ)]

by

C. Hawkes and T. H. Coaker

DEPARTMENT OF APPLIED BIOLOGY, UNIVERSITY OF CAMBRIDGE,
DOWNING STREET, CAMBRIDGE CB3 9LU, UK

The behavioural responses of insects to host-plant odours have proved difficult to interpret but correct interpretation is required to distinguish the stimuli and how they operate.

Host-plant odour increases the activity of cabbage root flies, a feature which may be used to assess the biological activity of odours. However, this approach does not distinguish between attractants and repellents. Release and recapture of ^{32}P marked flies in the vicinity of a host crop provided information on orientation, only mated gravid females moving upwind towards the crop. Age, sex and mated state did not, however, influence receptor potential in neurophysiological tests using host-plant odours, suggesting a CNS 'block' of behavioural responses. Tests in a large wind tunnel have confirmed that only mated gravid females orientate to host-plants, the odour causing anemotaxis. Orientated responses can be inhibited by 'disturbance' factors, physiological condition and low activity level of the flies, and diurnal periodicity. In order to identify odour components and categorize behavioural responses, the insect should be permitted to exhibit its potential behaviour. It appears that small olfactometers are unlikely to achieve this end.

INTRODUCTION

Responses of insects to host-plant odours have been studied using a variety of laboratory techniques. Attempts have been made to generalize from such results and to make suggestions about the influence of odours on host-plant finding under field conditions. Thorsteinson (1960) suggested that insects do not orientate to plants beyond a few metres, postulating that the principal effect of odour is to inhibit locomotor activity and thus act as an 'aggregant'. The difficulty in observing host-plant finding under field conditions has not permitted substantiation or contradiction of this concept. Fraenkel (1969) observed that the role of olfaction in host-plant finding has been a "very much neglected" subject.

The validity of present concepts depends largely, therefore, on the results of laboratory experiments and it appears that there is some difficulty in their interpretation. Firstly the apparatus used may modify the behaviour of the insect so that 'normal' responses are inhibited or restricted. Secondly when positive results are observed these may be open to several interpretations and may not indicate an orientated response (Kennedy, 1965). There is, therefore, some confusion about the use of terms such as attractant, arrestant, repellent and deterrent, and Fraenkel (1969) avoided this difficulty by classifying all stimuli as attractants or repellents. In behavioural studies it may be advantageous to face this difficulty and use a more detailed system of classification such as that of Dethier (1970). However, a broad classification must be considered only as a first step in a behavioural study for such a scheme is inadequate to describe the variety and complexity of the behaviour of insects in host-plant finding.

This paper reports work on the adult cabbage root fly [*Erioischia brassicae* (Bouché)] which has proceeded towards this first step. It illuminates some of the difficulties of observing behavioural responses to host-plant odour and single substances and shows how orientated responses may be overlooked.

OLFACTOMETER AND NEUROPHYSIOLOGICAL TESTS

Traynier (1967a) showed that host-plant odour increased the activity of gravid females but stimulated non-gravid females and males only slightly. An orientated response could not be demonstrated when using several types of olfactometer with moving air. It was suggested that unorientated activation of females enhanced the probability of making contact with host-plants in the field. Once at the plant stimulation of contact chemoreceptors, particularly by mustard oil glucosides, results in descent to the soil and the completion of the sequence of behavioural events which lead to oviposition (Traynier, 1967b).

Olfactometer tests were carried out to establish a method for assessing the biological activity of the components of host-plant odour (Coaker and Smith, 1968, 1969). Low levels of responsiveness to brassica odour were observed when counts were made of the flies at ventilated ports in a 'choice chamber'. In order to obtain a 50% response level, the design finally adopted exploited the positive phototaxis which *E. brassicae* displays in confined conditions. The biological activity of odour was assessed by observing how the response to light was modified, compounds being classified as excitatory or inhibitory.

Neurophysiological tests (electroantennograms — EAG's) were carried out in parallel with the olfactometer tests. The results of the two methods showed a general correlation when comparing the activity of some volatiles of cooked cabbage. Propionaldehyde and allylamine gave +ve EAG responses similar to those of cabbage leaf-juice and increased movement of females to light in the olfactometers. Allylbromide and propylamine gave —ve EAG's and no response in the olfactometer. Acetaldehyde gave +ve EAG's although smaller than those of cabbage leaf-juice and decreased responsiveness to light in the olfactometer.

Using the same methods, Wallbank (1972) obtained +ve EAG's and responses in the olfactometer with allylisothiocyanate and hexylacetate, naturally occurring volatiles from brassicas. However, when used in field traps only allylisothiocyanate increased captures.

It appears, therefore, that although there is a general correlation between the results of the three test methods at least two if not all three may have limitations. The EAG is the summated receptor potential of many cells which may behave in different ways and intracellular recordings would be preferable for a correct interpretation (Coaker and Finch, 1972). In the olfactometer tests a variety of types of responses are possible and any compound which increased activity could be classed as excitatory. Consequently,

86

these tests do not allow a distinction to be made between compounds which increase activity but act either as repellents or attractants. Trapping tests can also be interpreted in more than one way because increased captures could result for example from an increase in activity in the presence of odour or from an orientated response.

OBSERVATION OF HOST-PLANT FINDING IN THE FIELD

Flies marked with ^{32}P were released downwind of a host-crop and their behaviour was determined by recapturing with yellow water-traps (Hawkes, 1971, 1974). Large proportions ($> 80\%$) of mated and gravid females moved upwind from at least 24 m from the nearest brassica plants indicating an orientated response. Males and non-gravid and/or unmated females were not attracted although neurophysiological tests showed that age, sex and mated state did not influence receptor potential. There appears to be a CNS 'block' of the receptor input resulting from odour stimulation, which is removed by mating and the attainment of the gravid state.

The field observations did not permit the separation of odour and visual stimuli and so do not demonstrate conclusively that odour was causing orientation.

OBSERVATIONS IN A LARGE WIND TUNNEL

Detailed observations of behaviour were not easily obtained in the field and experimental conditions could not be readily reproduced. In order to carry out critical observations, experiments were conducted in a large wind tunnel to identify the stimuli which resulted in orientation. The $6 \times 2.3 \times 1.8$ m tunnel had a laminar air flow of 0.6 m/sec at 20 ± 0.5 °C and $62 \pm 4\%$ R.H. and was lit from above with fluorescent lighting. The floor and side walls up to a height of 0.9 m were covered with brown paper bearing a 0.5×0.5 m grid of black lines. The remaining areas were white.

Groups of 15 to 20 mated and gravid females were placed downwind of brassica plants in a $0.4 \times 0.3 \times 0.3$ m cage through which air was able to pass. Flies could leave the cage by either flying upwind or downwind but in more than 60% of the flights observed in 5 tests, the individuals did not leave the cage. Less than 10% of the flights were more than 0.5 m upwind. It appeared that although odour increased activity of the females the majority of flights were upwards rather than upwind. This behaviour is analogous to that observed in the olfactometer tests where host-plant odour could produce increased movement towards light.

Positive phototaxis is related to confinement. When females were released into the tunnel initially about 45% settled on the higher white sections of the side walls but after about an hour the distribution had changed with about 80% on the lower wall area covered in brown paper. Under field conditions flies were observed to stay largely within their 'boundary layer' and were rarely observed to fly higher than about 1 m (Hawkes, 1972).

The behaviour of individual females was observed in the wind tunnel by allowing them to settle on a piece of muslin placed over a cage; the muslin was removed and placed on the floor of the tunnel and then brassica plants were introduced upwind of the fly. When females were orientated across the wind at the start of the test two criteria of responsiveness could be used. The initial response was a turn upwind without forward movement. The second was an upwind flight taking the individual an average of about 0.7 m upwind. About a third of the females walked upwind 5–10 cm before they flew upwind. A response was considered to be the occurrence of both the initial upwind orientation and an upwind flight of at least 0.5 m. Of the 65 mated and gravid females tested 70% showed this response to the presence of brassicas upwind. No responses were observed in tests of 60 unmated and/or non-gravid females and 19 males.

The response of females to the brassica plants could have been produced by odour or visual stimuli or both. Allylisothiocyanate was used to assess whether odour alone could give an orientated response. Mated and gravid females showed the same behaviour on exposure to allylisothiocyanate as they did to brassica plants. The results suggested that the initial upwind oiientation was an odour induced anemotaxis.

CONCLUSIONS

There appears to be several factors which modify or inhibit the orientated response to odour:

1. Physical disturbance due to handling reduced the level of responsiveness.

2. Confinement increased positive phototaxis and thus interfered with the demonstration of orientated responses.

3. There was a positive correlation between the activity of females prior to exposure to odour and their response, indicating that a 'minimum' level of activity was required.

4. Females responded to odour only during a part of their life, after mating had occurred and when ovarian development was complete.

5. Ovipositional response to the host-plant follows a diurnal periodicity.

The results of tests in small olfactometers are clearly complicated by several of the factors listed above and the observation of orientated responses may be particularly difficult when such equipment is used. A problem of interpretation arises in any test where the insects can react to odour in a variety of ways i.e. where 'biological activity' encompasses heterogeneous behavioural responses. These difficulties can only be avoided if each type of response is studied individually.

The work in the large wind tunnel has so far demonstrated that a component of host-plant odour can act as an attractant for the cabbage root fly. Further work is required to asses whether odour stimulates anemotaxis after take-off or has other effects and whether visual stimuli modify the response to odour.

The implication of the present conclusions is that the behaviour of insects in host-plant finding can be misinterpreted through the artifactual effects of the test method used. When more species have been studied using apparatus more appropriate to the observation of orientated responses the current concepts of the role of olfaction in host-plant finding may have to be revised.

ACKNOWLEDGEMENT

We wish to acknowledge the Agricultural Research Council for financial
support.

REFERENCES

COAKER, T. H. and SMITH, J. (1968): Response to host plant odour. *Rep. natn.
Veg. Res. Stn. for 1967*, 73.
COAKER, T. H. and SMITH, J. (1969): The behaviour of adult cabbage root fly. *Rep.
natn. Veg. Res. Stn. for 1968*, 73–74.
COAKER, T. H. and FINCH, S. (1972): The association of cabbage root fly with its
food and host plants. In: van Emden, H. F. (ed.): *Insect/Plant Relationships*.
Blackwell, London.
DETHIER, V. G. (1970): Some general considerations of insect responses to chemicals
in food plants. In: Wood, D. L. et al. (eds): *Control of Insect Behavior by Natural
Products*. Academic Press, New York–London.
FRAENKEL, G. (1969): Evaluation of our thoughts on secondary plant substances.
Ent. exp. & appl. **12**, 473–486.
HAWKES, C. (1971): Studies on the adult cabbage root fly – field behaviour. *Rep.
natn. Veg. Res. Stn. for 1970*, 93.
HAWKES, C. (1972): The estimation of the dispersal rate of adult cabbage root fly
[*Erioischia brassicae* (Bouché)] in the presence of a brassica crop. *J. appl. Ecol.*
9, 617–63.
HAWKES, C. (1974): Dispersal of adult cabbage root fly [*Erioischia brassicae* (Bouché)]
in relation to a brassica crop. *J. appl. Ecol.* **11**, 83–93.
KENNEDY, J. S. (1965): Mechanisms of host plant selection. *Ann. appl. Biol.* **56**,
317–322.
THORSTEINSON, A. J. (1960): Host selection in phytophagous insects. *Ann. Rev.
Ent.* **5**, 193–218.
TRAYNIER, R. M. M. (1967a): Effect of host plant odour on the behaviour of the
adult cabbage root fly, *Erioischia brassicae. Ent. exp. & appl.* **10**, 321–328.
TRAYNIER, R. M. M. (1967b): Stimulation of oviposition by cabbage root fly, *Erio-
ischia brassicae. Ent. exp. & appl.* **10**, 401–412.
WALLBANK, B. E. (1972): Studies on the adult cabbage root fly – host plant volatiles.
Rep. natn. Veg. Res. Stn. for 1971, 61–70.

Symp. Biol. Hung. 16, pp. 91–94 (1976)

SMALL ERMINE MOTHS OF THE GENUS *YPONOMEUTA* AND THEIR HOST RELATIONSHIPS (LEPIDOPTERA, YPONOMEUTIDAE)

by

W. M. Herrebout, P. J. Kuijten and J. T. Wiebes

DEPARTMENT OF SYSTEMATIC ZOOLOGY AND EVOLUTIONARY BIOLOGY,
UNIVERSITY, LEIDEN, THE NETHERLANDS

Insect–host-plant relationships are often studied in cultivated species of economic importance. While the thorough knowledge acquired in modern plant-protection research is appreciable, the study of insects and plants under more natural conditions may add new data relevant to an understanding of the evolution of their relationship.

Among the small ermine moths of the genus *Yponomeuta* Latr., there are European forms feeding (as larvae) on plants cultivated in orchards (viz., *Malus* sp., *Prunus* spp.). On the other hand, there are also representatives feeding on plants growing in a more natural situation (e.g. *Euonymus* sp., *Salix* spp., *Sedum* sp.). Thus, the genus provides excellent opportunities to combine the knowledge of the applied aspect and data on more natural relationships. There already exists an important and extensive literature on both practical and theoretical research concerning *Yponomeuta*, on which some comprehensive reports were made by Parrott and Schoene (1912), Thorpe (1929, 1930), Pag (1959), Wiegand (1962). Moreover, an analogous investigation is that on the New World flies of the genus *Rhagoletis* by Bush (1966); we hope our study to become complementary to his.

The genus *Yponomeuta* has a wide, holarctic and palaeotropical, distribution. The European forms are listed in Table 1 of the present paper, with the genera and families of their host-plants. The main features of the life cycles are presented in Fig. 1.

The nine forms mentioned show different degrees of relationship. There are two species groups. Group A comprises two good species (viz., nos. 1, 7), next to a complex of five closely related forms (nos. 2–6). The status of these forms is of great scientific interest; they are discussed below. The second group, B, has two species (viz., nos. 8, 9).

Table 1 shows that there is no parallelism between the classification of the moths and that of their host-plants. There seems to be no fixed phylogenetic connection between the two groups of organisms (although the recurrent appearance of Celastraceae as host-plants, in several species-groups, may be of some phylogenetic significance). Instead, there is an apparent ecological relation, in that all host-plants but two (viz., *Salix* and *Prunus padus*), belong to one and the same type of vegetation.

Some of the forms (nos. 2–6) were indicated by Thorpe (1929, 1930) as host races, or incipient species. Friese (1960) could not detect any differential characters in their genitalia, and combined them in the complex species *Y. padellus*. Recently, Gershenson (1974) attributed to all forms the status of full species, even describing a new one from yet another host-plant (viz.,

TABLE 1

Nine European representatives of the genus Yponomeuta *and their principal host-plants*

	Salicaceae	Rosaceae		Crassulaceae	Celastraceae
		Pomoidea	Prunoidea		
Group A					
1. *Y. evonymellus* (L.)			*P. padus*		
2. *Y. cagnagellus* (Hübner)					*Euonymus*
3. *Y. malinellus* Zeller	*Salix*	*Malus Crataegus*			
4. *Y. padellus* (L.)			*P. spinosa* *P. domestica*		
5. *Y. rorellus* (Hübner)					
6. *Y. mahalebellus* Guenée			*P. mahaleb*		
7. *Y. irrorellus* (Hübner)					*Euonymus*
Group B					
8. *Y. vigintipunctatus* (Retz.)				*Sedum*	
9. *Y. plumbellus* (Schiff.)					*Euonymus*

Y. rhamnellus Gershenson from *Rhamnus cathartica*, Rhamnaceae). The forms are mainly recognized by their host-plants, and thus all but one (*Y. padellus* s. str.) are monophagous. As a rule it is not possible to rear larvae taken from one host on plants of another species, or the development on the unnatural host is greatly retarded and the moths finally emerging are much smaller than normal. Our investigations of the acceptability of several host-plants will focus on the possibility that the preference of the larvae is being determined by the first bite (of bark) they take while still under the cover of the aggregated egg shells. Thorpe (1930) already induced females of *Y. malinellus* and *Y. padellus*, from *Malus* and *Crataegus*, respectively, to oviposit on the other plant. Unfortunately, no food preferences of the larvae were tested, but in the experiments the females preferred their own specific host for oviposition.

The members of the *padellus*-complex partly overlap in their distribution, and they are also synchronous (see Fig. 1). Yet their populations seem to keep separate in the field. Several attempts to interbreed the forms were made. Wildbolz and Riggenbach (1965) were able to rear both combinations of *Y. malinellus* and *Y. padellus* until the third generation. In our institute, A. Breure (unpublished work) obtained viable offspring (first instar larvae) from each of all possible combinations of nos. 2, 3, 4, and 5 (see Table 1). Among these, marked differences were found in the total number of eggs laid per female. There also was some variation in the number of fertile eggs laid, the greatest number (ca. 100–180) being produced by females of *Y. cagnagellus* in all combinations. Under laboratory conditions, females of *Y. rorellus* only laid ca. fifty eggs when paired with their own males, and

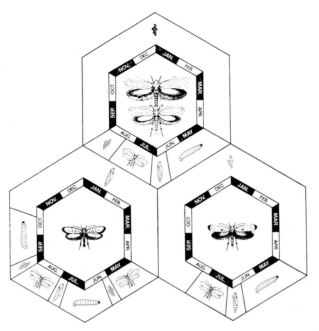

Fig. 1. Diagrammatic representation of the life cycles of the species listed in Table 1
Upper part: group A, depicted are nos. 1 and 3. Lower part: group B, left no. 8 and
right no. 9.
The species of group A hibernate as first instar larvae under egg-shells, which are
combined so as to form a common protective shield. *Y. plumbellus* hibernates in the
egg stage, whereas *Y. vigintipunctatus* has a pupal diapause in its second generation.
All larvae begin their subsequent development in buds or in leaf-mines, except for
Y. plumbellus which in its first instar is a borer in young shoots. From the second
instar on, all live more or less gregariously in conspicuous webs

even less (and then mostly sterile) when combined with males of the other
forms. *Y. malinellus* and *Y. padellus* were found to be intermediate in this
respect.

Neither the data on oviposition preferences, nor those on hybridisation
success, can form conclusive arguments for or against the specific identity
of the forms. The possible rôle of sex pheromones as signals by which the
forms may keep apart, was investigated in the following preliminary experiment. In an area where *Crataegus* predominates and several *Euonymus*-
trees occur (and where no infestations of other possible food plants were
present), traps were baited with virgin females of five forms (viz., nos. 1 to
5 inclusive). The series of traps baited with females of *Y. padellus* caught a
large number of males, whereas traps with females of *Y. cagnagellus* attracted
some males. The traps baited with females of *Y. evonymellus* (no. 1), *Y. malinellus* (no. 3), or *Y. rorellus* (no. 5), did not attract any males, which indicates that "forms" nos. 3 and 5 behaved as did the "good species" no. 1.
In the case of conspecificity of the "forms", one would have expected nos.
3 and 5 to attract some males.

Another series of traps were provided with mixtures of chemical compounds known to be sex attractants to other Lepidoptera. In these experi-

ments an abandoned apple-orchard with a heavy infestation of *Y. malinellus* was included. In general, *Y. malinellus* and *Y. cagnagellus* were found to react to the same compounds, while *Y. padellus* was attracted to an altogether different group of compounds. These results will be dealt with more fully in a forthcoming paper.

In conclusion, the results obtained up to now show that the "host races" or "forms" of the small ermine moths to some extent do behave as full species. The identification of the proximate factors separating their populations, and the measuring of their reproductive isolation, may give some clue to the origin and the evolution of the forms and their host relationships.

REFERENCES

BUSH, G. (1966): The taxonomy, cytology, and evolution of the genus *Rhagoletis* in North America (Diptera, Tephritidae). *Bull. Mus. Comp. Zool.* **134** (11), 431–562.

FRIESE, G. (1960): Revision der paläarktischen Yponomeutidae unter besonderer Berücksichtigung der Genitalien (Lepidoptera). *Beitr. Ent.* **10**, 1–131.

GERSHENSON, Z. S. (1974): Yponomeutidae, Argyresthiidae. *Fauna Ukraini*, **15** (6), 1–132.

PAG, H. (1959): *Hyponomeuta*-Arten als Schädlinge im Obstbau. *Z. ang. Zool.* **46**, 129–189.

PARROTT, P. J. and SCHOENE, W. J. (1912): The apple and cherry ermine moths. *Techn. Bull. New York Agric. Expt. Sta.* **24**, 1–40.

THORPE, W. H. (1929): Biological races in *Hyponomeuta padella* L. *J. Linn. Soc. Zool.* **36**, 621–634.

THORPE, W. H. (1930): Further observations on biological races in *Hyponomeuta padella* L. *J. Linn. Soc. Zool.* **37**, 489–492.

WIEGAND, H. (1962): Die deutschen Arten der Gattung *Yponomeuta* Latr. *Ber.* 9. *Wandervers. dtsch. Ent.*, *Tagungsber.* **45**, 101–120.

WILDBOLZ, TH. and RIGGENBACH, W. (1965): Beobachtungen und Versuche über Unterschiede zwischen der Apfelgespinstmotte und der Zwetschgengespinstmotte. *Schweiz. Ztschr. Obst-Weinbau* **101** (74), 105–111.

Symp. Biol. Hung. 16, pp. 95–99 (1976)

CHEMICAL AND BEHAVIORAL FACTORS INFLUENCING FOOD SELECTION OF *LEPTINOTARSA* BEETLE

by

T. H. Hsiao

DEPARTMENT OF BIOLOGY, UTAH STATE UNIVERSITY,
LOGAN, UTAH 84322, USA

Oligophagous insects are well suited for the study of interactions between plant chemicals and insect behavior because of their high degree of host plant specificity. Our studies of several Solanaceae feeders belonging to the genus *Leptinotarsa*, especially *L. decemlineata*, have demonstrated that both primary and secondary plant chemicals are indispensable in the regulation of feeding behavior (Hsiao and Fraenkel, 1968a,b,c,d; Hsiao, 1969). Primary plant chemicals such as sugars, amino acids, phospholipids, etc. are common to all plants and serve as essential phagostimulants to elicit initial feeding response and to maintain continuous feeding on the host-plant (Hsiao, 1972). Many secondary plant chemicals (e.g. alkaloids in Solanaceae) have now been shown to exert repellent, deterrent, and toxic effects, and their presence provides the basis of insect resistance in the majority of plants (Hsiao, 1974).

The role of secondary plant chemicals as host-specific token stimuli is less well understood and sometimes rather difficult to demonstrate experimentally. Considerable confusion exists in the terms currently used to describe the effects of these compounds on the feeding behavior of various phytophagous insects (Beck, 1965; Fraenkel, 1969; Schoonhoven, 1968). We have recently initiated a series of behavioral tests with larvae of several *Leptinotarsa* species to assess the influence that larval conditioning might have on the subsequent feeding preference for such host. Several experiments were also conducted on artificial diets fortified with leaf extracts of the test plant to define the roles of various chemical stimuli in the regulation of feeding behavior of these oligophagous insects. All our experiments were carried out in incubators maintained at a constant temperature of 26.5 °C and a photoperiod of 18 h light and 6 h darkness.

Rearings of newly hatched larvae of *L. decemlineata*, *L. texana*, and *L. haldemani*, on different food plants or on an artificial diet either before or after they had consumed their own egg shells, indicate that the larvae deprived of feeding on egg shell invariably initiated feeding more rapidly on their respective diets, but required a longer overall feeding period to reach the 2nd instar (Fig. 1). In comparison, larvae that were allowed to feed on their own egg shells grew faster and had a better survival rate, especially when reared on less acceptable food plants. The egg shells apparently provide the larvae with some nutrients and enhance subsequent feeding and growth. Wardojo (1969a) previously reported that larvae of the Colorado potato beetle when deprived of feeding on egg shells grew slower and produced smaller adults than normally reared individuals. In

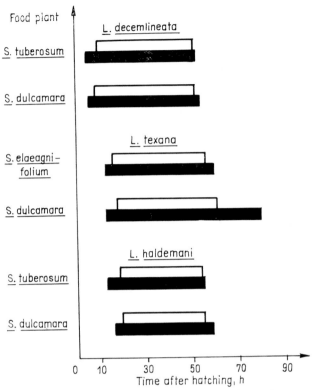

Fig. 1. Influence of feeding habits on rate of growth of 1st instar larvae of *Leptinotarsa* species fed on different *Solanum* plants. Data represent average from rearings of 50 larvae on each plant. Upper and lower bars represent, respectively, larvae that have and have not consumed their own egg shells before feeding on the plant. Each bar shows the time of initiation of feeding on the food plant and the time of moulting into 2nd instar

another experiment, larvae of *L. decemlineata* that had consumed their egg shells were offered the potato leaf or the artificial diet as food and both groups were observed to initiate feeding at about the same time (Fig. 2). However, the initiation of feeding was significantly delayed when a less acceptable plant such as *S. elaeagnifolium* was offered. The fact that newly hatched larvae were equally eager to feed, regardless of their prior feeding experiences, tends to suggest that the initial biting and feeding responses of the larvae are manifested as a consequence of an internal state of hunger rather than the influence of external chemical stimuli present in the host-plant. In other words, host-specific token stimuli apparently are not necessary to regulate feeding at this stage of larval development.

Influence of larval conditioning of a particular host on subsequent feeding preference was investigated with larvae of *L. decemlineata* (Fig. 3). First instar larvae that had not been exposed to any food source showed a definite preference for *S. tuberosum* over *S. dulcamara*, *S. elaeagnifolium*, or *Hyoscyamus niger* when offered a two-way choice of plants. Rearings of larvae on a particular host tended to induce a slight feeding preference for

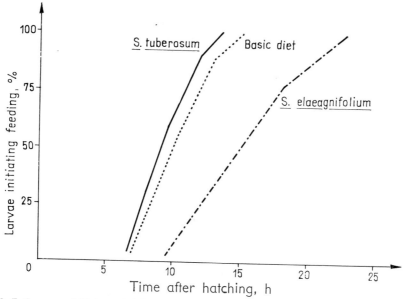

Fig. 2. Influence of diets on initiation of feeding of 1st instar larvae of *L. decemlineata*. Each curve represents response of 50 larvae that have consumed their own egg shells before feeding on the diets

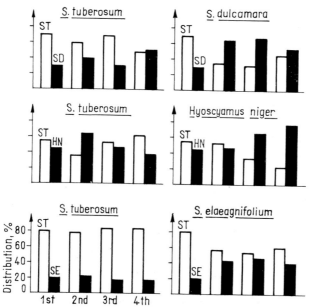

Fig. 3. Effects of larval conditioning on feeding preference of *L. decemlineata* larvae. Each plant species name refers to the host that the test larvae were reared from. Each data set represents the average response of 5 replicates of 10 larvae each conducted in a two-way choice experiment between *S. tuberosum* and one other plant. First instar larvae were not exposed to food plants before the test

TABLE 1

Growth and development of newly hatched 1st instar larvae reared on artificial diets fortified with potato leaf fractions

Diets	Initial No.	No. reaching 2nd instar	Development time (h)	
			Mean ± S.E.	Range
Basic diet (B.D.)	30	26	71.2±1.33	66–91
B.D. + 5% leaf powder	30	30	56.1±0.54	51–61
B.D. + 5% acetone extracted leaf powder	30	30	58.0±0.59	53–61
B.D. + 5% ethanol extracted leaf powder	30	30	59.6±0.70	50–66
Potato leaves	30	30	51.9±0.66	47–59

that plant. But the results of larval conditioning were not strong enough to demonstrate a major shift in feeding preference among the solanaceous plants tested, indicating that induction of feeding preference did not appear to play a significant role in host selection of the larvae. A similar conclusion was reached in the work of de Wilde et al. (1960) and Bongers (1970) who studied the effects of larval conditioning on adult feeding preference of *L. decemlineata* among *Solanum* species. However, our tests conducted with *L. decemlineata* reared on *S. elaeagnifolium*, a less acceptable host-plant, showed a gradual increase in tolerance to this plant with progressive larval instars. This finding suggests that sensory accommodation to deterrent compounds may be a more important factor in explaining the slight alteration in feeding preferences exhibited by larval conditioning on the various food plants tested rather than to infer the existence of an induced preference to any host-specific token stimuli.

An artificial diet (Wardojo, 1969b) that is adequate to induce larval feeding and growth was used to detect the presence of phagostimulants in the host-plants. Larval feeding preferences were determined with choice experiments on artificial diets with or without the addition of leaf extracts. First instar larvae of *L. decemlineata* that had not been exposed to food plants showed a significant preference for diets fortified with potato leaf powder, and leaf powders extracted with petroleum ether or acetone. After further extraction of these leaf powder fractions with alcohol, the resulting alcohol extract or residue portion were incorporated into the basic diet and both were preferred over the basic diet alone. This result indicates that more than one phagostimulant is present in the host plant. Growth experiments were also conducted with diets fortified with leaf fractions that had been extracted by acetone or ethanol. Table 1 shows substantial improvement in larval feeding and the duration of development on the leaf fraction fortified diets when compared to performance on the basic diet. It appears that nutritional factors from the host-plant may also be involved in promoting larval feeding and growth. The above findings are exactly the same as results obtained previously from 4th instar larvae reared on potato leaves (Hsiao, 1972). From this evidence, it would seem that larval responses to phagostimulants of the host-plant are innate, and are not a consequence of induced preference from rearing on the host. Since several phagostimulants and growth factors seem to be required to produce optimal larval response in feeding and growth and since the distribution of these compounds is not exclusive to solanaceous plants, it can be assumed that none of these substances can be considered as host-specific token stimuli for *L. decemlineata*.

From the available evidence, it may be concluded that the selection of food plant by larvae of *Leptinotarsa* species is influenced mainly by the qualitative and quantitative differences in deterrent and repellent chemicals of the plant. It is also evident that larvae of these species are highly discriminatory in their feeding habits and their feeding responses are initiated only by the presence of phagostimulants, both primary and secondary plant chemicals. These compounds act synergistically to induce initial feeding and to maintain continuous feeding on the food plants. Although larval responses to host-plant phagostimulants are innate, there is no experimental evidence to indicate that host-specific token stimuli from Solanaceae play significant roles in larval feeding behaviour. Our findings agree with the view of Jermy (1961, 1966) that the differences in sensitivity of chemoreceptors to deterrent chemicals may be the key factors in explaining the basis of oligophagy among phytophagous insects.

REFERENCES

BECK, S. D. (1965): Resistance of plants to insects. *Ann. Rev. Ent.* **10**, 207–232.

BONGERS, W. (1970): Aspects of host-plant relationship of the Colorado beetle. *Meded. Landbouwhogesch. Wageningen* **70** (10), 1—77.

FRAENKEL, G. (1969): Evaluation of our thoughts on secondary plant substances. *Ent. exp. & appl.* **12**, 473–486.

HSIAO, T. H. (1969): Chemical basis of host selection and plant resistance in oligophagous insects. *Ent. exp. & appl.* **12**, 777–788.

HSIAO, T. H. (1972): Chemical feeding requirements of oligophagous insects. In: Rodrigez, J. G. (ed.): *Insect and Mite Nutrition*. Amsterdam, North-Holland Publ. Co., 225–240.

HSIAO, T. H. (1974): Chemical influence on feeding behavior of *Leptinotarsa* beetles. In: Browne, L. B. (ed.): *Experimental Analysis of Insect Behaviour*. Berlin, Springer-Verlag, 237–248.

HSIAO, T. H. and FRAENKEL, G. (1968a): The influence of nutrient chemicals on the feeding behavior of the Colorado potato beetle, *Leptinotarsa decemlineata* (Coleoptera: Chrysomelidae). *Ann. ent. Soc. Am.* **61**, 44–54.

HSIAO, T. H. and FRAENKEL, G. (1968b): Isolation of phagostimulative substances from the host plant of the Colorado potato beetle, *Leptinotarsa decemlineata* (Say). *Ann. ent. Soc. Am.* **61**, 476–484.

HSIAO, T. H. and FRAENKEL, G. (1968c): The role of secondary plant substances in the food specificity of the Colorado potato beetle, *Leptinotarsa decemlineata* (Say). *Ann. ent. Soc. Am.* **61**, 485–493.

HSIAO, T. H. and FRAENKEL, G. (1968d): Selection and specificity of the Colorado potato beetle for solanaceous and nonsolanaceous plants. *Ann. ent. Soc. Am.* **61**, 493–503.

JERMY, T. (1961): On the nature of the oligophagy in *Leptinotarsa decemlineata* Say (Coleoptera: Chrysomelidae). *Acta Zool. Acad. Sci. Hung.* **7**, 119–132.

JERMY, T. (1966): Feeding inhibitors and food preference in chewing phytophagous insects. *Ent. exp. & appl.* **9**, 1–12.

SCHOONHOVEN, L. M. (1968): Chemosensory bases of host plant selection. *Ann. Rev. Ent.* **13**, 115–136.

WARDOJO, S. (1969a): Some factors relating to the larval growth of the Colorado potato beetle, *Leptinotarsa decemlineata* Say (Coleoptera: Chrysomelidae), on artificial diets. *Meded. Landbouwhogesch. Wageningen* **69** (16), 1–76.

WARDOJO, S. (1969b): Artificial diet without crude plant materials for two oligophagous leaf feeders. *Ent. exp. & appl.* **12**, 698–702.

DE WILDE, J., SLOOF, R. and BONGERS, W. (1960): A comparative study of feeding and oviposition preference in the Colorado beetle (*Leptinotarsa decemlineata* Say). *Meded. Landbouwhogeschool Opzoekingssta., Gent* **25**, 1340–1346.

Symp. Biol. Hung. 16, pp. 101–108 (1976)

INTERACTIONS BETWEEN THE HOST-PLANT AND MATING UPON THE REPRODUCTIVE ACTIVITY OF *ACANTHOSCELIDES OBTECTUS* FEMALES (COLEOPTERA, BRUCHIDAE)

by

J. HUIGNARD

LABORATOIRE D'ÉCOLOGIE EXPÉRIMENTALE, UNIVERSITÉ FRANÇOIS RABELAIS, PARC GRANDMONT, 37200 TOURS, FRANCE

In the absence of beans, the virgin females of *Acanthoscelides obtectus* do not lay eggs, but retain 30 to 40 oocytes in their lateral oviducts. The presence of beans induces oviposition in the majority of females though the oocytes are deposited at random in the rearing-box. The host-plant stimulates oogenesis but the magnitude of this stimulating effect changes in subsequent generations. Mating also induces egg-laying but its influence changes with strains selected according to their behaviour in the presence of the host-plant. Interactions between host-plant and mating are complex and this work shows the importance of the variations of the influence of an external stimulus in laboratory experiments.

In a certain number of phytophagous insects, the host-plant, which has no direct trophic action throughout imaginal life, controls, however, the reproductive activity of females. In effect, it can facilitate sex meeting (Riddiford and Williams, 1967; Rahn, 1968; Labeyrie, 1970, 1971), or induce egg-laying (Pass, 1967; Benz, 1969; Deseő, 1970; Labeyrie, 1970), thus allowing oviposition at a site favourable to the growth of larvae. Besides, there is ample evidence that a second factor: mating, can also regulate the reproductive function by acting upon fecundity or oogenesis (Engelmann, 1970). This regulation is very important in the bean weevil (*Acanthoscelides obtectus*) (Huignard, 1974). The author studied the interactions between these two factors in the bean weevil.

Two strains, selected by Labeyrie (1968) were used in the course of the experiments. With females of strain II, mating alone cannot induce egg-laying and the presence of beans is always necessary. On the other hand, females of strain III that have mated can lay their eggs even without beans. The mating alone induce egg-laying and the presence of host-plant is not necessary. This selection shows that the determinism of this behavioural aspect is not the result of ethological plasticity alone.

INFLUENCE OF THE HOST-PLANT ON VIRGIN FEMALES

Virgin females of both strains reared for 16 days in the absence of beans do not lay eggs, but retain 30 to 40 mature chorioned oocytes in their lateral oviducts. These oocytes are produced during the early days of imaginal life without any external stimulations, then oogenesis stops (Fig. 1). They represent the basic production that seems to have nothing to do with strain, and changes very little in the course of generations (Table 1).

When virgin females are reared in the permanent presence of beans for

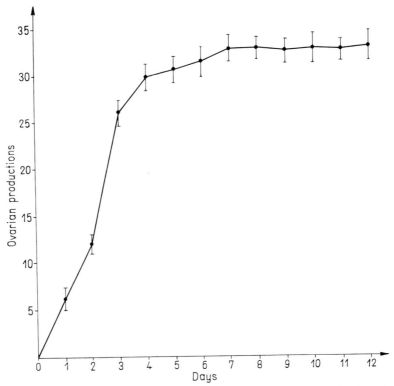

Fig. 1. Evolution of ovarian production of virgin females during the imaginal life in the absence of beans

TABLE 1

Evolution of basic production in the two strains over several generations

Generations	Strain II		Strain III	
	Fecundity	Ovarian productions	Fecundity	Ovarian productions
1	0	29.5 ± 1.1	0	30.2 ± 0.6
4	0	32.2 ± 1.4	0	33.2 ± 1.5
7	0	35.6 ± 1.6	0	35.8 ± 1.1
14	0	35.6 ± 1.3	0	32.2 ± 1.2
24	0	38.1 ± 3	0	34.1 ± 0.5

16 days, most of them lay eggs. Therefore, the presence of the host-plant allows oviposition, but the results found vary considerably. When examining throughout generations the distributions of virgin females of strain II according to their fecundity (Fig. 2) we can point out three cases:

a) Most females have a reduced fecundity, but a few of them react in the presence of the host-plant and all the produced oocytes are laid progressively.
b) The distribution is bimodal; thus the population consists of two types of females according to their reaction in the presence of the host-plant.

102

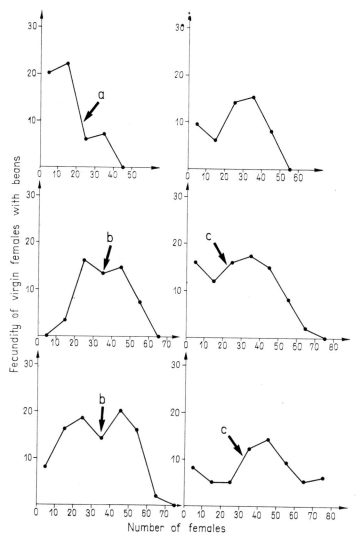

Fig. 2. Distribution of virgin females according to their fecundity in the presence of beans over several generations

c) Most females react in the presence of the host-plant and lay all the oocytes they produced; the distribution is unimodal.

Oviposition in virgin females always starts after significant delay (10 to 24 hours after the introduction of beans) whatever the age of insects or the state of retention of lateral oviducts. The fecundity, which is very low during the early days following the introduction of beans, increases progressively, then decreases at the end of the experiment (Fig. 3). The eggs laid are never clustered and seem to be deposited at random in the rearing-box.

The presence of the host-plant also stimulates oogenesis; as a matter of fact, the ovarian production of virgin females (number of eggs laid + num-

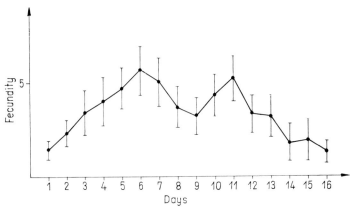

Fig. 3. Evolution of the daily fecundity of the virgin females

ber of retained oocytes) is higher than the basic production. The differences are always significant (Table 2). In the course of laboratory rearings, we have been able to observe in the two strains an increased magnitude of the stimulating effect due to the presence of host-plant throughout the generations. We may only speculate about the causes of these variations.

TABLE 2

Ovarian production of virgin females in the presence (HP+) or in the absence (HP—) of the host-plant over several generations

Strain II: generations 1,5,7,9,12,17
Strain III: generations 1,4,8,14,17,24

	Strain II			Strain III		
	Ovarian productions			Ovarian productions		
	HP—	HP+	t test	HP—	HP+	t test
Generations	29.1 ± 1.1	30.8 ± 0.7	3 (+)	28.2 ± 0.6	31 ± 0.9	2.8 (+)
	34.9 ± 1.9	38.4 ± 1.2	3 (+)	33.2 ± 1.5	39 ± 1.8	3.1 (+)
	34.6 ± 1.2	40 ± 2.7	3.8 (+)	34.2 ± 1.6	48 ± 3.7	4.2 (+)
	35.6 ± 1.3	53.2 ± 2.8	7.2 (+)	36.9 ± 3.2	56.3 ± 3.2	6.3 (+)
	37.6 ± 3.2	63.2 ± 2.9	8.6 (+)	36.3 ± 3.2	55.4 ± 3.7	5.4 (+)
	36.4 ± 3.4	62.1 ± 4.2	8.4 (+)	35.4 ± 2.6	56.3 ± 3.2	6.4 (+)

INFLUENCE OF THE HOST-PLANT ON FEMALES AFTER MATING

The experiments have been carried out in females of strain III, for it is possible to separate the influence of the host-plant, and the influence of copulation. After mating the females were isolated in the presence or absence of beans for 16 days.

a) *Isolated females without beans*

Mating strongly stimulates oogenesis (Table 3): ovarian production is significantly higher than in virgin females (control) kept under the same conditions (the difference is significant, $t = 18.9$).

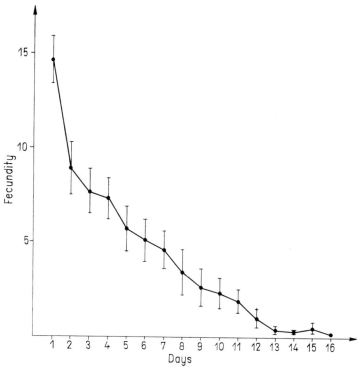

Fig. 4. Evolution of the daily fecundity of the strain III females after one mating in the absence of beans

TABLE 3

Ovarian production in the absence (HP—) or presence (HP+) of beans on the 16th day, in females of the same generation

	HP—		HP+	
	Number of females	Ovarian productions	Number of females	Ovarian productions
Mated females	73	61.5 ± 1.1	67	73.1 ± 2.1
Virgin females (control)	75	35.8 ± 1.1	62	56.3 ± 3

On the other hand, mating induces oviposition after a delay of 10 to 12 hours. Fecundity reaches its maximum 24 hours after mating (Fig. 4) then it gradually decreases till the end of the experiment. The eggs are clustered, generally stuck together in groups of 10 or 15 and stuck to the side of the rearing-box.

b) *Isolated females in the presence of beans*

In that case, their ovarian production is higher than without beans (the difference is clearly significant: $t = 7.7$). However, the stimulating effects due to the host-plant and to mating do not cumulate completely when they

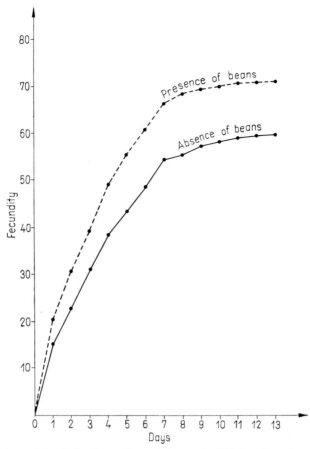

Fig. 5. Evolution of the daily fecundity of the strain III females after one mating in the absence or in the presence of beans (cumulative curve)

operate simultaneously. The ovarian production should be 82 (basic production + mating influence + bean influence) if the effects were totally cumulated, but it is only 73.1 (Table 4). The simultaneous presence of these two factors reduces the magnitude of their stimulating effects. Besides, we must take into account a limiting factor due to the female which produces only a definite number of mature oocytes probably depending on the quantity of fat-body reserves.

TABLE 4

Study of the simultaneous influence of host-plant and mating

Theoretical ovarian production: 82

1. Basic production
   +
2. Influence of mating } Ovarian production of mated females HP– *61.5*
   +
3. Influence of beans: 56.3–35.8 = *20.5*
   (virgin females with beans — basic production)

Experimental ovarian production: 73.1

106

The presence of beans reduces the delay between oviposition and mating (now it is only 3 or 4 hours) but the egg-laying rhythm is slightly altered (Fig. 5). The eggs are always clustered and are laid preferably under beans.

DISCUSSION

The various experiments show the respective rôles of the host-plant and of mating upon the reproductive activity of females. The host-plant can stimulate oogenesis and induce oviposition after a relatively long delay. Pouzat (1970, 1974) has shown that in *A. obtectus*, the signals coming from the bean are picked up by sensory organs and they probably regulate the reproductive function through the nervous system and haemolymph. Mating also effects oviposition but it has a more important and quicker action since a maximum fecundity period follows 24 hours later. It seems that the dilatation of bursa copulatrix due to the presence of spermatophore accounts for this action. The influence of mating upon oogenesis is mainly exerted through the haemolymph; in effect, certain male secretions contained in the spermatophore get into the haemolymph and can stimulate oogenesis (Huignard, 1974). This type of regulation should be determined precisely by research that is being carried out. Finally, mating influences the rhythm of egg-laying and allows the formation of egg-clusters; the process of this action, that seems to be related to the presence of spermatozoa in female genital tract (Huignard, 1974), is not yet known.

Finally, under normal conditions complex interactions must develop between the host-plant and mating when these two factors operate simultaneously. They provide a better adaptation of reproductive function to environmental changes and thus allow oviposition at a site favourable for larval development. The magnitude of the action of these two factors changes in different generations. These variations are probably in relation with the mode of selection of the strains in *A. obtectus*. Actually, we can suppose that the females with high fecundity giving a greater number of descendants are probably selected throughout generations. These experiments show the importance of these variations in laboratory rearings and the necessity of studying the influence of an external factor over several generations.

REFERENCES

BENZ, G. (1969): Influence of mating, insemination and other factors on oogenesis and oviposition in the moth *Zeiraphera diniana*. *J. Insect Physiol.* **15**, 55–71.
DESEŐ, K. V. (1970): The effect of olfactory stimuli on the oviposition behaviour and egg-production of some microlepidopterous species. *Colloque International du C.N.R.S.* **189**, 163–174.
ENGELMANN, F. (1970): *The physiology of insect reproduction*. Pergamon Press. Oxford.
HUIGNARD, J. (1974): Influence de quelques stimulations externes dues à la copulation sur la fonction reproductrice des femelles chez *Acanthoscelides obtectus*. *Ann. Sci. Nat.* (in press)
LABEYRIE, V. (1968): Longévité et capacité reproductrice des lignées d'*Acanthoscelides obtectus* sélectionnées en fonction de la réponse aux stimuli de ponte. *C.R. Soc. Biol.* **162** (12), 2203.
LABEYRIE, V. (1970): Influence déterminante du lieu de ponte sur la rencontre des sexes chez *Acanthoscelides obtectus*. *C.R. Acad. Sci. Paris* **271**, 1578–1581.

LABEYRIE, V. (1971): Trophic relations and sex meetings in insects. *Acta Phytopath. Acad. Sci. Hung.* **6**, 229–234.

PASS, B. C. (1967): Observations on oviposition by the alfalfa weevil. *J. econ. Ent.* **60**, 288.

POUZAT, J. (1970): Rôle des organes sensoriales céphaliques dans l'ovogenèse et l'émission chez la bruche du haricot (*Acanthoscelides obtectus* Say). *Colloque International du C.N.R.S.* **189**, 381–400.

POUZAT, J. (1974): Comportement de la bruche du haricot femelle, *Acanthoscelides obtectus*, soumise à différente stimuli olfactifs. *C.R. Acad. Sci. Paris*, **278**, 2173–2176.

RAHN, R. (1968): Rôle de la plante-hôte sur l'attractivité sexuelle chez *Acrolepia assectella*. *C.R. Acad. Sci. Paris*, **266**, 2004–2006.

RIDDIFORD, L. M. and WILLIAMS, L. M. (1967): Volatile principle from oak leaves: role in sex life of the *Polyphemus* moth. *Science*, **155**, 589–590.

Symp. Biol. Hung. 16, pp. 109–113 (1976)

INSECT—HOST-PLANT RELATIONSHIP — CO-EVOLUTION OR SEQUENTIAL EVOLUTION?

by

T. Jermy

RESEARCH INSTITUTE FOR PLANT PROTECTION, H–1525 BUDAPEST, PF. 102, HUNGARY

Co-evolution is the generally accepted theory for the evolution of insect—host-plant relationships, however, it can be shown that its main premises are inadequate: (1) most phytophagous insects have very low population densities compared to the biomass of their host-plants, therefore, they can hardly be important selection factors for the plant; (2) insect—host-plant interactions are not necessarily antagonistic: mono- and oligophagous insects, if their number is fairly high, may ideally regulate the abundance of their host-plants (mutual advantage); consequently, (3) resistance to insects is not a general necessity in plants and it cannot explain the presence of secondary plant substances; (4) parallel evolutionary lines of plants and insects which should result from co-evolutionary interactions are rare, while many closely related insects feed on botanically very distant plant taxa — a relationship which cannot be related to co-evolution.

Therefore, the theory of *sequential evolution* is proposed: the evolution of flowering plants propelled by selection factors (e.g. climate, soil, plant–plant interactions, etc.), which are much more potent than insect attacks, creates the biochemically diversified trophic base for the evolution of phytophagous insects, while the latter do not appreciably influence the evolution of plants.

A CRITICISM OF THE CO-EVOLUTIONARY THEORY

The insect—host-plant relationship represents a challenging problem for evolutionary considerations. The most generally accepted co-evolutionary theory has been expounded in detail by Ehrlich and Raven (1964) claiming that the trophic relations of phytophagous insects result from a very tight evolutionary interaction between plants and insects in which, on the one hand, the selection pressure represented by insect attacks induces the appearance of resistance mechanisms (mostly secondary substances) in the plants against the insects, while on the other hand, some insects succeed in overcoming this resistance by adapting themselves to these substances which may become feeding stimulants (token stimuli in host plant recognition). It is also supposed that such insects enter a new adaptive zone where they are free to diversify largely in the absence of other competing phytophagous insect species. The co-evolutionary theory is thought to be proved mainly by the fact that often closely related insect species feed on closely related plant species characterized by the presence of a group of specific plant chemicals (e.g. the lepidopterous tribe Pierini on cruciferous plants containing mustard oil glucosides). Fraenkel (1959) even asserts that the secondary substances in plants exist solely for the purpose of repelling and attracting insects.

In connection with the premisses of the co-evolutionary theory postulated by Ehrlich and Raven (1964) as well as implicitly or explicitly by many other authors, the following questions arise:

(1) Are phytophagous insects really important selection factors determining plant evolution?

(2) Are trophic plant–insect interactions always antagonistic from the evolutionary aspect, i.e. are the insects' attacks always "noxious" to the plant as a species?

(3) Is the plant–insect interaction the principal or perhaps the single reason for the existence of secondary plant substances?

(4) Are parallel evolutionary series (lines) proving co-evolutionary processes common in nature?

(5) Is the interspecific competition between phytophagous insects considerable?

Let us discuss these questions.

(1) To answer the first question one has to consider that most of the phytophagous insect species are rare in natural communities. Their population densities are most often extremely low compared to the available food resources. Bates (1958) is right to point out that "the great bulk of plant material never goes through the animal part of the energy-food cycle so neatly diagrammated in the textbooks." This applies the more so to phytophagous insects in natural communities where most of them are found only on a fraction of the individual specimens of their host-plant species. Since they affect only a very small part of the host-plant's population, their rôle as selection factor must be negligible compared to such potent environmental factors like soil, climate, plant–plant interactions within the community, diseases, etc. affecting regularly the whole or at least the major part of the population of a given plant species in a biotope.

It is also well known that those plant groups which are seldom attacked by phytophagous insects, like ferns, umbelliferous plants, etc., do not appear in great biomasses in most biocoenoses, while those groups which serve as a food for many insect species like graminaceous, leguminous plants, common forest trees, etc. are most abundant in natural communities disproving a noticeable population controlling effect of phytophagous insects.

The fact that introduced weeds have been successfully controlled in several cases by introduced phytophagous insect species is often referred to as a proof of the insects' population-regulating effect on plants. However, it has to be pointed out that both the introduced weed and its introduced phytophagous consumer are "strangers" in the given community which evolved without them. Therefore, students dealing with biological weed control rightly emphasize that phytophagous insects are effective control agents against weeds only because they are introduced without their natural enemies (Zwölfer, 1973). In the new environment they can develop high population densities when compared to the biomass of their host plant, which they can never reach in the community where they evolved. Thus, Wilson (1950) rightly concludes that if an introduced insect successfully controls an introduced weed, this does not mean that it is also an effective factor controlling the plant species's population density in its old environment. So the cases of successful biological control of weeds do not prove the insects being potent density controlling factors, i.e. selection factors for the plants.

110

(2) As far as the question of evolutionary antagonism between insect and plant species is concerned I should like to make reference to Rubtsov (1965) who pointed out that in the animal kingdom there is no indication of the development of a resistance against metazoan parasites. Rubtsov assumes that fatal parasitism is a built-in regulatory mechanism preventing overcrowding of the host which would result in exhausting the host's resources, i.e. a specialized parasite is the surest regulator of host population density.

It is common knowledge that in many plant species normal development is impeded by high population densities, therefore, it is logical to suppose that phytophagous insects may serve as specific regulators of host-plant abundance. In such cases the insects adapted to a given plant species serve as natural thinning-out factors promoting normal development of the remaining plant individuals, therefore, a resistant mutant of the plant would be handicapped in further spread as compared to the sensitive ones.

Such plant–insect relationships are mutualistic instead of antagonistic, i.e. they are advantageous for both parties, and they do not result in developing resistance against the specific insect partner.

Naturally, this applies only to insect species having high enough population densities as compared to the abundance of their host plant since rare species are not able to exert even such advantageous effects on the population of a common plant.

(3) Concerning the third question, it follows from the foregoing that the appearance of secondary plant substances cannot be solely the result of insect–plant interactions. In recent years it became evident that these substances are not waste products but intermediary steps of plant metabolism since they are continuously synthetized and decomposed (Zenk, 1968). They also turned out to be important information carriers in plant–plant interactions (allelopathy), and probably even more important as factors of resistance against diseases (Whittaker and Feeny, 1971). Therefore, resistance to insect attacks cannot be the "raison d'être" of these substances.

By the way, if we try to find by all means some important purpose for the existence of all secondary plant substances, we seem to forget that not all characters of an organism are involved in evolutionary selection mechanisms. There are characters which are neither advantageous nor disadvantageous for survival. It is very unlikely that, for instance, slight differences in the shape of the leaves of oak species had adaptive advantages or disadvantages affecting survival rate. The same must apply to many morphological characters by which plant or animal species can be distinguished. Why could we not suppose then that secondary plant substances represent, at least in part, only different possible pathways of metabolism which are in the same way species specific like many morphological characters, therefore, regarding evolution, they are also not necessarily advantageous or disadvantageous but often neutral. Lubischew (1969) was right in pointing out that natural selection is "a great destroyer, but a very feeble creator", which means that many characters were not created by selection. One can logically suppose also that many characters created by mutations are not destroyed by selection not only if they are advantageous but also if they are *neutral* for survival, like many morphological characters, and possibly also biochemical ones.

(4) If there were an effective mutual selection pressure between plants and insects decisively determining the evolution of both groups of organisms, then parallel evolutionary lines, i.e. series of closely related plant species attacked by closely related insect species should exist. However, such cases seem to be very rare (*Laspeyresia* moth species feeding on different rosaceous fruit trees can be mentioned as one of the very few examples). On the contrary, in most cases closely related insect species feed on botanically very distant plant species [e.g. *Rhagoletis cerasi* (Diptera, Trypetidae) on cherry, while the difficult to distinguish *Rh. berberidis* on *Berberis vulgaris*]; or closely related insect species live on the same plant species [e.g. *Phyllotreta* spp. (Coleoptera, Chrysomelidae) on the same cruciferous plant species, etc.]. Both latter types of relationships, which most frequently occur in nature, cannot be the result of insect and plant co-evolution as supposed by Ehrlich and Raven (1964), and others.

(5) To answer the question of the importance of interspecific competition in the evolution of phytophagous insects, one has to consider the already mentioned fact that in natural communities the population densities of most phytophagous insect species is always very low compared to the copiousness of available food. Inhomogeneous distribution as well as differences in the phenology of insect species further reduce the probability of competition for food. Consequently, competition must be negligible as a factor in the evolution of phytophagous insects. Thus, adaptation to a new host-plant cannot be regarded as an evolutionary "evasion" caused by interspecific competition.

THE THEORY OF SEQUENTIAL EVOLUTION

Müller (1975) has reported in this volume on a typical example of closely related insect taxa living on very different host-plants: This is most common in natural communities. Aphids, leafhoppers, fruitflies, bruchids, sawflies and many other groups of insects can be mentioned the species of which are sometimes difficult to distinguish by morphological characters but very well by their feeding habits (host-plants).

It is therefore logical to suppose that in many cases mutational changes in feeding behaviour represented enough isolation for sympatric speciation, thus, the flowering plants with their immense biochemical diversity serve as a base for the evolution of phytophagous insects, without an evolutionary feedback mechanism, i.e. in most cases without an appreciable selection effect exerted by the insects on the plants. This means that the evolution of phytophagous insects *follows* the evolution of plants, the latter being one of the most important selection factors in the evolution of insects.

This is the essence of the proposed theory of *sequential evolution* of phytophagy. This theory seems to be suitable for the evolutionary interpretation of all existing insect—host-plant relationships.

Finally, one more question arises, namely, what is the adaptive advantage of narrow host specificity for phytophagous insects? The question is the more logical since narrow specificity in feeding behaviour very often results in high mortality caused by the exhaustion of food resource, e.g. by total defoliation of the individual host-plant before larval development is completed, thus, mono- and oligophagy reduces the probability of survival, therefore, selection for polyphagy must be strong.

Kennedy (1953) emphasized that "the host plant is not merely something fed on, it is something lived on". It can be added that plants are the most sensitive bioindicators of a whole complex of ecological factors being effective in a given biotope, therefore, the choice of a plant species most often means also the choice of a well-defined combination of biotic and abiotic factors. Thus, the insect's chemoreceptors, being tuned to a plant species's biochemical profile consisting of primary and secondary plant substances, represent a reliable clue not only to a suitable source of food but also to a specific ecological situation to which the insect is adapted in all its requirements for development and reproduction. This must be the advantage of food specialization.

REFERENCES

BATES, M. (1958): Food-getting behavior. In: Roe, A. and Simpson, G. G. (eds.): *Behavior and evolution*. New Haven, 206–223.
EHRLICH, P. R. and RAVEN, P. H. (1964): Butterflies and plants: a study in coevolution. *Evolution* **18**, 586–608.
FRAENKEL, G. S. (1959): The raison d'être of secondary plant substances. *Science* **129**, 1466–1470.
KENNEDY, J. S. (1953): Host plant selection in Aphididae. *Trans. IXth Int. Congr. Ent.* **2**, 106–113.
LUBISCHEW, A. A. (1969): Philosophical aspects of taxonomy. *Ann. Rev. Ent.* **14**, 19–38.
MÜLLER, F. P. (1975): Hosts and non-hosts in subspecies of *Aulacorthum solani* (Kaltenbach) and intraspecific hybridizations (Homoptera: Aphididae). *Symp. Biol. Hung.* **16**, 187–190.
RUBTSOV, I. A. — Рубцов, (1965) И. А.: Антагонистические и мутуалистические отношения между хозяином и паразитом в природе. *Журн. общ. биол.* **26**, 166–175.
WHITTAKER, R. H. and FEENY, P. P. (1971): Allelochemics: chemical interactions between species. *Science* **171**, 757–770.
WILSON, F. (1950): Biological control of weeds. *New Biology* **8**, 51–74.
ZENK, M. H. (1968): Biochemie und Physiologie sekundärer Pflanzenstoffe. *Ber. Dtsch. Bot. Ges.* **80**, 573–591.
ZWÖLFER, H. (1973): Possibilities and limitations in biological control of weeds. *Bull. OEPP* **3** (3), 19–30.

Symp. Biol. Hung. 16, pp. 115–119 (1976)

THE ROLE OF VOLATILE OIL COMPOSITION FOR TRUNK PEST RESISTANCE IN CONIFEROUS PLANTS. EXPERIMENTS ON LUMBER

by

V. S. Karasev

LABORATORIYA ZASHTSHITY DREVESINY, UKRNIIMOD, KIEV–6, USSR

It has been found that the damage caused by *Callidium violaceum* L. in unbarked coniferous lumber stored within a year is in the average: in spruce 95%, in pine 60% and in fir 30% of its length. Some of fir lumber was not damaged at all. The amount of volatile oil (VO) in fir phloem was 7–8 times as much as in spruce. In laboratory tests the composition of VO (by gas-liquid chromatography method) and its toxicity (lethal concentration 50 by probit method) as well as the toxicity of terpenes obtained from the investigated species were determined. The VO toxicity in undamaged fir was 3.2 times as high as in spruce, and it was 1.5 times as high as in little damaged fir. Larval toxicity of substances present in the wood decreases in the following order: bornil acetate, Δ3-carene, alpha-pinene, VO of undamaged fir, VO of little damaged fir, beta-pinene, limonene, VO of spruce. The toxicity of VO in undamaged fir (per 1 mg of insect's weight) for young larvae was 34—39 times as high as for 4th—5th instar larvae.

The resistance of fir lumber to *C. violaceum* was found to increase with the increase of bornil acetate contents which proved to be the most toxic substance. Its amount when no damage occurred was 18.7%, in case of little damage (2 larvae per m) it was 5%, while in badly damaged spruce lumber (20 larvae per m) the substance was absent.

Thus, if oleoresin flow is absent which can be observed not only in timber, but also in weakened trees growing under unfavourable conditions, trunk pest resistance depends on the amount of VO and its composition. The latter determines toxicity of VO for insects.

Trunk pest resistance was found to increase with improving growth conditions (Rudnev, 1962). This may be explained by the different amounts of protective substances in plants, i.e. the galipot terpenes in coniferous trees (Rudnev, 1962, 1968).

Some authors showed that pest resistance in coniferous trees depends on the properties and toxicity of the galipot terpenes, which are determined by growth conditions (Rudnev and Smelyanets, 1969; Rudnev et al., 1969, 1970; Smelyanets and Akimov, 1969). There are data for some conifers (spruce) proving that the composition of galipot terpenes and their toxicity do not significantly change with the change of growth conditions and with the physiological state of the plants, where trunk resistance to pests is achieved by the increase of oleoresin flow, which pours over the attacking pests (Vasetschko, 1969, 1972).

In growing trees most probably both factors are available, but it is not known which has the primary effect. Tests have been made to observe the effect of terpenes with the second factor (oleoresin flow) being absent. For this purpose pest resistance in conifers (spruce, pine and fir) was studied on unbarked logs, which were attacked to different degrees during a one-year storage by *Callidium violaceum* (L.).

8*

It was shown that this insect infested mostly spruce lumber. 95% of the unbarked surface of the spruce was damaged by larvae whose number fluctuated between 10 and 20 per m length of lumber, reaching 40 in some cases. In pine lumber the damage was 60% and in fir 30% with 2 larvae per meter in the latter. Most of the fir lumber was not damaged at all though they were laying close to the badly damaged spruce timber.

The concentration of volatile oils (terpenes) in badly damaged spruce yielded 0.05% of fresh weight, and it was 7–8 times lower than that in fir (Table 1). The higher content of volatile oils in damaged fir (0.41%), compared to undamaged ones (0.35%) can be explained by the fact that the larvae damaged the inner part of the bark (phloem) where oleoresin blisters are absent, which resulted in the increase of terpenes in the uneaten rest of the bark.

The results of tests carried out in the laboratory by probit analysis for determining the lethal concentration evoking the mortality of 50% insects (LC–50) showed that volatile oil toxicity in spruce was 3.2 times lower than in undamaged fir. In fact the effect of volatile oil in spruce is much lower due to its low concentration in the bark. Volatile oil toxicity in damaged fir was found to be 1.5 times lower than in undamaged fir.

Analysis of terpenes, extracted from the bark of spruce, of slightly damaged fir, and of undamaged fir showed that with the increase of volatile oil toxicity the concentration of bornil acetate and Δ3-carene increased. In the badly damaged spruce bornil acetate was absent (Table 2). Results of gas-liquid chromatography analysis were in conformity with the data concerning the high concentration of bornil acetate in volatile oils in different species of fir (Kurth, 1952; Ivanov, 1961).

Laboratory tests on some substances present in volatile oils of the wood showed (Table 3) that bornil acetate and α-terpineol were the most toxic substances to *C. violaceum* larvae. That corresponded to the data of some authors (Smelyanets, 1967; Vasetschko and Kuznetsov, 1969; Vasetschko et al., 1970) about the high toxicity of these substances to coniferous pests (*Aradus cinnamomeus* Panz., *Ips typographus* L., *I. amitinus* Eichh.) as well as to *Tribolium destructor* Uytt. The toxicity values of the two former were followed by those of Δ3-carene (which was toxic to *A. cinnamomeus* and *I. typographus*). α-Pinene was less toxic than Δ3-carene, but was still more toxic than volatile oils of undamaged fir, β-pinene and limonene.

TABLE 1

Volatile oil content in bark of spruce and fir
lumber and their toxicity for 4th–5th instar larvae
of Callidium violaceum

Species of wood	No. of larvae per linear metre of lumber	Content of volatile oil in bark, %	LC-50 toxicity of volatile oil (microliters per 1 cm³ Petri dishes)	
			per 1 larva	per 1 mg of weight
Spruce	20	0.05	0.8992	0.006423
Fir	2	0.41	0.4656	0.003554
Fir	0	0.35	0.1690	0.001988

TABLE 2

Composition of volatile oils in bark of stored spruce, pine and fir lumber

Species of wood	X_1	α-Pinene	Cam-phene	β-Pinene	Δ3-Carene	Limo-nene	X_2	Bornil acetate
Spruce	0.1	54.1	8.4	19.2	3.7	14.3	0.2	0
Pine	0	90.5	1.2	0.8	2.15	4.95	0.4	0
Slightly damaged fir	0.05	51.25	10.25	18.3	4.9	10.25	0	5.0
Unattacked fir	0.6	46.1	11.5	3.7	5.1	5.9	3.4	18.7
Fir balsam	0	0	0	0	1.0	4.1	0	94.9

X_1 and X_2 unidentified terpenes.

TABLE 3

Toxicity of volatile oil components in studied coniferous lumber and of insecticides for 4th–5th instar larvae of Callidium violaceum

Studied substances	LC-50 (microlitres per 1 cm³ Petri-dishes)	
	per 1 larva	per 1 mg of weight
α-Pinene	0.1652	0.001636
β-Pinene	0.2630	0.002104
Δ3-Carene	0.1483	0.001107
Limonene	0.5226	0.004020
Bornil acetate	0.1129	0.000836
Volatile oil of fir balsam	0.1045	0.000757
α-Terpineol	0.0974	0.000964
Gamma-BHC	0.0019	0.000014
Dipterex	0.0060	0.000040

Larval toxicity of the substances studied decreased in the following order: lindane, dipterex, volatile oil of fir balsam, bornil acetate, α-terpineol, Δ3-carene, α-pinene, volatile oil of undamaged fir, β-pinene, volatile oil of slightly damaged fir, limonene, volatile oil of spruce.

Considering the fact that pest resistance of coniferous lumber can be achieved by terpenes which are more toxic than the volatile oil in undamaged fir, it should be concluded that the increase of concentration of bornil acetate, Δ3-carene and α-pinene enhanced biological resistance, but β-pinene or limonene concentrations decreased the same. α-Terpineol, which can be found in some pine species (Kurth, 1952) increases pest resistance.

Thus, high resistance of fir to *C. violaceum*, when oleoresin flow was absent, can be explained by high concentrations of terpenes in its bark due to bornil acetate, the most toxic substance to this pest.

The effect of protective system in the fir becomes apparent when we consider the fact that the resin ducts in fir are concentrated mainly in the outer bark, where a dense layer of oleoresin blisters is formed. Thus upon hatching the larvae are facing protective bar. Considering also an essentially (34–39 times) higher toxicity of volatile oil to young larvae compared to L_4–L_5 larvae tested (Table 4), it can be assumed that larvae die

117

TABLE 4

Volatile oil toxicity in fir for 1st–3rd, 4th and 5th instar larvae of Callidium violaceum

Instar of larvae	Average weight of larvae	LC-50 (microlitres per 1 cm³ Petri-dishes)	
		per 1 larva	per 1 mg of weight
1st–3rd	35	0.0209	0.000059
4th	85	0.1690	0.001986
5th	183	0.4247	0.002321

while boring into the bark. Although the toxicity of terpenes is much lower than that of lindane or dipterex, their high concentration in the outer bark makes the intrusion of *C. violaceum* larvae into fir lumber impossible.

Thus when oleoresin flow is absent which can be found not only in cut trees but also in those growing under unfavourable conditions, the most resistant ones for trunk pests are species with highly toxic terpenes. With the improving of growing conditions the increase of oleoresin flow in damaged areas intensifies its toxic effect on insects.

Such species as the spruce, the terpenes of which have a low toxicity irrespective of the physiological state, can resist trunk pests only if these trees grow under favourable conditions ensuring intensive oleoresin flow.

Confirming the opinion of Rudnev (1968) about the cardinal role of protective substances in plant resistance, our data showed that their effect on trunk pests can be different according to different species of trees and their physiological state.

ACKNOWLEDGEMENT

I am indebted to Dr. G. I. Vasetschko for his helpful comments on the manuscript.

REFERENCES

IVANOV, L. A. — Иванов, Л. А. (1961): *Биологические основы добывания живицю.* Гослесбумиздат, Москва—Ленинград, 1—290.

KURTH, E. F. (1952): The volatile oils. In: Wise, L. E. and John, E. C. (eds.). *Wood chemistry.* **1,** 548–589.

RUDNEV, D. F. — Руднев, Д. Ф. (1962): Влияние физиологического состояния растений на массовое размножение вредителей леса. *Зоол. журн.* **41** (3), 313—329.

RUDNEV, D. F. — Руднев, Д. Ф. (1968): Причины снижения устойчивости насаждений и условия образования очагов вредителей в лесах Украины. *Бюл.н.-т. информации УкрНИИЗР,* 39—59.

RUDNEV, D. F. (1972): Causes of reproduction of forest pests and the nature of resistance in planting. *XIIIth Int. Congr. Entomol., Moscow, 1968. Proceedings,* Leningrad, **3,** 86–87.

RUDNEV, D. F. and SMELYANETS, V. P. — Руднев, Д. Ф. и Смелянец, В. П. (1969): О природе устойчивости древесных насаждений к вредителям. *Зоол. журн.* **48** (12), 1802–1810.

RUDNEV, D. F., SMELYANETS, V. P., AKIMOV, YU. A. and LISHTVANOVA, L. N. — Руднев, Д. Ф., Смелянец, В. П., Акимов, Ю. А. и Лиштванова, Л. Н. (1969): Значение защитных веществ в устройчивости сосны против вредителей. *Лесное хозяйство,* **12,** 51–53.

RUDNEV, D. F., SMELYANETS, V. P., AKIMOV, YU. A. and LISHTVANOVA, L. N. — Руднев, Д. Ф., Смелянец, В. П., Акимов, Ю. А. и Лиштванова, Л. Н. (1970): Причины различной устойчивости сосны к ее вредителям. *Лесное хозяйство*. **12,** 68–72.

SMELYANETS, V. P. — Смелянец, В. П. (1967): Устойчивость сосен крымской и обыкновенной к вредителям на юге Украины. *Автореф. канд. дис.*, Киев, 1–22.

SMELYANETS, V. P. and AKIMOV, YU. A. — Смелянец, В. П. и Акимов, Ю. А. (1969): Устойчивость сосны к вредителям и биологически активные вещества. *Труды у Всесоюзного совещания по иммунитету растений*, Киев, 58–62.

VASETSCHKO, G. I. — Васечко, Г. И. (1969): Значение токсичности живицы в устойчивости против короеда-типографа. *Труды у Всесоюзного совещания по иммунитету растений*, Киев, 70–73.

VASETSCHKO, G. I. — Васечко, Г. И. (1972): Зависимость между смоляным давлением и устойчивостью ели против короедов. *XIIIth Int. Congr. Entomol., Moscow, 1968. Proceedings*, Leningrad, **3,** 104–105.

VASETSCHKO, G. I. and KUZNETSOV, N. V. — Васечко, Г. И. и Кузнецов, Н. В. (1969): Токсичность терпеноидов для короедов ели. *Химия в сельском хозяйстве*, **12,** 33—34.

VASETSCHKO, G. I., KUZNETSOV, N. V., SMELYANETS, V. P. and GUZENOK, N. H. — Васечко, Г. И., Кузнецов, Н. В., Смелянец, В. П. и Гузенок, Н. Х. (1970): Инсектицидные особенности некоторых компонентов эфирных масел. *ДАН УССР, Б*, **3,** 275–278.

Symp. Biol. Hung. 16, pp. 121–123 (1976)

HOST-PLANT FINDING BY FLYING APHIDS

by

J. S. Kennedy

IMPERIAL COLLEGE FIELD STATION, SILWOOD PARK,
ASCOT, BERKS. SL5 7PY, UK

As far as we know flying aphids show only a very limited amount of selective orientation toward plants, and what they do show is visual rather than olfactory (Kennedy and Stroyan, 1959). That is to say selection of the appropriate host plant depends primarily on arrestant stimuli received after landing. This gives rise to some real problems which have seldom been considered. If selection does indeed depend on random landings most of which, therefore, occur on non-host plants, then it seems the aphids must make repeated landings and take-offs if they are to carry out enough "scanning" (sampling) of the flora. In fact, Moericke (1955) described just such a behaviour pattern long ago under the name *Befallsflug* [translated into English as "attacking flight" (Johnson, 1958) or "alighting flight" (Kennedy et al., 1961)]. This characterizes the behavioural phase which follows the phase of *Distanzflug*.

In *Befallsflug*, landings and take-offs have both been observed as *active* events in the aphids that have been watched. In order to land properly it is first necessary for a flying animal, like any aircraft, to minimize its motion relative to the fixed landing point (its ground speed) before touch-down. This means, especially for a weak flier like an aphid, minimizing wind drift. Fortunately, close to the ground, which is where the plants to be sampled by the flying aphid are to be found, there is always a layer of more slowly moving air [Taylor's (1960, 1974) "boundary layer"] within which the wind speed is not greater than the airspeed of the flying aphids (something less than one metre per second). Given such a layer of retarded air movement, the aphids could in principle resist wind drift and regain control of their own displacement over the ground and thus make controlled, visually-aimed landings. But they could not do this unless they reacted to the wind in two ways: (i) orienting into the wind, and (ii) adjusting their airspeed according to the wind speed. These reactions would have the combined result of minimizing the wind drift to which the air-borne insect is inevitably subject.

Field observations made on *Myzus persicae, Aphis fabae* and *Brevicoryne brassicae*, and also, more conclusively, laboratory experiments on *Aphis fabae,* have demonstrated that when flying near the ground these aphids do, in fact, react to wind drift in those two ways (Kennedy and Thomas, 1974). Moreover, the laboratory experiments, using a wind tunnel with a movable striped floor, showed that the two "station-keeping" reactions, orientation upwind and regulation of the airspeed according to the wind speed, can be elicited in still air when the wind drift is simulated by moving the visible substrate. In other words these reactions to wind are *optomotor* reactions.

There is no evidence as yet as to whether optomotor reactions also give these aphids some control of their height of flight above the ground. Some such control is suggested by the observation that aphids engaged in *Befalls-flug* do confine themselves, at least for a few minutes at a time, to their boundary layer. Anyway, altitude control is clearly a second and indeed prior problem for a flier that has been travelling on the wind above the vegetation. It must now somehow descend into, and stay flying for a while within, the boundary layer. Part of the solution to this problem, too, seems to be visual.

Moericke's (1955) idea was that a migrant aphid was, in the first place, attracted upwards by the predominantly short-wavelength light from the sky, but later on, after flying for some time, its phototactic response switched over, becoming negative to the sky light but now positive to the longer-wavelength light reflected from the plants and soils below. This was such an attractive hypothesis that it has often been quoted as fact (Kennedy and Stroyan, 1959; van Emden et al., 1969; Wigglesworth, 1972), although it is no more than a reasonable inference from field observations and there has been no experimental evidence to support it. We now have some direct evidence (unpublished, but referred to in part by Kennedy and Fosbrooke, 1972) from experiments in a "flight chamber". This is essentially a vertical wind tunnel, the working section of which is lit through the central area of its roof screen so that the aphid flies upwards toward the light but is held below it by the air current blowing downwards (Kennedy and Ludlow, 1974). When the floor screen of the working section is black, flying aphids do eventually become negative to the light from the central window above. But, even then, they very rarely descend toward the floor; instead they fly horizontally away from the top light toward the dark walls. If, however, a patch in the floor of the working section is coloured orange, then eventually the aphids fly actively downwards toward this patch.

This seems to confirm Moericke's idea of a phototactic switch-over after a certain amount of flying. But there is a further problem here. If the fliers were indeed "pulled" phototactically first upwards, and then downwards, as he originally suggested, then the aphids could never indulge in any horizontal *Befallsflug* among the vegetation but would rather dive into it. It was therefore postulated (Kennedy et al., 1961) that there was a phototactic "pull" both upwards toward the sky, and downwards toward the earth, simultaneously, throughout flight. The "switch-over" seemed likely to be a matter of the downward pull increasing in strength relative to the upward one, rather than a simple switch-over from one to the other. Such a system would permit both descent into the boundary layer and, at times the striking of some sort of temporary balance between the two opposite "pulls", thus holding the height of flight roughly steady at one level, for instance within the boundary layer. Flight chamber experiments (Kennedy and Ludlow, unpublished) have now confirmed that there is such a balance between simultaneous upward and downward phototactic "pulls", at any rate while an aphid is flying upwards. Thus an orange-reflecting floor patch reduces an aphid's rate of climb towards the top light, as compared with its rate of climb when the floor is entirely black. We are still engaged in experi-

122

ments to establish whether the top light continues to pull an aphid upwards even after the aphid has begun to fly downwards. The preliminary results make this appear likely, but more data are needed.

Those, then, are some of the behavioral mechanisms (doubtless there are others) which enable these aphids to exercise selection among plants and these mechanisms must, therefore, be counted as part of the host-selection process. Without them, the insects would have no choice. Other flying insects, even if they do not rely solely on arrestants for host selection, will often require similar mechanisms to ensure effective approaches to and landings upon their host-plants.

REFERENCES

van Emden, H. F., Eastop, V. F., Hughes, R. D. and Way, M. J. (1969): The ecology of *Myzus persicae*. *Ann. Rev. Ent.* **14**, 197—270.

Jonhson, B. (1958): Factors affecting the locomotor and settling responses of alate aphids. *Anim. Behav.* **6**, 9—26.

Kennedy, J. S., Booth, C. O. and Kershaw, W. J. S. (1961): Host finding by aphids in the field. III. Visual attraction. *Ann. appl. Biol.* **49**, 1—21.

Kennedy, J. S. and Fosbrooke, I. H. M. (1972): The plant in the life of an aphid. *Symp. R. ent. Soc. Lond.* **6**, 129—140.

Kennedy, J. S. and Ludlow, A. R. (1974): Co-ordination of two kinds of flight activity in an aphid. *J. exp. Biol.* **61**, 173—196.

Kennedy, J. S. and Stroyan, H. L. G. (1959): Biology of aphids. *Ann. Rev. Ent.* **4**, 139—160.

Kennedy, J. S. and Thomas, A. A. G. (1974): Behaviour of some low-flying aphids in wind. *Ann. appl. Biol.* **76**, 143—159.

Moericke, V. (1955): Über die Lebensgewohnheiten der geflügelten Blattläuse (*Aphidina*) unter besonderer Berücksichtigung des Verhaltens beim Landen. *Z. ang. Ent.* **37**, 29—91.

Taylor, L. R. (1960): The distribution of insects at low levels in the air. *J. Anim. Ecol.* **29**, 45—63.

Taylor, L. R. (1974): Insect migration, flight periodicity and the boundary layer. *J. Anim. Ecol.* **43**, 225—238.

Wigglesworth, V. B. (1972): *The principles of insect physiology*. London, Chapman and Hall.

Symp. Biol. Hung. 16, pp. 125–127 (1976)

COLONIZATION SITES OF SCALE INSECTS (HOMOPTERA: COCCOIDEA) ON DIFFERENT PLANT SPECIES AND VARIETIES

by

F. Kozár

RESEARCH INSTITUTE FOR PLANT PROTECTION, H–1525 BUDAPEST, PF. 102, HUNGARY

Three groups of scale insects living on wooden plants can be distinguished considering the parts of plants on which they can reproduce. *Sphaerolecanium prunastri* and *Epidiaspis leperii* infest branches only, while *Parthenolecanium corni* (L_1) attacks also the leaves. The San José scale lives on all parts of the plants except the roots. Significant differences have been found in the intensity of San José scale infestation between trunks, branches and fruits of susceptible apple varieties (Húsvéti Rozmaring, Jonathan, Starking, Golden Winter Parmen).

These differences show that the susceptibility of all main parts of the plant should be tested.

In scale insects food selection is practically reduced to the selection of feeding site by the mobile first instar larvae; even this is very limited due to the short range of larval mobility. Suitability of a host plant can be measured by the population densities (number of scales per plant surface unit) resulting from the rate of development, fertility, mortality, sexual ratio and other factors.

The scale insects can be grouped according to their feeding sites on the host plant. The species *Sphaerolecanium prunastri* and *Epidiaspis leperii* infest only the branches of the tree. *Parthenolecanium corni* colonizes the twigs and leaves, and the San José scale, *Quadraspidiotus perniciosus*, settles down on every parts of the tree. The studies of plant susceptibility have to be carried out according to these features as the investigations in the first two types can be restricted to the study of plant parts while in the third group — represented by the San José scale — simultaneously *all* plant parts, i.e. the trunk, branches, twigs, leaves, and fruits have to be considered for establishing host plant susceptibility. In case of *Sphaerolecanium prunastri* and *Parthenolecanium corni* significant differences have been found in the susceptibility of peach varieties by studying the individual densities on the branches (Kozár, 1972).

By studying the infestation on branches and on the trunk, in the literature the different apple varieties were grouped into "susceptible" and "moderately susceptible" categories (Prints, 1964; Prints, 1971a,b; Thiem and Schetters, 1958; Zagaynyi, 1964). According to these studies the apple varieties: Starking, Húsvéti Rozmaring, Jonathan and Golden Winter Parmen have to be ranged into the category "susceptible" for the San José scale.

The infestation of these four varieties was surveyed in 1972, in a 20-year-old commercial orchard, showing a medium San José scale infestation.

From each variety 20 trees were examined in order to establish the degree of infestation on the bark (10 times 1 sq.cm), on the two-year-old twigs (10 times 10 cm) and on 25 fruits of each tree selected at random (Table 1).

TABLE 1
Colonization density of the San José scale

Part of plant	Apple varieties			
	Húsvéti Rozmaring	Starking	Jonathan	Golden Winter Parmen
Trunk (scale/sq.cm)	0.71	0.05[a]	0.05	0.20[c]
Fruit (scale/fruit)	0.54	0.72[a,b]	1.54	1.16[c]
Twig (2nd year) (scale/10 cm)	1.16	1.32[b]	1.10	0.85

[a] $P = 5\%$; [b] $P = 1\%$; [c] $P = 5\%$.

In the course of the investigations significant differences were recorded between the varieties and the plant parts. The heaviest trunk infestation was on Húsvéti Rozmaring, in contrast to Starking, whereas the fruits of Jonathan trees showed the highest population densities.

In the Golden Winter Parmen variety the fruits were more infested than the trunk; similar differences were noted in the case of Starking, but not only regarding fruit and trunk infestation but also fruit and twig infestation. In Jonathan and Húsvéti Rozmaring no significant differences were observed, in accordance with the data of Prints (1973) who made his surveys in Moldavia.

These observations showed that in case of the San José scale it is very difficult to establish "susceptibility" of a variety in general, as the evaluation considers rarely the different plant parts, which may have important practical and theoretical consequences. Owing to this fact earlier data on susceptibility may be used only with restrictions, in the case of plants showing an overall infestation. It is well known that plant- and variety-susceptibility change considerably according to different growing, climatical, and geographical conditions and due to the same factors also the infestation on different plant parts may be different affecting also the development of insect pests. As it is well known, in the case of the San José scale the sex ratio differs on the different plant parts, indicating also the necessity of differentiated evaluation methods.

It may be noted, however, that even with the San José scale it is not important to consider always all plant parts in each plant species or variety. So, for example, in most peach varieties the hairy fruits are not colonized by the larvae, nor are the rough, cracked surface and necrotized parts of the bark tissues.

The causes of differences in scale colonization are not fully understood. Some data indicate the importance of morphological and physiological differences in plant parts (Prints, 1971b).

Summarizing it I should like to draw the conclusion that in evaluating plant susceptibility the relation of different scale species to their host plant has to be considered. If plant susceptibility against the San José scale is assessed, it is advisable to make independent surveys on all parts of the trees, since on the same level of varietal susceptibility there could be considerable differences between the density of the scales on the different parts of the trees.

REFERENCES

KOZÁR, F. (1972): Susceptibility of peach varieties to infection by scale, with special regard to San José scale. *Acta Phytopath. Acad. Sci. Hung.* **7** (4), 409–411.

PRINTS, E. YA. — Принц, Е. Я. (1971a): Растения повреждаемые калифорнийской щитовкой в Молдавии, *Энтомофауна Молдавии* 90–96.

PRINTS, E. YA. — Принц, Е. Я. (1971b): Характер повреждения побегов и устойчивость сливы и яблони к калифорнийской щитовке. *Энтомофауна Молдавии* 82–85.

PRINTS, E. YA. — Принц, Е. Я. (1973): Реакция различных сортов яблони на повреждении плодов калифорнийской щитовкой. *Фауна и Биология Насекомых Молдавии,* Кишинев, 54–62.

PRINTS, YA. I. — Принц, Я. И. (1964): Изучение устойчивости яблони и сливы и разработка методов борьбы с калифорнийской щитовкой и задача науки. *Сборник работ по калифорнийской щитовке,* Кишинев, 27–28.

THIEM, H. and SCHETTERS, C. (1958): Die Vermehrung der San José-Schildlaus (*Aspidiotus perniciosus* Comst.) in Abhängigkeit vom Zustand der Wirtspflanzen und der Stärke ihrer Erstbesiedelung. *Pflanzenschutz* München **10** (12), 138–142.

ZAGAYNYI, S. A. — Загайный, С. А. (1964): Кормовые растения калифорнийской щитовки. *Сборник работ по калифорнийской щитовке,* Кишинев, 52–59.

Symp. Biol. Hung. 16, pp. 129–132 (1976)

THE INFLUENCE OF LINOLEIC AND LINOLENIC ACIDS ON THE DEVELOPMENT AND ON THE LIPID METABOLISM OF THE COLORADO POTATO BEETLE (*LEPTINOTARSA DECEMLINEATA* SAY)

by

J. Krzymańska

INSTITUTE FOR PLANT PROTECTION, 60318 POZNAŃ, POLAND

The influence of linoleic and linolenic acids on the physiology of Colorado potato beetle was studied. The larvae and the beetles were fed with food supplemented with linoleic and linolenic acids. Observations were made on fertility, mortality, activity and feeding habits. The metabolism of fatty acids was studied, too.

All results showed that unsaturated fatty acids are very important factors in the physiology of Colorado potato beetle. The conclusion was made that linoleic acid possesses greater physiological value for Colorado potato beetle than linolenic acid. It was established that the composition of the diet directly influenced the composition of fatty acids in the insect's lipid.

INTRODUCTION

It is well known that deficiency of polyunsaturated fatty acids in the animal food causes certain disturbances in the animal's physiology. In general this deficiency is reflected by a low survival rate among the newborns by subnormal growth, skin lesions, and impaired ability to reproduce (Wagner and Folkers, 1964). Abnormal distribution of polyenoic acids, an increase of metabolic rate and marked increase in water intake are also connected with polyunsaturated fatty acid deficiency (Nunn and Smedley-Mac Lean, 1938). The disorder in the fatty acid metabolism caused by this deficiency is indicated by an increase of oleic, palmitoleic and trienoic acid content in organ and depot fat and in plasma lipids. A simultaneous decrease in linoleic, arachidonic, pentaenoic and hexaenoic acids in the plasma and erythrocytes were also observed (Holman, 1960; Greenberg and Moon, 1961). According to the results of recent research, polyunsaturated fatty acids are the precursors of prostaglandins, compounds possessing a great biological activity. It is very probable that among others these also act against arteriosclerosis (Ramwell and Shaw, 1971; De Cicco, 1971).

The question of polyunsaturated fatty acid requirement by insects is not a simple one. There are remarkable differences in this respect among the different species of insects. In some cases they are not necessary for growth and development, but most species show requirement for these acids, i.e. for linoleic and linolenic acids (Vanderzant et al., 1958).

Generally speaking, the quantity and quality of lipids in food and their influence on the metabolism is of fundamental significance in insect physiology.

Fat is the main source of energy especially in the critical periods of insect development, as in hibernation, starvation and metamorphosis (Langhlin, 1956; Villeneuve and Lemonde, 1963), as well as in reproduction.

9

In previous investigations (Krzymańska, 1962) the quantity of total lipids and unsaturated fatty acids in Colorado potato beetle food has been determined. It was found that the content of total lipids, as well as the content of linoleic and linolenic acid increases and simultaneously the content of oleic acid decreases during the vegetation period. The content of linolenic acid was higher than that of linoleic acid.

The aim of the present work is to examine and compare the effect of linoleic and linolenic acid on the physiology of the Colorado potato beetle.

MATERIALS AND METHODS

In the biological experiments potato leaves supplemented with linoleic and linolenic acids were used as insect diet. In each treatment 500 L_1 instar larvae were used at the beginning of experiment. After reaching L_4 instar one part of larvae was taken for biochemical analyses and the others were left to pupate. The rearing experiments were continued with the newly hatched adults. The beetles were reared in special isolators — 10 females and 5 males in each. For all experimental treatments 4 repetitions (4 isolators) were used. Potato leaves given as a food were sprayed with the ether solutions of fatty acids. The quantities of added fatty acids were twice higher in the supplemented food than in the natural food.

The development of larvae and beetles and their activity, mortality, feeding behaviour, and chiefly the reproduction of beetles were regarded as biological criteria for the influence of unsaturated fatty acids on the physiology of Colorado potato beetle.

In the biochemical investigation the total lipid content and composition of fatty acids was determined in L_4 larvae and in adults.

The content of the total lipid was determined by the method of Folch et al. (1957) using the mixture of chloroform and methanol (2 : 1) as a solvent for extraction. The composition of fatty acids was examined by gas-liquid chromatography (Krzymańska, 1962).

RESULTS AND DISCUSSION

Results obtained from the experiment regarding the length of time of the adult's activity showed that there were no significant differences between the different treatments. There were some differences in activity time between particular years, but there were no differences in the activity of beetles from treatments with supplemented food and control. It can be asserted with high probability that the examined compounds had no effect on the duration of beetle activity.

The mortality of beetles was different in particular treatments and in particular years. However, we have not found any correlation between the enriched food and the mortality of beetles.

The mortality occurred mainly in the final period of the rearing time and it had no influence on the fecundity of the adults, which was the main criterion of the biological effect of the essential fatty acids on the insect.

130

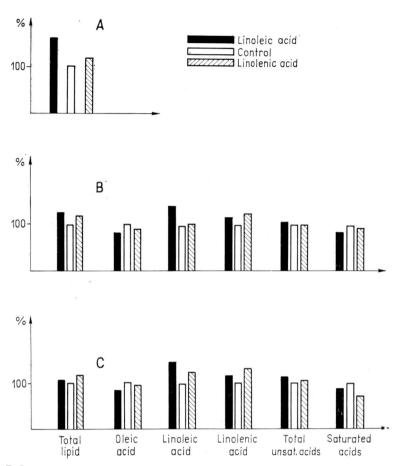

Fig. 1. Influence of linoleic and linolenic acids on the reproduction and lipid content
of the Colorado potato beetle
A: Reproduction of beetles in experiments with food supplemented with linoleic and
linolenic acid
B: Content of total lipid and fatty acids in L_4 larvae
C: Content of total lipid and fatty acids in the beetles

On the basis of observations made regarding the quantity of the consumed food it may be stated that the addition of the examined compounds
did not affect feeding behaviour either. Also the amounts of leaves eaten
by adults were almost equal for all rearing treatments, so in fact beetles
have consumed different quantities of the examined substances.

As it was mentioned above, the reproduction of the adults was taken
as a main biological criterion in our experiments, and in this regard the
most important differences took place. The results are shown in Fig. 1A,
where the amount of eggs laid by the females from the control rearings is
taken as 100%, the amount of eggs from experimental rearings with linoleic,
and linolenic acid, were 167% and 121%, respectively. Comparing these
results it can be stated that essential fatty acids — chiefly linoleic acid —
affect directly the fecundity of the Colorado potato beetle.

On the basis of biochemical investigations it can be stated that by enriching the food with the essential fatty acids an increase in the unsaturated fatty acid content in the lipids of larvae and adults occurs. The direct dependence between the composition of diet fat and that of insect fat was observed. Results are given in Figs 1B and 1C.

By comparing the analytical results in respect to lipid compounds in larvae and in beetles, it was established that the total lipid content was higher in the adults, but higher content of unsaturated fatty acids was found in larval lipid. This suggests that in the Colorado potato beetle, like in other insects, the ingestion of large quantities of unsaturated fatty acids in larval stage results in proper pupal development.

In summing up it can be stated that both unsaturated fatty acids examined, but mainly linoleic acid, belong to the compounds of great biological activity for the Colorado potato beetle. This is manifested in their influence on fatty acid metabolism, which is reflected in the fecundity of the insect.

REFERENCES

DE CICCO, A. (1971): Physiological action of polyunsaturated fatty acids. *Acta Vitaminol. Enzymol.* **25**, 13.

FOLCH, J., LESS, M. and SLOANE, S. G. H. (1957): A simple method for the isolation and purification of total lipids from animal tissues. *J. Biol. Chem.* **226**, 497.

GREENBERG, L. D. and MOON, H. D. (1961): Alterations in the blood fatty acids in single and combined deficiencies of essential fatty acids and vitamin B_6 in monkeys. *Arch. Biochem. Biophys.* **94**, 405.

HOLMAN, R. T. (1960): The ratio of trienoic : tetraenoic acids in tissue lipids as a measure of essential fatty acid requirement. *J. Nutrit.* **70**, 405.

KRZYMAŃSKA, J. (1962): Badania nad lipidami lisci ziemniaka i ich wplywem na rozwój stonki ziemniaczanej (*Leptinotarsa decemlineata* Say). *Prace Nauk IOR* **4**, 53–100.

LANGHLIN, R. J. (1956): Storage and utilization of reserves by the garden chafer (*Phyllopertha horticola* L.). *J. Exp. Biol.* **33**, 566.

MEAD, J. F., SLAFON, W. H. and DECKER, A. B. (1956): Metabolism and essential fatty acids. *J. Biol. Chem.* **218**, 401.

NUNN, L. C. and SMEDLEY-MACLEAN, (1938): Studies of the essential unsaturated fatty acids in their relation to the fat-deficiency disease of rats. *Bioch. J.* **32**, 2178.

RAMWELL, P. W. and SHAW, Y. E. (1971): The biological significance of the prostaglandins. *Ann. N. Y. Acad. Sci.* **180**, 10.

VANDERZANT, E. S., KERUR, D. and REISER, R. (1958): The role of dietary fatty acids in the development of the pink bollworm. *J. econ. Ent.* **50**, 606.

VANDERZANT, E. S. and RICHARDSON, C. D. (1964): Nutrition of the adult boll weevil: lipid requirements. *J. Insect Physiol.* **10**, 267.

VILLENEUVE, J. L. and LEMONDE, A. (1963): Variations des lipides en cours de la métamorphose de *Tribolium confusum* Duval (Coleoptère: Tenebrionidae). *Arch. Int. Physiol. Biochim.* **71**, 143.

WAGNER, A. and FOLKERS, K. (1964): *Vitamins and Coenzymes*. New York.

Symp. Biol. Hung. 16, pp. 133–136 (1976)

THE IMPORTANCE OF THE CO-EVOLUTIVE POINT OF VIEW IN THE INVESTIGATION OF THE REPRODUCTIVE RELATIONS BETWEEN INSECTS AND HOST-PLANTS

by

V. Labeyrie

LABORATOIRE D'ÉCOLOGIE EXPÉRIMENTALE, UNIVERSITÉ
FRANÇOIS RABELAIS, PARC GRANDMONT, 37200 TOURS, FRANCE

In many cases, like feeding, reproduction is connected with the trophic relations. Signals produced by trophic basis induce different reproduction sequences. For example, the production of pheromones by *Acrolepia assectella* female is induced by leek leaves. On the other hand, leek leaves stimulate ovarian activity and oviposition.

In some cases an efficient signal for meeting of sexes is not produced by the food itself, but by an other element of the environment actually associated with the trophic basis. For example, both sexes of *Philophylla heraclei* are attracted by trees near the celery field. The chemical attraction on insects or the chemical action upon their reproductive physiology, is generally produced by allelochemics developed by plant during the process of evolution. The value of these signals is not the same for all the species of consumers. In a general way, it is a barrier, and only some oligophagous or monophagous species evolved to surmount it. Only a co-evolutive process makes them transgress these protective barriers of the plant. By this way the same chemical signal is attractive on some species and repulsive on other ones. So is it as far as leeks are concerned.

We have the ability to use this selective evolution to protect the plants against pests by the introduction of a specific chemical signal in another trophic chain.

First, it is necessary to state my concept of co-evolution. There is co-evolution when several species evolve with mutual relations, particularly when there are trophic relations. They evolve together. With this definition we do not need to know if there are mutualistic advantages or antagonistic relations.

Besides, the advantageous or antagonistic notions have not the same content at the individual level as at the population level. For example, when a weak plant is destroyed selectively by a consumer, for instance when *Pinus* trees are destroyed by *Scolytidae*, this is a dramatic event for these trees, but the disappearance of these phenotypes is perhaps an advantageous thing for the *Pinus* population. On the other hand, at the population level "advantageous" or "antagonistic" are relative terms.

The presence of consumers, a density-dependent factor of the host in the system, may, when the system is functioning well, promote homeostasis of the host population. Thus, Pimentel and Soans (1970) underline that the two species profit from an economy of stable production and consumption.

Also the dispersion of the hosts, a very efficient protection against consumers, is, to a large extent, advantageous for the consumers too, since it promotes displacements and facilitates panmixis.

Hence each factor may be advantageous or mutualistic within a certain range and antagonistic outside these limits.

The second concept which is necessary to define is that of adaptation. In this paper there is only the question of genetic adaptation, because physiological or ethological plasticity are to a large extent genetically defined. For example, for more than ten years we have been working with selected strains of *Acanthoscelides obtectus* (Coleoptera, Bruchidae) in connection with their responsiveness to the egg laying stimuli provided by the host (Labeyrie, 1960). We have a strain unable to oviposit in the absence of *Phaseolus vulgaris* seeds, and another strain able to oviposit in empty petri dishes. The comparison of these strains has permitted (Huignard, 1970) to precise the influence of various host-stimuli upon oogenesis and to determine (Pouzat, 1970) their influence upon the value of the signals received by the sensory receptors.

Thus, adaptation is never the result of an isolated selective pressure. Each pressure is a part of the overall selective pressure of a certain environment.

For these reasons in a given environment a particular selective pressure may induce a directed selection if this factor is, at a particular time, the dominant selective factor. When there are repetitions of this situation, this factor is able to direct the selection.

For example, in plants it may promote the occupation of habitats that are unfavourable in many aspects, but advantageous because they reduce the selective pressure from a too heavy consumer. Such stations act like shelters and favour the species survival and renewed colonization of the forsaken habitats.

Also consumers may forsake a usable host when its occurrence is too irregular, for example, *Cassia grandis* consumed by Bruchidae (Janzen, 1973). In these conditions a new host, more regular in occurrence, will become the new principal host. This is the case with many insects that use preferentially cultivated imported crops in an ecosystem. The absence of crop rotation favours this new trophic relation.

In this context of complex pressures the host-consumer co-evolution can produce real evolutionary escalation: During this escalation specialized structures and specific compounds of the host eliminate many consumers. But, at the same time, some consumers evolve by using these structures and these compounds. To understand this mechanism it is necessary not to forget that efficient policemen produce efficient thieves and that efficient thieves produce efficient policemen.

In plants, such structures like spines, or trichomes, or waxes make up barriers against consumers in general but *Opuntia* spines are used by *Cactoblastis cactorum* for oviposition (Monro, 1967).

In the experimental ecology laboratory we study the attractive substances of *Allium porrum*, for *Acrolepia assectella (Lepidoptera, Plutellidae)*. But the complex of chemicals produced by the wounding of a leaf is, at the same time, a repellent material for the other phytophagous insects, and even toxic for several species. For example a piece of leek leaf in *Drosophila* cage kills them. These substances possess also a bactericidal action against pathogens, potential invaders of the wound, and even against soil bacteria. Perhaps, the same substances favour the germination of leek seeds and inhibit the germinaceous seeds, competitors of leek.

This evolutionary co-specialization has some mutualistic aspects for the plant and the consumer.

It reduces interspecific competition for the specialized and victorious consumer. And, if the trophic chains were actually isolated, it might favour a total homeostasis with regular population fluctuations and total density dependence.

If we study the consumer's side of the co-evolutionary process, we observe that it affects many aspects of its biology. Earlier I underlined the importance of trophic relations in the sexual behavior of insects by reference to two kinds of mechanism. One of them arises when the host-plant induces pheromone production and so meeting of the sexes. This phenomenon has been observed in *Antheraea polyphemus* (Riddiford and Williams, 1967) and in *Acrolepia assectella* (Rahn, 1968).

The second mechanism, where both sexes are simultaneously attracted by the plant, which thus plays the role of "lieu de rendez-vous" described by Prokopy in 1968 in *Rhagoletis pomonella*, and observed in *Urophora siruna-seva* (Zwölfer, 1969), exists also in Coleoptera like *Acanthoscelides obtectus* (Labeyrie, 1970a).

But this co-evolutionary sexual behaviour is not necessarily induced by host signals. The consumer does not react only to the host signals. It reacts to all the signals of the host environment. It reacts to a complex integration of signals. Perhaps, this aspect allows to understand the choice of oviposition site when the trophic level ability has not yet appeared (Labeyrie and Huignard, 1973). Maybe it is important for an understanding of the choice of oviposition site by *Locusta migratoria*.

Lewi (in press) confirms this hypothesis at the level of sexual behaviour, in the case of *Philophylla heraclei* (Diptera, Trypetidae). In this species, mating is not carried out upon the host-plant, *Sepium graveolens* (Umbelliferae) but under the trees near the host-plant. The adults feed upon honey dew of Aphids and Coccids upon extra floral nectaries. But *Sepium graveolens* grows upon wet soils characteristic of forests. So, the presence of trees is a permanent element of the natural host-plant environment. It is in connection with signals from this linked element that the co-evolution develops, and that this kind of sexual behaviour appeared.

I will not indicate the other aspects of the action of the host-plant upon the other sequences of reproductive biology. A previous symposium (C.N.R.S., 1970; Labeyrie, 1970b) has brought many examples of these ecological relations.

To conclude, crop protection is an applied aspect of population dynamics, and population dynamics has an evolutionary aspect, but this evolution is difficult to understand without the examination of trophic relations.

It is possible to use these co-evolutionary relations in order to protect the crop, not only by using natural pesticides produced by the plant, but also by using lures. These lures are natural mimetic structures or chemical compounds. It is possible to use them to thwart the co-evolutionary relations between host and consumers.

Thus Jones et al. (1970) proposed using natural attractants for *Heliothis zea*, produced by *Zea mays*, to induce oviposition by this insect in a place where larval growth is impossible.

On the opposite side Robert (1975) proposes to use extracts of *Castanea sativa* to inhibit the oviposition of *Scrobipalpa ocellatella* upon *Beta*.

These are only particular examples and I think it is possible to increase the ways of efficient and clean pest control by using the co-evolutionary aspects of population dynamics.

REFERENCES

C.N.R.S. (1970) *Colloque International du C.N.R.S.* **189**, *L'influence des stimuli externes sur la gamétogenèse des insectes.* Tours, 1969.

HUIGNARD, J. (1970): Analyse expérimentale de certain stimuli externes influençant l'ovogenèse chez *Acanthoscelides obtectus* Say (Coléoptère, Bruchidae). In: *Colloque International du C.N.R.S.* **189**, *L'influence des stimuli externes sur la gamétogenèse des Insectes,* 357–380.

JANZEN, D. H. (1973): Tropical agroecosystems. These habitats are misunderstood by the temperate zones, mismanaged by the tropics. *Science,* **182**, 1212–1219.

JONES, R. L., BURTON, R. L., BOWMEN, M. C. and BEROZA, M. (1970): Chemical inducers of oviposition for the corn earworm, *Heliothis zea* (Boddie). *Science* **168**, 856—857.

LABEYRIE, V. (1960): Action de la présence de grains de haricot sur l'ovogenèse d'*Acanthoscelides obtectus* Say. *C. R. Acad. Sci. Paris* **250**, 2626–2628.

LABEYRIE, V. (1970a): Influence déterminante du lieude ponte sur la rencontre des sexes chez *Acanthoscelides obtectus. C. R. Acad. Sci. Paris* **271**, 1578–1581.

LABEYRIE, V. (1970b): Signification adaptative de l'intégration des signaux fournis par le milieu extérieur lors de l'ovogenèse des insectes. In: *Colloque International du C.N.R.S.* **189**, *L'influence des stimuli externes sur la gamétogenèse des Insectes,* 21–43.

LABEYRIE, V. and HUIGNARD, J. (1973): Relations trophiques et comportement reproducteur des insectes. *Ann. Soc. Royale Zool. Belgique* **103** (1), 43–51.

MONRO, J. (1967): The exploitation and conservation of resources by populations of insects. *J. Anim. Ecol.* **36**, 531–547.

PIMENTEL, D. and SOANS, O. (1970): Animal populations regulated to carrying capacity of plant host by genetic feed-back. *Proc. Adv. Study Inst. Dyn. Numb. Popul. Osterbeck.*

POUZAT, J. (1970): Rôle des organes sensoriels céphaliques dans l'ovogenèse et l'émission chez la Bruche du Haricot, *Acanthoscelides obtectus* Say. In: *Colloque International du C.N.R.S.* **189**, *L'influence des stimuli externes sur la gamétogenèse des Insectes,* 381–400.

PROKOPY, R. J. (1968): Visual responses of apple maggot flies *Rhagoletis pomonella* (Diptera, Tephritidae): orchard studies. *Ent. exp. & appl.* **11**, 403–422.

RAHN, R. (1968): Rôle de la plante hôte sur l'attractivité sexuelle chez *Acrolepia assectella. C. R. Acad. Sci. Paris* **266**, 2004–2006.

RIDDIFORD, L. M. and WILLIAMS, L. M. (1967): Volatile principle from oak leaves: role in sex-life of Polyphemus moth. *Science,* **158**, 139–141.

ROBERT, P. CH. (1975): Inhibitory action of chestnut-leaf extracts (*Castanea sativa* Mill.) on oviposition and oogenesis of the sugar beet moth (*Scrobipalpa ocellatella* Boyd.; Lepidoptera, Gelechiidae). *Symp. Biol. Hung.* **16**, 223—227.

ZWÖLFER, H. (1969): *Urophora siruna-seva* (Diptera, Trypetidae) a potential insect for the biological control of *Centaurea solstitialis* in California. *Techn. Bull. Com. Inst. Biol. Contr.* **11**, 105–155.

136

Symp. Biol. Hung. 16, p. 137 (1976)

FINDING OF FEEDING AND EGG-LAYING SITES BY THE MIGRATORY LOCUST, *LOCUSTA MIGRATORIA* L.

by

J. R. LE BERRE and H. LAUNOIS-LUONG

LABORATOIRE D'ENTOMOLOGIE ET D'ÉCOPHYSIOLOGIE EXPÉRIMENTALES,
FACULTÉ DES SCIENCES, UNIVERSITÉ PARIS-SUD, 91405, ORSAY, FRANCE

(Summary only)

Many authors have investigated the search and finding of host-plants by insects; the majority of the mechanisms studied referred to the vision and olfactorial orientation.

In the migratory locust (solitary phase) two principal groups of behavioural reactions have been established:

— search for feeding and egg-laying sites;

— host-plant selection and search for sites of egg-laying inside these regions.

In the present short report we have dealt only with the first group of behavioural reactions.

— In a biotope of mosaic type the females are not distributed at random. During the sexual maturation and pre-oviposition they prefer the environment with medium humidity (mesophily). After mating the fertilized females move to humid or even very humid habitats (hygrophily).

— The composition of the vegetation can be decisive in some cases, but there is no strict relationship between the presence of preferred plant species and the distribution of females. The knowledge of microclimate (phytoclimate) seems to be more important.

— The actographic data are influenced by the physiological state of the females collected in the Nature. Especially starving and the presence of eggs ready to be laid bring about a remarkable increase in locomotive activity.

It is thus a whole series of behavioural reactions which help the females to find their feeding and egg-laying sites in which also their specific preferences are displayed.

Symp. Biol. Hung. 16, pp. 139–151 (1976)

MOUTH PARTS AND RECEPTORS INVOLVED IN FEEDING BEHAVIOUR AND SUGAR PERCEPTION IN THE AFRICAN ARMYWORM, *SPODOPTERA EXEMPTA* (LEPIDOPTERA, NOCTUIDAE)

by

WEI-CHUN MA

INTERNATIONAL CENTRE OF INSECT PHYSIOLOGY AND ECOLOGY,
P.O. BOX 30772, NAIROBI, KENYA

In lepidopterous larvae, receptor systems known thus far to be involved in olfaction and taste perception are situated on the antennae, the maxillae, and the posterior surface of the labrum. Experiments with larvae of the African armyworm, *Spodoptera exempta*, employing the amputation of various mouth parts suggest the existence of additional chemoperception sites in some other parts of the buccal cavity. These sites probably include sugar-sensitive receptors and must be responsible for the food discriminative capability of the larvae after removal of the antennae, maxillae, and labrum together. Results of amputation and electrophysiological experiments indicate that sugar-sensitive receptors are also located in the medial and lateral maxillary sensilla styloconica and in the preoral sensilla coeloconica of the armyworm. In addition, each of the medial styloconic sensilla contains a sensitive receptor responding to inositol, but not to sugars. Among various sugars tested in feeding experiments, D-fructose and sucrose appear as the most effective phagostimulants for the armyworm. Other active sugars can be arranged in the following order of decreasing stimulatory effectiveness: D-fructose ≥ sucrose > raffinose > maltose ≥ D-glucose > melibiose > > D-galactose. L-rhamnose and, more unexpectedly, the cyclohexitol *m*-inositol are unable to positively influence feeding activity. The functional significance of the chemo-sensory information from the different mouth part receptors in regulating the normal food intake behaviour as well as the implications of the observations in understanding the strictly graminivorous behaviour of the armyworm are discussed.

INTRODUCTION

In the relationship between the African armyworm, *Spodoptera exempta* Walker, and its host-plants the egg-laying adult moth contributes little to the determination of the type of food used by the next larval generation. According to Hattingh (1941) and our own observations, the female moth is unselective with regard to the species of plant or other substrate on which the eggs are deposited, as long as it has a relatively smooth surface. Conversely, only plants belonging to the families of Gramineae and Cyperaceae are acceptable as food for the larval stages of the armyworm (review by Brown, 1962). The neonate larvae escape from unsuitable substrates by hanging from a silken thread, which after breaking gives them a certain buoyancy in air. In the typical rangeland-like habitats of the armyworm the wind-borne larvae have a good chance of landing on the wild grasses that form their natural food.

The later instars are able to show a more active type of food selection. Apart from damaging rangeland, groups of fifth- or sixth-instar larvae may invade plots of maize and other cereal crops. During displacement activity in the field the larvae are commonly observed nibbling and subsequently rejecting forbs and weeds that they encounter. The behaviour suggests that avoidance of non-food plants is less important in food selection than actual probing behaviour. Taste perception is probably of crucial importance in determining the strictly graminivorous behaviour of the armyworm. The present paper concerns some observations on the functional significance of the different mouth parts and sensory receptors concerned with feeding. Attention has especially been focussed on the feeding stimulatory effect of sugars and on the sites of sugar-sensitive chemoreceptors. However, before considering these aspects it will be useful to introduce some of the regulatory mechanisms that are known to control the various components of normal feeding behaviour of lepidopterous larvae and the types and distribution of the receptor systems involved.

OLFACTION AND TASTE

For the armyworm, an insect with a biting-chewing feeding habit, the complete sequence of normal feeding behaviour comprises orientation, palpation, biting, nibbling, and continuous feeding. Literature related to the rôle of olfaction and taste in influencing the diverse behaviour pattern in lepidopterous larvae has recently been reviewed by Schoonhoven (1972) (see also Ma, 1972 and Dethier, 1973).

Olfactory receptors associated with the sensilla basiconica on the apices of the antennae (Dethier and Schoonhoven, 1969) guide the larva in short-range orientation toward the plant (e.g. Dethier, 1941; Ishikawa et al., 1969). Odours emanating from the plant surface seem to facilitate feeding responses rather than to influence orientation behaviour when perceived by some of the eight sensilla basiconica located on the apices of the maxillary palps (Ishikawa et al., 1969). The remaining sensilla basiconica on the maxillary palp have been suggested to act as gustatory organs (Schoonhoven and Dethier, 1966). Behaviour studies of the silkworm *Bombyx mori* by Ishikawa et al. (1969) have shown that palpal sense organs are involved in the perception of morin and beta-sitosterol, which promote feeding in this insect. The exact gustatory function of the palpal receptors is still little understood. In the pierid *Pieris brassicae* it has been concluded that palpal receptors are neither involved in the animal's ability to perceive sugars nor in its biting responses to mustard oil glycosides (Ma, 1972). In various studies it has been found that the palps of lepidopterous larvae exert a spontaneous inhibitory influence on biting activity, in contrast to other parts of the maxilla (Ishikawa et al., 1969; Ma, 1972).

Apart from sites on the palps, sensory organs with a contact chemosensory function are located on the maxillary galeae as well as on the posterior surface of the labrum. In certain species of caterpillars also the hypopharynx has been implicated as a region bearing chemosensory organs (Dethier, 1937). However, the identification of gustatory organs by morphological and anatomical studies and by electrophysiological recording has thus far been limited to two different types:

    (i) *sensillum styloconicum*, relatively short and thick-walled pegs situated as one pair on each of the maxillary galea (Ishikawa and Hirao, 1963; Schoonhoven and Dethier, 1966), and

    (ii) *sensillum coeloconicum*, pit pegs located on the epipharyngeal wall of the preoral cavity (Ma, 1972).

Epipharyngeal sensilla coeloconica may be absent in some species, but in most lepidopterous larvae they seem to be present in one or more pairs. They are innervated by three bipolar neurons with chemosensory function (Fig. 1b). Figure 1a shows the distribution of the coeloconic sensilla on the epipharyngeal surface of the labrum of *S. exempta*. As in other species the sensilla occur here in association with a pair of campaniform sensilla that register strains and pressures generated in the cuticle during the ingestion of food. The styloconic sensilla serve both chemical and mechanical sensory modalities. Each organ is innervated by four chemosensory neurons and

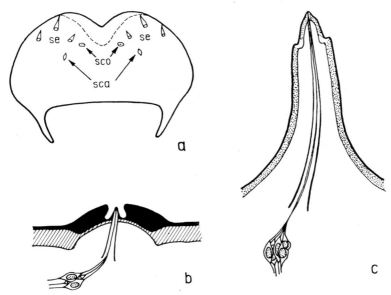

Fig. 1. Two types of sensilla involved in caterpillar taste perception: sensillum coeloconicum (b); sensillum styloconicum (c). Part (a) shows the epipharyngeal wall of the labrum of *S. exempta* with the distribution of tactile setae (se), sensilla coeloconica (sco) and sensilla campaniformia (sca)

one mechanoreceptor. The latter monitors deflections of the distal papilla in its flexible socket during contact with food particles, while the main shaft of the sensillum is not provided with a flexible socket (Fig. 1c). Extensive analyses have been made recently on the interspecific variation in action spectra of the various chemoreceptors of the maxillary sensilla styloconica (Schoonhoven, 1968; Ma, 1972; Dethier, 1973).

The integrative nerve centres for primary responses from the maxillary receptors are situated in the suboesophageal ganglion, which contains motor centres for biting and locomotory activity. Axons from the preoral sense organs are connected to the anterior stomodeal nervous system via the labrofrontal nerves to the tritocerebral lobes of the brain. Quantitative behaviour analyses have led to the supposition that the sensory input from the maxillary receptors controls biting movements and locomotion, while gustatory information from the preoral receptors (obtained during nibbling, i.e. the biting and chewing of small food particles without ingestion) mainly governs the swallowing actions necessary for the actual ingestion (Ma, 1972). According to this concept stimuli that evoke biting reactions ("feeding incitants", *sensu* Beck, 1956), and at the same time arrest locomotion, primarily act on the maxillary receptors, whereas stimuli ("feeding stimulants") which in addition tend to promote ingestional movements, thus leading to continuous feeding, also act on the receptors residing in the preoral or oral cavity. Furthermore, specific deterrent compounds or adverse concentrations of other stimuli may either inhibit biting reactions or inhibit ingestional responses or both.

SUGARS AS PHAGOSTIMULANTS FOR THE ARMYWORM

The feeding experiments were conducted with maize-bred last (sixth) instar larvae weighing between 90 and 140 mg and reared under crowded conditions at 27 °C. The larvae were starved for 2 hrs prior to the experiment and one larva was transferred to each PVC vial of 4.2 cm diameter containing a block of agar-cellulose diet (4 per cent w/w for each component). The diet contained 0.1 M of a sugar compound. This concentration was chosen because it is very close to the optimum stimulus concentration for sugar receptors in lepidopterous larvae (Ma, 1972), while it is also approximately the concentration of sucrose found in most grasses (Waite and Boyd, 1953). The vials were incubated at 30 °C constant and at 12L:12D light regime and the diet refreshed every 24 hrs. In order to minimize effects due to variability in health condition of groups of test animals two control tests were run parallel to each experiment: one control consisting of basic diet without sugar and one of diet containing 0.1 M sucrose. In addition, only larvae that lived for three days on the experimental diets were considered in the evaluation of the results.

The phagostimulatory effect of the active sugars is expressed as a value (consumption index or C.I.) relative to the effect of sucrose and basic diet (Table 1). As a second response parameter the number of larvae showing no positive response (C.I. < 1) was determined. From the original data it could be calculated that L-rhamnose and the cyclohexitol m-inositol do not induce responses which are significantly different from those on the basic control diet (Student's t-test). As shown in Table 1, D-fructose and sucrose induced the highest average feeding rate among the seven active sugars. This is correlated with positive responses (C.I. > 1) in 100 per cent of the test larvae. The attractiveness of these two sugars for the armyworm larvae is also expressed by the fairly constant daily intake over the whole period (three days) of each experiment, which stands in contrast to the behaviour on the moderately active sugars. The initially low intake on raffinose, maltose, glucose, melibiose, and galactose during the first day is followed by a 3 to 5-fold increase during the 2nd day, which again is at least in part correlated with an increase in the per cent larvae showing some positive (C.I. > 1) reaction. Under the conditions of the experiment the feeding responses of the larvae on the moderately active sugars thus may reveal a temporal lowering of the acceptance threshold.

The possible inhibitory effect of the two inactive compounds, i.e. inositol and rhamnose, was assessed in further tests with diets containing a low

142

TABLE 1

Feeding stimulative effectiveness for S. exempta *of sugars relative to the effect of sucrose. C.I. = consumption index; s.e. = standard error of the mean. C.I. < 1 signifies the per cent larvae showing a consumption index smaller than one*
(See further text)

	Day 1		Day 2		Day 3		N
	C.I. ± s.e.[2]	C.I.<1 (%)	C.I. ± s.e.	C.I.<1 (%)	C.I. ± s.e.	C.I.<1 (%)	
Sucrose[1]	100.00	0	100.00	0	100.00	0	
Fructose	112.23 ± 13.96	0	135.03 ± 14.88	0	123.24 ± 10.12	0	21
Raffinose	33.84 ± 8.34	32	95.51 ± 16.21	0	101.61 ± 15.05	5	19
Maltose	13.10 ± 3.09	26	51.45 ± 7.94	11	80.94 ± 8.84	0	19
Glucose	13.67 ± 5.20	54	63.45 ± 10.27	14	69.78 ± 11.72	18	28
Melibiose	4.89 ± 1.42	50	18.19 ± 3.64	19	34.77 ± 7.38	25	16
Galactose	5.68 ± 1.94	40	17.87 ± 4.15	20	28.98 ± 6.21	20	15
Rhamnose	4.46 ± 1.49	53	4.73 ± 2.14	53	5.76 ± 2.76	67	15
Inositol	5.01 ± 3.73	70	5.86 ± 1.93	50	8.57 ± 3.25	55	20

[1] Mean production per larva on sucrose: 10.40 mg (day 1); 7.16 mg (day 2) and 7.92 mg (day 3) dry fecal pellets; N = 180. On basic diet without sugars added: 1.76 mg (day 1); 1.37 mg (day 2); 0.84 mg (day 3); N = 39.

[2] C. I. = mg dry fecal pellets $\frac{\text{test compound} - \text{basic diet}}{\text{sucrose} - \text{basic diet}} \times 100$; a sucrose control and basic diet control run for each test.

concentration of sucrose in admixtures with inositol or rhamnose. The results, shown in Table 2, confirm the suggestion that both compounds are unable to influence the larval feeding responses in a significant manner. In view of the finding that the larva possesses sensitive maxillary inositol receptors (see below), the effect of inositol was investigated in more detail. Table 3 shows that inositol does not have any positive interaction with sucrose, but that at the relatively high concentration of 0.1 M it may exert a slightly inhibitory effect on the response to low concentrations of sucrose. Table 3 also shows that sucrose at a concentration of 0.004 M still induces a significant positive feeding response in the larvae. The experiments further show that the low concentrations of sucrose do not cause a much higher intake on the 2nd day than on the first day of the period of observation, which is contrary to what might have been expected from the results shown in Table 1. Although a certain post-ingestional feed-back effect could be involved, an exact physiological explanation of this interesting difference cannot be given at this moment.

In other species of lepidopterous larvae inositol synergizes the feeding response to sucrose (see Discussion). The failure of inositol to enhance feeding in the armyworm places the functional significance of the inositol receptors associated with the medial pair of maxillary sensilla styloconica open to question. As yet it remains unknown whether the receptors mediate rejectance or whether the observed inhibitory effects might be ascribed to inhibition of the sugar stimuli at the sugar receptor sites. This problem has to be investigated electrophysiologically. Thus far we have used the glass capillary electrode technique (after Hodgson and Roeder, 1956) only in a preliminary survey of loci bearing sugar receptors. From these experiments it was concluded that sugar-sensitive receptors reside in both pairs of maxillary sensilla styloconica as well as in the epipharyngeal sensilla

TABLE 2

Average dry fecal pellets (mg) ± S.D. per larva of S. exempta on diets containing sucrose in admixtures with L-rhamnose and m-inositol. [The significance level of differences (Student's t-test) at 90 per cent or higher confidence level are indicated]

Test medium	Day 1	Day 2	Day 3	N
0.02 M sucrose + 0.1 M rhamnose	3.31 ± 2.04 } <0.05	3.16 ± 1.52	3.88 ± 2.51	14
0.02 M sucrose	4.20 ± 2.20 } <0.05	4.18 ± 1.69 } <0.1	4.80 ± 2.36 } <0.1	10
Control	2.87 ± 0.77	3.11 ± 1.28	3.16 ± 1.59	14
0.02 M sucrose + 0.1 M inositol	3.06 ± 2.07 } <0.05	2.88 ± 2.43	3.51 ± 2.61	16
0.02 M sucrose	3.41 ± 2.32 } <0.02	3.34 ± 2.41 } <0.001	3.00 ± 2.56 } <0.001	17
Control	1.61 ± 0.56	0.16 ± 0.28 } <0.001	0.29 ± 0.57 } <0.001	12

10

TABLE 3

Average dry fecal pellets (mg) ± S.D. per larva of S. exempta on diets containing various concentrations of sucrose and m-inositol.
(Significance of differences at 90 per cent or higher confidence level are indicated)

Exp.	Test medium	Day 1	Day 2	Day 3	N
Ia	0.1 M sucrose + 0.1 M inositol	10.01 ± 5.10	8.30 ± 3.54	5.61 ± 2.52	18
	0.1 M sucrose	6.98 ± 4.38 } <0.001 } <0.001	7.65 ± 4.93 } <0.001	5.89 ± 3.70 } <0.001 } <0.001	18
	Control	1.47 ± 0.60	2.06 ± 0.82	1.65 ± 1.03	17
Ib	0.01 M sucrose + 0.1 M inositol	1.74 ± 1.85	2.89 ± 1.50	2.89 ± 2.55	16
	0.01 M sucrose	4.32 ± 2.60 } <0.01 } <0.02	5.14 ± 1.31 } <0.01 } <0.001	5.47 ± 1.95 } <0.001	12
	Control	1.47 ± 0.61	2.06 ± 0.82	1.65 ± 1.03	17
Ic	0.004 M sucrose + 0.1 M inositol	0.69 ± 0.25	0.64 ± 0.51	1.02 ± 0.55	12
	0.004 M sucrose	1.13 ± 0.28 } <0.001 } <0.01	1.11 ± 0.57 } <0.05 } <0.05	1.33 ± 0.56 } <0.05	12
	Control	0.55 ± 0.37	0.67 ± 0.42	0.89 ± 0.49	13
IIa	0.02 M sucrose + 0.01 M inositol	3.86 ± 2.77	5.46 ± 4.38	5.91 ± 4.05	19
	0.02 M sucrose	5.47 ± 2.32 } <0.01	6.29 ± 3.45 } <0.001 } <0.02	5.79 ± 2.76 } <0.001 } <0.01	19
	Control	2.54 ± 0.98	1.13 ± 0.46	0.69 ± 0.48	8
IIb	0.02 M sucrose + 0.001 M inositol	2.01 ± 1.16	2.99 ± 2.55	4.36 ± 3.59	16
	0.02 M sucrose	3.39 ± 2.32 } <0.05 } <0.1	4.83 ± 2.98 } <0.1 } <0.001 } = 0.01	5.95 ± 3.68 } <0.01 } <0.01	18
	Control	1.94 ± 0.72	0.82 ± 0.55	1.06 ± 1.02	11

coeloconica. Receptors responding to stimulation with solutions of inositol were only found in the medial pair of styloconic sensilla. This receptor type generated maximal impulse frequencies at 90–100 impulses per second during the first second of stimulation with concentrations of 10^{-3} M inositol. The response threshold of the receptor lied at approximately 10^{-6}–10^{-5} M concentration. Details of the electrophysiological studies will appear elsewhere.

AMPUTATION EXPERIMENTS

The experiments carried out on the armyworm involved the removal of the antennae, the maxillae (amputation at the base of the galea), and the labrum (amputation at the clypeo-labral suture). Last-instar larvae six to ten hrs after moulting were immobilized on ice and the mouth parts removed with iridectomy scissors under a stereomicroscope. After drying of the wounds under a desk lamp the larvae were kept on maize leaves at 25 °C until the feeding experiments 18 hrs later. Post-operative mortality, i.e. due to the effect of the operation and not to other factors, was never higher than 15 per cent in all cases. The feeding experiments were conducted with the techniques mentioned above.

Figure 2 gives an idea of the effect of the various amputations on the feeding capability of the larva on its natural food. In this experiment prepupae were formed around 90 hrs after the operation irrespective of the type of amputation and hence of the feeding rate and total intake. Normal pupae, though of different average weights, were formed in all cases. A steady increase in feeding rate can be observed in the first three days in all groups. Except for a slower start during the first day with labrectomized larvae the daily intake is not much different between maxillectomized and labrectomized larvae. In both groups the daily intake is lower than in the non-operated control group, but much higher than in larvae having both the maxillae and the labrum removed. Interference with the mechanics of normal food intake behaviour as well as a lack of adequate sensory information necessary for the performance of normal feeding could have caused the impediment of normal feeding activity. The partial recovery of feeding activity in the treated larvae suggests that motor behaviour during feeding shows a certain plasticity with regard to control by the various mouth parts and their receptors.

Larvae with antennae, maxillae and labrum removed together and presented with a non-food plant such as cassava leaves show short moments of nibbling followed by total rejectance. In the absence of essential physical differences between the non-food plants and the normal food, it may be concluded that loci with chemosensory organs are present either in the hypopharynx or further down the buccal cavity of the armyworm. This supposition was further tested in experiments concerning the ability of sugar perception (see below).

The question arises as to whether there exists any difference at all between the functional significance of the external maxillary receptors and that of the internal buccal receptors in the larval rejectance behaviour. This problem, which obviously cannot be resolved by choice tests, was elucidated to some extent in experiments in which larvae were presented

146

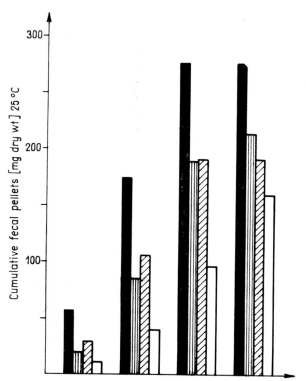

Fig. 2. Effect of various treatments on daily food intake in one last-instar larva of *S. exempta*: non-operated (black columns); maxillectomy (striped columns); labrectomy (oblique-striped); combined maxillectomy and labrectomy (white columns). Means of five larvae per treatment; all larvae originated from the same egg-batch and treatments were performed shortly after the fifth larval moult. Food: sorghum leaves. Temperature: 25 °C constant

with cassava leaves in a no-choice situation. More than 75 per cent of maxillectomized larvae will ultimately start feeding and continue growing until prepupa formation. Of labrectomized larvae, however, only about 10 per cent succeeds to feed in a sufficient amount as to maintain growth and development. In several observations it was noted that the armyworm is potentially quite able to grow and develop on cassava. It may be concluded that presence of deterrents or an inadequacy of the food to stimulate feeding results in a stronger inhibition in the presence of the maxillary receptors (including those on the palps) than when only perceived by the internal buccal receptors in the absence of the maxillae. The sensory control of the initial triggering of biting activity is likely to be of great importance in determining the animal's food specificity.

The answer as to what extent the larvae with amputated mouth parts retain their taste sensitivity to sugars was sought through feeding experiments using the two most effective sugars, fructose and sucrose, as feeding stimulants. Since it was not well feasible to design the experiments in complete blocks including all treatments the observations were made on separate groups, each including response measurements on diet with sugar

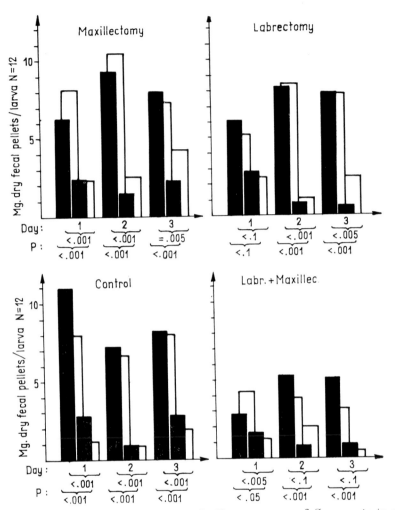

Fig. 3. Effects of various treatments on feeding responses of *S. exempta* to sucrose (left column of each pair of black columns) and to fructose (left column of each pair of white columns). The right column of each pair of columns gives the average response to diet without any sugar added. P means the probability that there is no difference between responses on diets with and without sugar (Student's *t*-test)

added and diet without sugar. The treatment means presented in Fig. 3 corroborate the conclusion from the electrophysiological observations concerning the existence of sugar receptors in the maxillary and the epipharyngeal sensilla. The positive responses to the presence of sugar in larvae with maxillae and labrum removed together suggest that sites of sugar-sensitive receptors are situated in the buccal cavity in addition to those on the epipharyngeal surface of the labrum. It may be argued that the observed difference in intake could have been caused by post-ingestional feed-back effects on the larval feeding activity. Although the present experimental approach does not allow to rule out this possibility the strength of the response often seemed to be too great to be conceivably accounted for by a mere feed-back effect on the animal's activity. Moreover, in assessing the data only those larvae have been considered which were actively moving about throughout the three days period of the experiment. However, to resolve this problem additional behaviour observations are needed as well as morphological and electrophysiological studies of the suggested sensory organs.

DISCUSSION AND CONCLUSION

The feeding behaviour of lepidopterous larvae involves at least two known types of olfactory organs and three known types of gustatory organs. The work carried out thus far suggests the existence of more than four loci bearing chemosensory organs. The armyworm is equipped with sites of sugar-sensitive receptors on the maxillae, the epipharyngeal surface of the labrum, and most likely also on as yet unidentified sites in other parts of the buccal cavity. The present feeding experiments, designed in such a way as to exclude the problem of orientation, indicate some of the relative importance of the receptors in the feeding behaviour of the African armyworm. Sugars such as D-fructose and sucrose seem to induce biting reactions in the armyworm as well as swallowing responses, and, therefore, can be considered as true feeding stimulants.

Dietary sugars are important as a readily available energy source and, in herbivorous animals, at the same time act as phagostimulants. In phytophagous insects (reviewed by Schoonhoven, 1968) usually the sugars commonly occurring in green plants are the most effective in stimulating feeding. Sugars to which the armyworm responds can be arranged in the following order of decreasing phagostimulatory effectiveness: fructose $\geq$ sucrose $>$ raffinose $>$ maltose $\geq$ glucose $>$ melibiose $>$ galactose. While essential as phagostimulants dietary sugars cannot be responsible for the strictly graminivorous behaviour of the armyworm. Apart from differences in relative order the same sugars elicit feeding responses in the polyphagous cotton leafworm *Spodoptera littoralis*. According to Meisner et al. (1972) the preference order for *S. littoralis* is: sucrose $>$ raffinose $>$ maltose $>$ fructose $>$ melibiose $>$ glucose, while Khalifa et al. (1974) established the following sequence: sucrose $>$ fructose $>$ maltose $=$ cellobiose $>$ lactose $>$ galactose $=$ glucose.

In combined thin-layer and paper chromatography of water extracts from dry leaf powder of various species of wild (e.g. *Chloris gayana*, *Cynodon dactylon*, *Pennisetum clandestinum*, *P. purpureum*) and cultivated (maize, sorghum) grasses we found considerable quantities of fructose and glucose in the free form to be present, with glucose the most dominant. Pentoses are present only in very small quantities. Other saccharides like maltoses and sucrose are found in various amounts; they are especially well-represented in sorghum leaves. Inositol seems to be present in various quantities while oligosaccharides only occur in negligible amounts. In general the taste sensitivity of the armyworm to sugars thus seems to be well-adapted to the occurrence of free sugars in the natural food plants. In this respect inositol appears as an exception. The inertness of this compound in influencing feeding behaviour in spite of the presence of sensitive inositol receptors is not readily explainable. In other species of lepidopterous larvae, e.g. the tobacco hornworm *Manduca sexta* (Yamamoto and Fraenkel, 1960) and the silkworm *B. mori* (Ishikawa et al., 1969), inositol is inactive when tested alone but synergizes the effect of sucrose. In such cases it may be assumed that inositol-sensitive receptors are absent in the buccal cavity. The maxillary inositol receptors in these species (Ishikawa, 1963; Schoonhoven, 1968), when stimulated adequately, will then trigger biting activity, but in the absence of feeding stimulants no appreciable ingestion will take place. Consequently, in the armyworm the input from the maxillary inositol receptors is probably not translated by the CNS into biting behaviour.

The adaptive value of the possession of taste receptors with no apparent functional meaning can only be a matter of speculation. From a neurophysiological viewpoint it may be conceived that a progressive reduction of the larval taste sensitivity to compounds with a very wide distribution is accompanied by a greater acuity to compounds with a more limited occurrence. This could provide one of the several mechanisms leading to a higher food plant specificity of the animal. One concrete example is given by the chemoperceptive properties of *P. brassicae* larvae. This oligophagous species possesses receptors sensitive to mustard oil glycosides (Schoonhoven, 1967) in conjunction with specific deterrent-sensitive receptors (Ma, 1972). At the same time the larvae, by exception, lack inositol-sensitive receptors and are behaviourally completely unresponsive to the presence of fructose (Ma, 1972). Thus, *P. brassicae* performs better as an oligophagous insect than it would have done by responding to such ubiquitous plant compounds as fructose and inositol.

Insofar as the African armyworm is concerned it remains unknown whether the strictly graminivorous behaviour is primarily based on its ability to detect some specific chemical cue or whether it is mostly prevented from feeding on plants other than grasses by the action of deterrent factors. Especially the latter possibility seems to be of general applicability in mono- and oligophagous insects (Jermy, 1966). However, the present results may provide the first step to a detailed knowledge of the complexity of mechanisms underlying the food discriminative behaviour of the African armyworm.

REFERENCES

BECK, S. D. (1956): Resistance of plants to insects. *Ann. Rev. Ent.* **10**, 207–232.

BROWN, E. S. (1962): The African Armyworm *Spodoptera exempta* (Walker) (Lepidoptera, Noctuidae): A review of the literature. *Commonwealth Institute of Entomology*, London.

DETHIER, V. G. (1937): Gustation and olfaction in lepidopterous larvae. *Biol. Bull.* **72**, 7–23.

DETHIER, V. G. (1941): The function of the antennal receptors in lepidopterous larvae. *Biol. Bull.* **80**, 403–414.

DETHIER, V. G. (1971): Electrophysiological studies of gustation in lepidopterous larvae. I. Comparative sensitivity to sugars, amino acids and glycosides. *Z. vergl. Physiol.* **72**, 343–363.

DETHIER, V. G. (1973): Electrophysiological studies of gustation in lepidopterous larvae. II. Taste spectra in relation to food-plant discrimination. *J. comp. Physiol.* **82**, 103–134.

DETHIER, V. G. and SCHOONHOVEN, L. M. (1969): Olfactory coding by lepidopterous larvae. *Ent. exp. & appl.* **12**, 535–543.

HATTINGH, C. C. (1941): The biology and ecology of the armyworm *(Laphygma exempta)* and its control in South Africa. *Sci. Bull. Dep. Agric. S. Afr.* **217**, 50.

HODGSON, E. S. and ROEDER, K. D. (1956): Electrophysiological studies of arthropod chemoreception. I. General properties of the labellar chemoreceptors of Diptera. *J. Cell. Comp. Physiol.* **48**, 51–75.

ISHIKAWA, S. (1963): Responses of maxillary chemoreceptors in the larva of the silkworm, *Bombyx mori*, to stimulation by carbohydrates. *J. Cell. Comp. Physiol.* **61**, 99–107.

ISHIKAWA, S. and HIRAO, T. (1963): Electrophysiological studies on taste sensation in the larvae of the silkworm, *Bombyx mori*. Responsiveness of sensilla styloconica in the maxilla. *Bull. Sericult. Exp. Sta.* **18**, 297–357.

ISHIKAWA, S., HIRAO, T. and ARAI, N. (1969): Chemosensory basis of hostplant selection in the silkworm. *Ent. exp. & appl.* **12**, 544–554.

JERMY, T. (1966): Feeding inhibitors and food preference in chewing phytophagous insects. *Ent. exp. & appl.* **9**, 1–12.

KHALIFA, A., SALAMA, H. S., AZMY, N. and EL-SHARABY, A. (1974): Taste sensitivity of the cotton leafworm, *Spodoptera littoralis*, to chemicals. *J. Insect Physiol.* **20**, 67–76.

MA, WEI-CHUN (1972): Dynamics of feeding responses in *Pieris brassicae* Linn. as a function of chemosensory input: a behavioural, ultrastructural and electrophysiological study. *Mededelingen Landbouwhogeschool, Wageningen*, **72** (11), 162 pp.

MEISNER, J., ASCHER, K. R. S. and FLOWERS, H. M. (1972): The feeding response of the larva of the Egyptian cotton leafworm, *Spodoptera littoralis* Boisd., to sugars and related compounds. I. Phagostimulatory and detterent effects. *Comp. Biochem. Physiol.* **42A**, 899–914.

SCHOONHOVEN, L. M. (1967): Chemoreception of mustard oil glucosides in larvae of *Pieris brassicae*. *Proc. K. Ned. Akad. Wtsch.* (C), **70**, 556–563.

SCHOONHOVEN, L. M. (1968): Chemosensory bases of host plant selection. *Ann. Rev. Ent.* **13**, 115–136.

SCHOONHOVEN, L. M. (1972): Plant recognition by lepidopterous larvae. *Symp. Roy. Ent. Soc. London* **6**, 87–99.

SCHOONHOVEN, L. M. and DETHIER, V. G. (1966): Sensory aspects of hostplant discrimination by lepidopterous larvae. *Arch. Néerl. Zool.* **16**, 497–530.

WAITE, R. and BOYD, J. (1953): The water-soluble carbohydrates of grasses. 1. Changes occurring during the normal life-cycle. *J. Sci. Food Agric.* **4**, 197–203.

YAMAMOTO, R. T. and FRAENKEL, G. (1960): Assay of the principal gustatory stimulant for the tobacco hornworm, *Protoparce sexta*, from solanaceous plants. *Ann. ent. Soc. Am.* **53**, 499–503.

Symp. Biol. Hung. 16, pp. 153–155 (1976)

SYSTEMIC DAMAGE CAUSED
BY *TRIOZA APICALIS* ON CARROT

by

M. MARKKULA, S. LAUREMA and K. TIITTANEN

AGRICULTURAL RESEARCH CENTRE, DEPARTMENT OF PEST
INVESTIGATION, 01300 TIKKURILA, FINLAND

The carrot psylla (*Trioza apicalis* Först.) is an important pest principally only in Northern Europe. This is one reason why, until recently, little information was available on the mechanism and effect of the damage to the plant.

The damage is evident in the deformation of carrot leaves. It is easily recognized, being best described as a parsley-like appearance of the leaves. In addition, the growth rate of the whole plant and especially of its roots decreases. In the worst cases the yield is almost nil.

SYSTEMIC TOXAEMIA

The mechanism and effects of the injury to the plant have been studied in Czechoslovakia (Laska, 1964, 1974) and in Finland (Markkula and Laurema, 1971). The damage seems to be a form of systemic toxaemia. Laska (1964) was the first to demonstrate that the leaf the carrot psylla is feeding on remains undamaged. The parsley-like deformation appears only in the youngest leaf which is not yet visible at the time that the psylla is feeding.

The degree of injury depends closely on the length of time the psyllas stay on the plants. At our Institute the severity of the injury was assessed according to a scale from 0 to 5 (where 0 = undamaged and 5 = completely injured). When one overwintered female visited a plant, the average damage rating was 0.3 for one hour, 1.9 for 4 hours and 2.9 for 24 hours. These figures are average for 10 females. The plants were small carrot seedlings with only two growing leaves. During the greater part of the visit on the plant (over 70%), the psylla females were in a sucking position. The time of actual food intake could not be established. The behaviour of the 10 females was observed for 2–7 hours.

On average, the injury became visible within two days (range of variation 1–5) from the moment the psyllas were set on the plants. If the animals were left on the plant, the growth of the leaves was inhibited and there was ample development of lateral roots. When the psyllas were removed, new, undamaged leaves appeared and the plants began to recover, but the deformed leaves remained parsley-like.

The damage caused by overwintered males, larvae and new generation females was significantly less serious than that caused by overwintered females. The damage caused by the larvae did not appear to be systemic,

153

as in the studies by Laska (1964), but the leaves on which larvae were feeding were slightly deformed.

EFFECT ON THE YIELD

In experiments performed over a number of years at our Institute, it has
been found that during years when psyllas were abundant, the yield was
almost nil in untreated experimental plots. The density of the psylla population need not be high. In the case of small carrot seedlings, one psylla per
plant is enough to cause an almost complete loss of the yield.

When one female or one male per plant was enclosed in a cage in the field
for the whole growing season, it was observed at the time of harvest that
the plants injured by a female had no main root and therefore gave no
yield. The roots of plants damaged by a male weighed on average 13 grams
(range 7.6–15.4 grams) and the roots of intact plants 55 grams (range
42.5–63.7 grams).

EFFECT ON THE CHEMICAL COMPOSITION OF THE YIELD

The damage affects the contents of carotene and vitamin C in the roots.
Two-kilogram samples of badly damaged (damage rating 4) and undamaged carrots were selected from the crop, and the contents of carotene,
vitamin C, sugar and starch were determined as milligrams per kilogram.
The damaged roots had over 30% less carotene but almost 30% more
vitamin C than the healthy roots. The damage had not affected the content of sugar and starch.

INJURIOUS SUBSTANCE IN PSYLLA SALIVA

Several facts, especially the rapid commencement and cessation of the
damage, which almost coincides with the visit of the psylla, indicate that
it is not caused by a virus or other phytopathogenic organism spread by
the pest. The cause of the damage is a substance secreted with the saliva
of the psylla, which disturbs the metabolism of the plant. Attention has
been concentrated on elucidating the nature of the chemical substance and
the physiological mechanism of its effect.

The first studies were devoted to determining whether the toxin is one
of phytohormones. The substance secreted by the psylla does not appear
to be indolyl-3-acetic acid or any similar phytohormone (Markkula and
Laurema, 1971). The psylla stunted the growth of the leaf stalks and veins.
In fact indolyl-acetic acid had an opposite effect by stimulating growth
of the stunted leaf stalks.

The ^{14}C-labelling method was used in further studies on the salivary
secretions of the psylla on an artificial nutrient substrate and in natural
conditions on carrot seedlings. Five compounds were distinguished in psylla
saliva collected from the artificial substrate: glucose, fructose, saccharose,
inositol and one unknown amine. Examination of the carrot seedlings that
had harboured the ^{14}C-labelled psyllas revealed a number of radioactive

small- and large-molecular compounds, some of which, at least, had been metabolized by the plant and had originated elsewhere than from the saliva of the psyllas.

Attention will now be directed to identifying the unknown amine in the psylla saliva and examining its role in the development of the toxaemia.

REFERENCES

LASKA, P. (1964): Příspěvek k bionomii a ochraně proti *Trioza apicalis* Först. (Triozidae, Homoptera). *Zool. Listy,* **13,** 327–332.

LASKA, P. (1974): Studie über den Möhrenblattfloh (*Trioza apicalis* Först.) (Triozidae, Homoptera). *Acta Scient. Nat. Acad. Scient. Bohemosl.* **8** (1), 1–44.

MARKKULA, M. and LAUREMA, S. (1971): Phytotoxaemia caused by *Trioza apicalis* Först. (Hom., Triozidae) on carrot. *Ann. Agric. Fenn.* **10,** 181–184.

Symp. Biol. Hung. 16, pp. 157–161 (1976)

INFLUENCE OF THE NECTARILESS CHARACTER IN COTTON ON HARMFUL AND BENEFICIAL INSECTS

by

F. G. MAXWELL,[1] M. F. SCHUSTER,[2] W. R. MEREDITH[3] and M. L. LASTER[4]

[1]DEPARTMENT OF ENTOMOLOGY, MISSISSIPPI AGRICULTURAL AND FORESTRY EXPERIMENT STATION; [2]DEPARTMENT OF ENTOMOLOGY; [3]USDA, ARS AND DELTA BRANCH EXPERIMENT STATION, MISSISSIPPI AGRICULTURAL AND FORESTRY EXPERIMENT STATION; MISSISSIPPI STATE UNIVERSITY, MISSISSIPPI STATE, MISSISSIPPI 39762, USA

Several years research in the Mississippi Agricultural and Forestry Experiment Station at Mississippi State University have shown that the nectariless trait reduces the incidence of boll rot, tarnished plant bugs, fleahoppers, and bollworms on cotton plants. The nectariless trait will be an important component in the development of pest management schemes for the future control of these pests in Mississippi and other areas of the cotton belt. No major breeding problems have been indicated in the use of the nectariless trait; therefore, its incorporation into existing germplasms and breeding programs is receiving high priority.

The nectar from cotton, *Gossypium hirsutum* (L.) has long been known as a primary or secondary source of food for a number of beneficial insects, as well as certain insect pests of cotton (Trelease, 1879; Tyler, 1908; Lukefahr and Rhyne, 1960; Butler et al., 1972). From an evolutionary standpoint, Tyler (1908) speculated that the attraction of ants and other beneficial insects was greater than the attraction of harmful insects. However, probably of greater evolutionary importance is the attraction of bees for cross-pollination and hybridization. Without bees, cotton would be essentially self-pollinated, and the chance of producing new recombinants and genetic changes would be greatly reduced.

The nectaries of cotton have been well described by Tyler (1908). Extrafloral nectaries are located on the main rib, and sometimes on other prominent ribs, of the underside of the leaf. Three outer involucral nectaries are located at the base of the involucral bracts, on the outside. Three inner involucral nectaries are located between the calyx and the involucre. A wild tetraploid species, *G. tomentosum* Nutt., native to Hawaii, has no extrafloral nectaries. Meyer and Meyer (1961) described this nectariless trait and its inheritance when it was transferred to tetraploid *G. hirsutum* L. types. The presence of two recessive genes, ne_1 and ne_2, results in the absence of the extrafloral nectaries on the leaf midribs and bracts and between the bracts and the calyx. The floral nectary located at the base of the inside of the calyx is not removed by these two genes. Thus honey bee activity inside the flower should not be reduced and perhaps even increased.

Mound (1962), in a review of research connected with nectaries of cotton, reported that the nectaries are composed of many closely packed multicellular papillae, the cells of which are small, with large nuclei and dense cytoplasm. No evidence was found of a direct vascular supply to a nectary. Mound (1962) reported that the main sugars were sucrose, glucose, and fructose. Butler et al. (1972) also reported that cotton nectar was composed mainly of glucose and fructose, with a small amount of sucrose, and that under Arizona conditions, honey bees removed an average of about 3.4 kg/ha of nectar per day from 'Pima S-1' (*G. barbadense* L.). Research by Hanny and Elmore (1974) indicated that cotton nectar is probably an important source of amino acids for insects. They reported that amino nitrogen constitutes an average of 0.04% of the extrafloral cotton nectar. Twenty-four amino acids were isolated from the nectar. The nectaries of cotton, therefore, are an important link in the food chain of many insects, supplying them with moisture, simple sugars, amino acids and certain minerals.

Several reports from outside Mississippi document the insect-suppression properties of nectariless cotton. Lukefahr and Rhyne (1960) reported that nectariless strains had much lower numbers of cotton leafworm, *Alabama argillacea* (Hűbner) and the cabbage looper, *Trichoplusia ni* (Hűbner) than normal nectaried varieties. Lukefahr et al. (1965) reported that nectariless cottons suppressed bollworm, *Heliothis zea* (Boddie) and tobacco budworm, *H. virescens* (Fabricius) egg deposition when the insects were confined to cotton growing in cages. In field experiments, Lukefahr et al. (1966) observed significantly lower populations of bollworms on a nectariless and glabrous strain than on nectaried and hirsute varieties. In cage studies, Davis et al. (1973) reported that *H. zea* moths laid only 45% as many eggs on nectariless 'Acala' strains as on nectariless companion lines. Benschoter and Leal (1974) showed that moths of *Bucculatrix thurberiella* Busck, the cotton leafperforator, lived significantly longer on nectaried cotton than on nectariless.

The remaining discussion on nectariless cotton concerns field research findings in Mississippi. This team research in the past few years has been an attempt to obtain a complete evaluation of the nectariless trait and to develop advanced breeder lines for use by seed companies in producing commercial varieties for the midsouth.

In 1969 and 1970, the influence of the presence or absence of nectaries on boll rot was evaluated. The incidence of boll rot was significantly less on nectariless cotton than on the normal type, as indicated in Table 1. Nectariless cotton would eliminate that type of boll rot caused by patho-

TABLE 1

Effect of the nectariless cotton character in BC_4F_4
and BC_5F_4 generations on boll rot and percent boll rot

Generation and variety	Boll rot loss, kg/ha		% boll rot	
	Normal	Nectariless	Normal	Nectariless
Stv 7A BC$_4$	175	138	5.19	4.40
DPL Sm leaf, BC$_4$	198	166	6.01	5.50
Dixie King, BC$_4$	198	140	6.60	5.64
Average	190	148	5.93*	5.18
Stv 7A BC$_5$	287	228	7.71	6.75
DPL Sm leaf, BC$_5$	278	243	7.87	6.93
Dixie King, BC$_5$	286	237	8.73	7.19
Average	283	236	8.11*	6.96

* Significantly more boll rot at the 0.01 level of probability.

gens entering through the nectaries. Also, since nectariless cottons reduce levels of some piercing type insects, boll rot transmitted through insect damage might be reduced.

Tarnished plant bugs, *Lygus lineolaris* (Palisot de Beauvois) are increasing as an important cotton pest in the Mississippi Delta. We have noted reductions of 43% to 65% in tarnished plant bug populations on nectariless cotton, as indicated in Table 2. Similar reductions of cotton fleahoppers, *Pseudatomoscelis seriatus* (Reuter), have also been observed, as shown in Table 3.

The most important insect pests in the Mississippi Delta are the bollworm, *H. zea*, and the tobacco budworm, *H. virescens*. Nectariless cottons in our field experiments have reduced the numbers of bollworm-damaged squares in each of three years (Table 4). Reductions of 15 to 58% were

TABLE 2

Effect of the nectariless cotton character on populations of tarnished plant bugs

Year	Reference	% Reduction[v]
1971	Scales (pers. comm.)	64.7*
1971	Laster and Meredith (1974)	43.2*
1972	Laster and Meredith (1974)	55.6*
1973	Schuster and Maxwell (1974)	63.6**

* and ** indicate significance at 10% and 1% probability levels, respectively.
[v] In comparison with populations on normal cotton plants.

TABLE 3

Effect of the nectariless cotton character on population of fleahoppers

Year	Reference	% Reduction[v]
1971	Meredith *et al.* (1973)	23.6
1973	Schuster and Maxwell (1974)	57.6*

* Indicates significance at the 5% probability level.
[v] In comparison with populations on normal cotton plants.

TABLE 4

Effect of the nectariless cotton character on number of damaged squares

Year	Reference	% Reduction[v]
1971	Laster and Meredith (1974)	15.7
1972	Laster and Meredith (1974)	22.5*
1973	Schuster and Maxwell (1974)	57.7*

* Indicates significance at the 10% probability level.
[v] In comparison with populations on normal cotton plants.

159

recorded. In the 1973 studies involving 10-acre replicated plots, reductions in bollworm eggs and larvae were also recorded. This agrees with the cage results of Davis et al. (1973) who obtained 45% reduction in bollworm egg numbers. These results indicate that the nectariless trait, where grown in large acreages, has an excellent potential of being a major contribution to bollworm control.

Our studies (Schuster and Maxwell, 1974; Laster and Meredith, 1974) also showed that the nectariless trait had no great deleterious effect on natural insect enemies.

The evaluation of any resistant trait is incomplete without an appraisal of its effect on yield, earliness of yield, and fiber properties. In Table 5, it has been reported (Meredith et al., 1973) that plants with the nectariless trait, as measured in small-plot tests, are approximately equal in yield and fiber properties to their nectaried recurrent parents. A nectariless genotype by variety interaction is also indicated in Table 5 for lint yield. Stoneville

TABLE 5

Average performance at six locations of three BC_6F_4 nectariless cottons and their recurrent parents for ten yield and quality traits

Variety	Total lint, kg/ha	Lint, 1st harvest	Lint, %	Boll size	Seed index	50% SL+	2.5% SL+	T_1	E_1	Micronaire values
Stv 7A var	1.115*	588	35.3**	6.08*	12.7	0.58*	1.18	18.9	6.5	4.98
Stv 7A ne	1.067	587	34.8	5.92	12.7	0.57	1.18	18.6	6.4	4.97
DPL SL var	1.091	532	35.8	5.84**	11.9	0.58	1.22*	18.8	8.6	4.80
DPL SL ne	1.133*	598**	36.1*	5.62	11.7	0.57	1.18	18.9	8.4	4.75
Dixie King var	1.036	551	32.7	7.40	15.2	0.57	1.18	18.5	6.9	4.37
Dixie King ne	1.030	568	33.0*	7.41	14.8	0.57	1.17	18.0	6.9	4.38
Average var	1.081	557	34.6	6.44	13.3	0.58*	1.19*	18.8	7.3	4.72
Average ne	1.077	584	34.6	6.32	13.1	0.57	1.18	18.5	7.2	4.70

* and ** indicate significantly higher than its nectariless or nectaried counterpart at 0.05 and 0.01 probability levels, respectively.
SL + = Span length.

7A's yield was significantly reduced (4.3%) by the nectariless trait, and conversely, Deltapine Smoothleaf's yield was increased significantly (3.8%). This indicates that breeders will need to search for the best genetic background in which to incorporate the nectariless trait. The genetic analyses indicate that no reduction in yield would be expected by using nectariless cottons. This agrees with the conclusions reported by Davis (1969) in New Mexico. Nectariless cottons also matured earlier than their nectaried counterparts. This is probably related to less early-season tarnished plant bug damage on the nectariless cottons.

Use of the nectariless trait in the Mississippi Delta will require that commercial seed companies quickly incorporate this trait into their breeding programs. In reviewing their research, it is obvious that they are now placing a high priority on the nectariless trait. Nectariless strains developed by Stoneville and Delta and Pine Land Company are already programmed by our research team for extensive field testing this season. It is quite probable that, within the next two to three years, nectariless commercial

160

varieties will be available on the open market and will be extensively grown primarily for insect and disease resistance in Mississippi and other areas of the cotton belt.

REFERENCES

BENSCHOTER, C. A. and LEAL, M. P. (1974): Relation of cotton plant nectar to longevity and reproduction of the cotton leafperforator in the laboratory. *J. econ. Ent.* **67** (2), 217–218.

BUTLER, G. D., LOPER, G. M., MCGREGOR, S. E., WEBSTER, J. L., and MARGOLIS, H. (1972): Amounts and kinds of sugars in the nectars of cotton (*Gossypium* spp.) and the time of their secretion. *Agron. J.* **64**, 364–368.

DAVIS, D. (1969): Agronomic and fiber properties of smooth nectariless 'Acala' cotton. *Crop Sci.* **9**, 817–819.

DAVIS, D. V., ELLINGTON, J. J. and BROWN, J. C. (1973): Mortality factors affecting cotton insects: I. Resistance of smooth and nectariless characters in Acala cotton to *Heliothis zea, Pectinophora gossypiella* and *Trichoplusia ni. J. Environ. Qual.* **2**, 530—535.

HANNY, B. W. and ELMORE, C. D. (1974): Amino acid composition of cotton nectar. *J. Agr. Food Chem.* **22**, 476—478.

LASTER, M. L. and MEREDITH, W. R., Jr. (1974): Influence of nectariless cotton on insect pest populations. *MAFES Res.* Highlights.

LUKEFAHR, M. J., COWAN, C. B., PFRIMMER, T. R. and NOBLE, L. W. (1966): Resistance of experimental cotton strain 1514 to the bollworm and cotton fleahopper. *J. econ. Ent.* **59**, 393–395.

LUKEFAHR, M. J., MARTIN, D. F. and MEYER, J. R. (1965): Plant resistance to five lepidoptera attacking cotton. *J. econ. Ent.* **58**, 516–518.

LUKEFAHR, M. J. and RHYNE, C. L. (1960): Effects of nectariless cottons on populations of three lepidopterous insects. *J. econ. Ent.* **53**, 242–244.

MEREDITH, W. R., Jr., LASTER, M. L., RANNEY, C. D. and BRIDGE, R. R. (1973): Agronomic potential of nectariless cotton (*Gossypium hirsutum* L.). *J. Environ. Qual.* **2**, 141–144.

MEYER, J. R. and MEYER, V. G. (1961): Origin and inheritance of nectariless cotton. *Crop Sci.* **1**, 167–169.

MOUND, L. A. (1962): Extrafloral nectaries of cotton and their secretions. *Emp. Cotton Grow. Rev.* **39**, 254–261.

SCHUSTER, M. F. and MAXWELL, F. G. (1974): The impact of nectariless cotton on plant bugs, bollworms, and beneficial insects. *Beltwide Cotton Production Research Conference Proceedings.* (n. p.)

TRELEASE, W. (1879): Nectar; what it is, and some of its uses. In: Comstock, J. H. (ed.) *Report upon Cotton Insects. U.S. Dept. Agric. Publ.* 319—343.

TYLER, F. J. (1908): The nectaries of cotton. Part V. *U.S. Dept. Agr. Plant Ind. Bull.* **131**, 45–54.

11

Symp. Biol. Hung. 16, pp. 163–172 (1976)

THE EFFECTS OF QUALITATIVE AND QUANTITATIVE CHANGES OF DIET ON EGG PRODUCTION IN *LOCUSTA MIGRATORIA MIGRATORIOIDES*

by

A. R. McCAFFERY

CENTRE FOR OVERSEAS PEST RESEARCH, COLLEGE HOUSE, WRIGHTS LANE, LONDON W8 5SJ, UK

A diet of poor quality *Agropyron repens* is often insufficient to initiate or maintain egg development in *Locusta migratoria migratorioides*. Locusts given lush grass show a normal somatic growth period followed by initiation and completion of egg development.

The provision of poor quality grass to maturing locusts (with vitellogenic oocytes) previously fed on lush grass reduces the rate of egg pod production and raises the rate of terminal oocyte resorption. On completion of their gonotrophic cycle these locusts do not reinitiate further vitellogenesis.

Quantitative studies show that the provision of gradually decreasing sub-optimal amounts of *Agropyron repens* produces corresponding decreases in the rate of egg production and increases in the rate of terminal oocyte resorption. Ingestion of less than 80 mg (dry wt.) of grass/female/day is insufficient to initiate egg development in locusts whose somatic growth period is normal. At levels of feeding at or above this threshold, total egg production is approximately proportional to the dry weight intake of grass.

It can be shown that the initiation of the first vitellogenic cycle occurs 48 hours before yolk sequestration by the oocytes is visible. The metabolic and endocrine significance of these results is briefly discussed.

INTRODUCTION

The resorptive breakdown of the developing terminal oocytes is one of the major factors limiting the egg production and consequently the population size of *Locusta migratoria* in the Middle Niger outbreak area (Farrow, 1972). This process and the rate of egg production may be affected by the nutritional state of the locust (Petavy, 1966). The experiments described here attempt to determine the extent to which feeding is a factor in the initiation and maintenance of oocyte development in *Locusta migratoria*.

RESULTS

A. Qualitative studies

Oogenesis is affected by the presence of males (Highnam and Lusis, 1962). In order to measure the food utilisation of females with males present, groups of 10 female *Locusta* were caged with 10 males which had their mouths sealed with wax so they were unable to eat. The males were starved for 48 hours before the experiment so that they produced no faeces and were replaced by fresh, similarly treated males, after 48 hours with the females. The females were fed on *Agropyron repens:* one group was given

11*

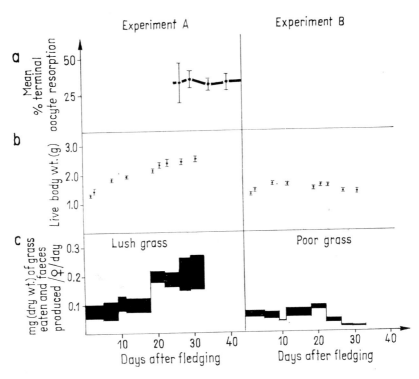

Fig. 1. a) The rate of terminal oocyte resorption, b) the live body weight, c) the leve of food intake (upper line) and faeces production (lower line) (shaded area represents quantity of food utilised) of adult female *Locusta migratoria* fed on lush or poor quality *Agropyron*

lush, green grass and a second group was fed on a poor quality diet of the grass which had moist, green central stems and browning leaf blades.

The females ate and utilised less of the poor grass, on a dry weight basis, at all stages of somatic and reproductive development (Fig. 1c), and no marked increase in ingestion occurred at any stage compared with lush-grass fed females. Poor-diet females gained weight normally during the first part of somatic growth but subsequently lost weight gradually. The live body weight of lush-grass fed females was significantly higher than that of the poor-grass fed females from day 12 after fledging ($t = 4.548$, $p < 0.001$) (Fig. 1b).

Mating behaviour was seen on day 21 in the lush-grass fed females followed by copulation on day 22 and the first oviposition on day 25. Subsequently egg-pods were produced continuously. In contrast, the poor-grass fed females exhibited no sexual activity and no egg-pods were laid. Dissection of these females showed that the oocytes had developed to a length of 1.5 mm during somatic growth, but no vitellogenesis had taken place.

The total blood protein concentration of the poor-grass fed females fell from day 13 whilst that of the lush-grass fed females rose (Table 1). The blood volume of poor-grass fed females was maintained until about day

24 after which it fell to a level below that of the lush-diet group. As a result the total blood protein of the poor-grass fed females fell sharply during the last week of the experiment (Table 1).

The failure to initiate egg development in poor-grass fed females may have been due to low food intake and utilisation which resulted in low haemolymph protein levels. Lack of water was a probable cause of later mortality but since no mortality occurred until some time after the normal initiation of vitellogenesis (about day 18) it is unlikely that this was a cause of the failure to initiate oocyte development. In order to verify these points a further experiment was carried out.

Cages of ten *Locusta* females were set up with non-eating males as described above. One group of females was given 1.5 g of lush *Agropyron*/female/ day (water content 81% equivalent to 280 mg dry weight/female/day). A second group was fed on poor quality *Agropyron* at a rate of 0.5 g/female/day (water content 43% equivalent to 280 mg dry weight/female/day) together with a lush grass supplement of 0.2 g/female/day (dry weight equivalent 40 mg/female/day). All cages were provided with fresh tap water daily in petri dishes.

Using this treatment no mortality occurred in the poor-diet females. Egg

TABLE 1

Blood volume, haemolymph protein concentration and total haemolymph protein of adult female Locusta migratoria *fed on lush and poor diets of* Agropyron repens *(n = 5)*

Age, days	Blood volume (μl)			
	lush-grass fed	poor-grass fed	t	p
4	231.5± 5.07	236.2±16.72	0.27	NS
13	256.8±20.1	250.8±16.8	0.23	NS
20	301.4±22.6	342.0±25.4	1.20	NS
22	329.6±48.2	372.0±60.2	0.55	NS
29	324.4±57.1	165.0± 3.8	2.09	NS

Age, days	Haemolymph protein conc. (g/100 ml)			
	lush-grass fed	poor-grass fed	t	p
4	3.91±0.29	3.84±0.14	0.22	NS
13	4.56±0.29	4.31±0.16	0.76	NS
20	4.74±0.13	2.88±0.38	4.67	<0.005
22	4.63±0.60	2.51±0.18	3.35	<0.01
29	5.80±0.46	1.60±0.38	6.23	<0.001

Age, days	Total haemolymph protein (mg)			
	lush-grass fed	poor-grass fed	t	p
4	9.06±0.19	9.06±0.13	0	NS
13	10.86±0.39	10.82±0.66	0.05	NS
20	18.85±1.44	12.54±1.15	3.42	<0.01
22	20.68±5.37	12.08±1.90	1.51	NS
29	23.86±5.22	3.52±0.92	5.22	<0.001

production in the lush-grass fed group of females was similar to that observed in the previous experiment but poor-grass fed females given lush grass supplements and water showed no signs of sexual activity, did not show vitellogenesis and did not lay any egg-pods.

To reduce any adverse effects of poor feeding during somatic growth, one group of ten females was fed on the lush-grass diet until at least the end of somatic growth when the protein synthetic apparatus is fully developed (day 22) and sexual activity begins. From day 23 the poor diet as described above was provided. A second group of ten females was given the reverse feeding regimes in which the poor diet was provided up to day 21 followed by the lush diet from day 22.

Before the change of diet the lush-grass fed females took in large amounts of food and the quantities utilised were similar to those observed during the first experiment although the absolute amounts ingested were higher (Fig. 2c). Despite a fairly high intake the rate of food utilisation in the initially poor-fed females decreased until faeces production virtually equalled food intake. Correlated with this the lush grass fed locusts gained weight steadily whereas the initially poor-grass fed females managed only to maintain their fledging weight (Fig. 2b).

Following the reversal of diet on day 23 the initially lush-grass fed females maintained a fairly high level of ingestion and utilisation of food for 10 days. After this time the intake and utilisation of food fell markedly (Fig. 2c). When initially poor *Agropyron* fed females were fed on a lush-grass diet the overall food intake, although similar to that just prior to the change of diet, resulted in a much increased level of food utilisation (Fig. 2c). The former animals maintained their body weight at the time the diet was changed whilst the latter gained weight following the change of diet at a rate of 0.065 g/day (lush-fed controls gained weight from fledging at a rate of 0.048 g/day) (Fig. 2b).

Sexual activity in the initially lush-grass fed females began on day 22, copulation occurred on day 25 and first egg-pods were laid on day 26. Initially poor-grass fed females showed sexual activity 5 days after the change of diet on day 26, copulation occurred on day 29 and the first egg-pods were laid on day 30. The latter females then laid pods continually until the experiment was terminated.

The whole group of ten females fed initially on lush grass followed by poor grass laid only four pods in total. The later that these pods were laid the higher was the observed rate of terminal oocyte resorption (Fig. 2a). Since the development of these clutches was not synchronous the rate of oocyte resorption and resultant clutch size varied according to the length of time that the female was able to eat lush grass. A further three females showed no signs of vitellogenic activity.

Females fed first on poor grass followed by lush grass laid their first egg-pods 9 days after the change of diet. High rates of terminal oocyte resorption were associated with the development of these first pods but later egg-pods had higher clutch sizes as a result of decreased resorption of oocytes (Fig. 2a). This level of oocyte resorption took two weeks to fall to a control level of about 30%. It will be noted that during these experiments rates of terminal oocyte resorption of 30–70% were found. Resorption rates outside these levels were not observed.

166

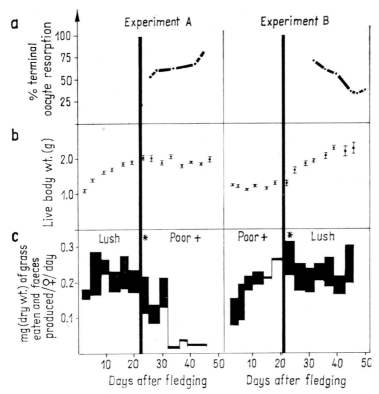

Fig. 2. Effects of changes* in *Agropyron* diet quality on: a) the rate of terminal oocyte resorption, b) the live body weight, c) the levels of food intake (upper line) and faeces production (lower line) (shaded area represents quantity of food utilisation) of adult female *Locusta migratoria*

The results of these experiments suggest that it is the *quantity* of utilisable food that is an important factor in initiating egg development in *Locusta* females.

B. Quantitative studies

Groups of ten *Locusta* females were fed 'ad lib', i.e. with a surplus of food, on lush *Agropyron repens* from fledging until day 15. Somatic growth proceeded normally and the females gained weight in the usual way. From day 16 individual groups were fed on various amounts of lush from 40 to 240 mg dry weight/female/day. Controls were continuously fed 'ad lib' on lush grass. Experimental females ate all the grass provided. All cages were provided with a source of water.

Females given 240 mg dry weight/female/day of lush grass gained weight in a similar fashion to the 'ad lib' fed controls. The provision of smaller quantities of food had a correspondingly greater effect on the reduction in fresh body weight (Table 2). Copulation took place on days 18–19 in all groups. First egg-pods were laid on day 19. There was no apparent correla-

TABLE 2

Fresh body weight of adult female Locusta migratoria *fed
on various amounts of* Agropyron *from day 16*
Mean live body weight (g)

Feeding regime: mg dry wt/female/day	Day 5	Day 14	Day 19	Day 26
40	1.49±0.07	2.21±0.05	2.01±0.09	1.90±0.08
80	1.65±0.06	2.46±0.07	2.32±0.07	2.22±0.07
120	1.72±0.03	2.33±0.06	2.32±0.06	2.48±0.07
160	1.77±0.06	2.38±0.07	2.46±0.06	2.60±0.10
200	1.73±0.04	2.21±0.06	2.34±0.05	2.70±0.09
240	1.74±0.04	2.27±0.05	2.37±0.06	2.99±0.12
'Ad lib'	1.75±0.04	2.29±0.07	2.55±0.05	3.04±0.08

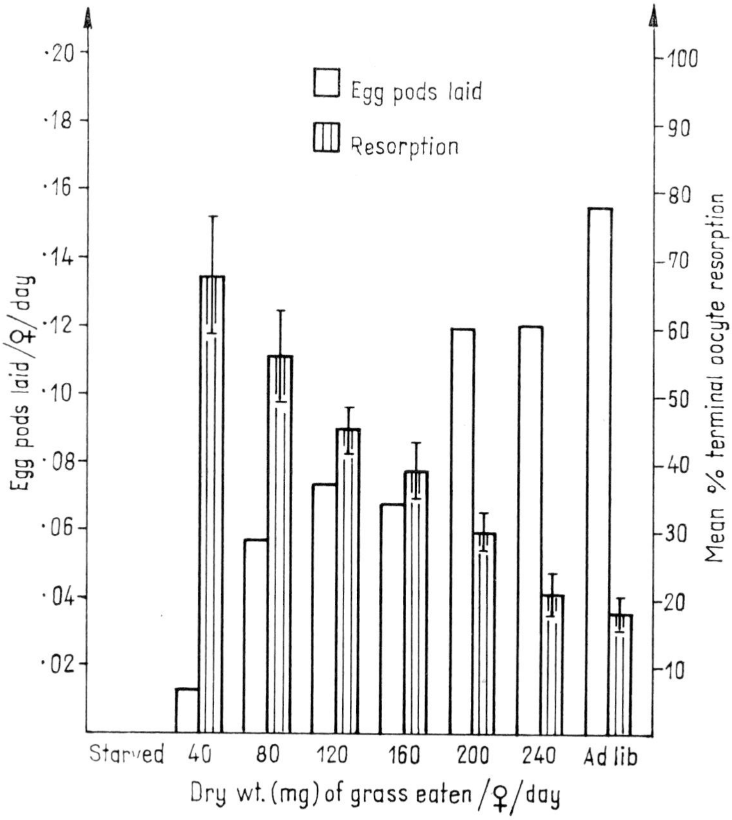

Fig. 3. The rate of egg-pod production and terminal oocyte resorption of adult female
Locusta migratoria fed on various quantities of lush *Agropyron*

tion between the amount of food available and the time of oviposition in
these experimental groups and from the time of the first oviposition continuous egg production occurred in all groups except those fed 40 mg dry
weight/female/day.

In this latter group only two egg-pods were produced during the experiment, both by the same female. Of the remaining females all but one showed signs of vitellogenic activity accompanied by small single resorption bodies on all the ovarioles due to total resorption of all oocytes. No further initiation of subsequent gonotrophic cycles took place. The single female able to produce two egg-pods must be considered as atypical. Since group feeding was adopted it may well have been possible for such a female to obtain sufficient food at the expense of others to initiate egg development.

'Ad lib' fed females had the highest rate of egg-pod production, whilst 200 and 240 mg dry weight/day fed females had somewhat lower rates. Further decreases in the quantity of food supplied resulted in corresponding reductions in the rate of egg-pod production (Fig. 3).

The number of ovarioles in the ovary was statistically similar in all the groups and females given 240 mg dry weight of grass/day produced mean clutches of 62.13 ± 2.41 eggs which was not significantly different (t $= = 1,303$) from the mean clutch size of 65.30 ± 1.23 found in 'ad lib' fed controls. Significant reductions in clutch size and increases in the rate of terminal oocyte resorption (Fig. 3) were found as the food provided was decreased

To confirm these results two groups of *Locusta* females were fed individually for 5 hours in glass jars of 1 litre capacity. After feeding, the females were returned to the cages in which fed males were present but in which no further food was provided. In this way the stimuli resulting from moderate crowding and the presence of males was maintained whilst individual feeding levels of the female were precisely monitored. The two groups of females were 'ad lib' fed on lush *Agropyron* from fledging to day 16. From day 17 restricted amounts of lush were placed in the individual feeding jars such that the amounts of grass consumed by the locusts were calculated to be 35 ± 3 and 77 ± 15 mg dry weight/female/day for the two groups. A third group was 'ad lib' fed throughout the experiment as a control. The egg production of the three groups was monitored for three weeks after the first egg-pods were laid.

The two groups of females eating reduced amounts of food laid their first egg-pods on days 15 and 16. In the first week after the change of diet,

TABLE 3

Analysis of individual ovarian development of female
Locusta migratoria *given reduced amounts of food*
at various times after fledging

Day of diet change	n	Previous egg development	Oviposited	100% resorption	% Vitellogenesis initiated
8	8	0	0	0	0
9	8	6	0	6	75
10	7	4	2	2	57
11	8	3	0	3	37.5
12	7	4	3	1	57
13	8	8	5	3	100
14	8	8	2	6	100
15	9	9	0	9	100
16	9	9	7	2	100

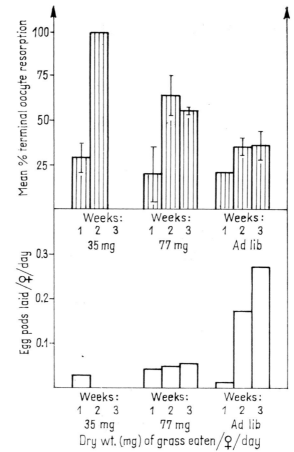

Fig. 4. The rate of egg-pod production and terminal oocyte resorption of adult female *Locusta migratoria* at various times after the beginning of sub-optimal feeding

the 35] mg dry weight/day fed females laid a few more egg-pods then their egg production stopped during the second week as a result of total oocyte resorption (100%) or the lack of initiation of further vitellogenesis (Fig. 4). The 77 mg dry weight fed females continued to lay egg-pods but the rate of oocyte resorption was higher than that seen in the 'ad lib' fed group of females.

These results together with those obtained previously suggest that successful initiation of egg development requires a certain minimum food intake.

Table 3 shows the results obtained when the time of the change over from 'ad lib' to a suboptimal diet was varied between days 8 and 16. In this experiment days 8 and 9 appeared critical for the initiation of egg development although a good deal of variation occurs in the timing of the initiation. In general the later the change from 'ad lib' feeding occurs the greater will be the number of females exhibiting vitellogenesis. The ultimate rate of terminal oocyte resorption at oviposition depends on how long the female has spent on a reduced diet since the initiation of oocyte growth.

DISCUSSION

The experiments described here show that both poor quality and low (suboptimal) quantities of food reduce or abolish the egg production of female *Locusta migratoria*.

Poor quality *Agropyron* diets are insufficient to initiate vitellogenesis. It has been shown in the past that green food plants are important for maximal egg production in the locusts *Dociostaurus maroccanus* (Merton, 1959) and *Schistocerca gregaria* (Cavanagh, 1963). The addition of small quantities of lush food and water to a poor *Agropyron* diet whilst decreasing mortality and markedly increasing ingestion fails to bring about increases in utilisation or egg development in *Locusta*. Qualitative factors of the diet may therefore be important in initiating and maintaining egg production in the female.

Vitellogenesis does not occur in poor *Agropyron* fed females despite the fact that during somatic growth the oocytes develop to a point at which the process normally occurs. Following a change from poor to lush grass feeding rapid oviposition occurs although the high rate of terminal oocyte resorption suggests that fat body metabolism does not fully recover for some weeks. Specific stimuli associated with certain levels of lush grass feeding appear therefore to be associated with the initiation of vitellogenesis.

Quantitative experiments reveal that continuous egg-pod production requires the ingestion of at least 75–80 mg dry weight of lush *Agropyron* each day and where initiation does occur the resultant level of oocyte resorption reflects the nutritional status during egg development. A subsequent change from optimal feeding to sub-optimal or poor quality feeding results in high resorption. If this subsequent diet is of poor quality or is less than 75 mg dry weight/female/day the next gonotrophic cycle is not initiated. Slight reductions in food quantity first affect the rate of egg-pod production. Further decreases affect both this rate and the rate of oocyte resorption. These observations agree with those of Petavy (1966) on *Locusta migratoria cinerascens* and artificial diets.

The timing of the initiation process for vitellogenesis occurs some seven days prior to oviposition. Although there is considerable individual variation this sensitive period is some 24–48 hours prior to visible yolk deposition and may correlate with the activation of the corpora allata (Johnson, 1972) whose secretion is required during vitellogenesis (Minks, 1967).

The process of vitellogenesis is hormonally controlled (Highnam et al., 1963a) and feeding may affect the release of hormonal factors from the corpora cardiaca (Highnam et al., 1966) in turn influencing the corpora allata (Girardie, 1967). Since resorption of terminal oocytes is considered as a result of competition between developing oocytes for vitellogenic protein and corpus allatum hormone (Highnam et al., 1963b) it is clear that feeding may affect the vitellogenic process either by altering the nutritional state of the haemolymph and/or by affecting the hormonal balance

171

of the insect. The influence of feeding and nutritional factors on egg development is therefore a complex phenomenon and is being investigated further.

REFERENCES

CAVANAGH, G. G. (1963): The use of the Dadd synthetic diet as a food for adult *Schistocerca gregaria* (Forsk.) and the effects of some additions and modifications to it. *J. Insect Physiol.* **9**, 759—775.
FARROW, R. A. (1972): The African migratory locust in its main outbreak area on the Middle Niger: quantitative studies of solitary populations in relation to environmental factors. 2 vols. *Ph. D. Thesis*, Reading.
GIRARDIE, A. (1967): Contrôle neuro-hormonal de la métamorphose et de la pigmentation chez *Locusta migratoria cinerascens* (Orthoptère). *Bull. biol. Fr. Belg.* **101**, 79–114.
HIGHNAM, K. C. and LUSIS, O. (1962): The influence of mature males on the neurosecretory control of ovarian development in the desert locust. *Q. J. microsc. Sci.* **103**, 73–83.
HIGHNAM, K. C., LUSIS, O. and HILL, L. (1963a): The role of the corpora allata during oocyte growth in the desert locust, *Schistocerca gregaria* Forsk. *J. Insect Physiol.* **9**, 587–596.
HIGHNAM, K. C., LUSIS, O. and HILL, L. (1963b): Factors affecting oocyte resorption in the desert locust *Schistocerca gregaria* (Forskal). *J. Insect Physiol.* **9**, 827–837.
HIGHNAM, K. C., HILL, L. and MORDUE, W. (1966): The endocrine system and oocyte growth in *Schistocerca* in relation to starvation and frontal ganglionectomy. *J. Insect Physiol.* **12**, 977–994.
JOHNSON, R. A. (1972): Activity of the corpora allata in the migratory locust *Locusta migratoria migratorioides* R. and F. *Ph. D. Thesis*, Sheffield.
MERTON, L. F. H. (1959): Studies in the ecology of the Moroccan locust (*Dociostaurus maroccanus* Thunberg) in Cyprus. *Anti-Locust Bull.* **34**, 123.
MINKS, A. K. (1967): Biochemical aspects of juvenile hormone action in the adult *Locusta migratoria*. *Archs. néerl. Zool.* **17**, 175–257.
PETAVY, G. (1966): Rapports entre alimentation et physiologie chez un insecte: *Locusta migratoria cinerascens* Fab. (Orthoptera, Acrididae). I. Utilisation de milieux nutritifs artificiels. *Annls. Nutr. Aliment.* **20**, 279–299.

172

Symp. Biol. Hung. 16, pp. 173–180 (1976)

THE VALUE OF ARTIFICIAL FEEDING TECHNIQUES IN APHID–HOST-PLANT STUDIES

by

T. E. MITTLER

DIVISION OF ENTOMOLOGY AND PARASITOLOGY, UNIVERSITY OF CALIFORNIA, BERKELEY, CALIF. 94720, USA

What can we learn about the association between aphids and their host plants by studying the behaviour, growth, and reproduction of these insects on artificial diets? This is a question that has frequently been asked. I will try to answer it from a variety of points of view, and try to show the applicability and value of this approach to our understanding of aphid–host-plant interactions.

The first attempts to feed aphids artificially were made by Hamilton in 1935 in relation to the transmission of plant viruses. Using plant epidermis, solutions of dyes and polonium, and as unsophisticated a piece of equipment as a gold-leaf electroscope, Hamilton demonstrated that such indicators were ingested by aphids and that they could be introduced into plants on which the aphids were subsequently allowed to feed.

Rubber membranes proved to be of value in the first attempts (Maltais, 1952) to feed aphids aseptically on chemically defined diets. While other materials (e.g. polyethylene sheeting) have also proved suitable for certain applications (such as for retaining pressurized diets), there is no doubt that the introduction of Parafilm* to aphid studies by Bradley (1956) has contributed considerably to the successful development of the various techniques now available for feeding aphids on artificial diets.

The simplest devices for so doing, the so-called sachets developed 10 years ago (Mittler and Dadd, 1964a) are quick to prepare and allow one to work with small volumes (fractions of a ml) of bacteriologically sterile diet (Dadd et al., 1967). However, considerably more complex feeding assemblies have also been employed successfully in particular situations (Akey and Beck, 1971). Some of the more recently developed systems (Harrewijn, 1973; Akey and Beck, 1975) even allow continuous or periodic diet changes to be affected automatically without dislodging the aphids from the membranes. Such systems hold promise also for rearing of other sucking insects, e.g. scales, that cannot be moved once settled.

While the original impetus for research on artificial feeding of aphids came from plant-virus workers (see review by Watson and Plumb, 1972) the potentials of this technique in other areas of aphid research have been recognized for some years. On the ecological side, it provides the means for testing how the susceptibility or resistance of plants to aphid infestation can be related to established differences in the plants' nutritive content

---

* Now marketed as Parafilm® "M" by the American Can Co., Neenah, Wisconsin.

(Auclair et al., 1957; van Emden and Bashford, 1971; Harrewijn, 1972). Auclair's paper in this symposium is particularly pertinent in this regard.

Qualitative and quantitative data on certain plant constituents can be correlated with the performance of aphids on the analysed plants. But while such correlations can be of great practical value, the causal effects must be determined under strictly defined dietary conditions. These are difficult enough to achieve with synthetic media (Mittler, 1972), but compared to plants, where one difference may merely reflect others, the use of artificial diets holds out considerable hope for the elucidation of this important area. Thus the numerous nutritional studies to date on food utilization and the dietary requirements of a number of aphid species for amino acids, sugars, minerals, sterols, and vitamins, are of interest not only in themselves. Their predictive ecological value may hopefully be applied in plant breeding programs. The metabolic implications of the nutritional findings have recently been reviewed by Dadd (1973) in the framework of insect dietetics in general.

The study of aphid nutrition is especially fascinating and challenging since it deals with an insect-symbiote complex. The artificial feeding technique has made it possible to demonstrate that the symbiotes may be caused to degenerate completely by nutritional means, such as through trace-mineral deficiencies (Ehrhardt, 1966) or by the feeding of antibiotics (Ehrhardt and Schmutterer, 1966). Subsequent changes in the aphids' nutritional requirements, e.g. a dependence on a dietary supply of all essential amino acids (Mittler, 1971) or an inability to synthesize sterols (Ehrhardt, 1968a), could be taken advantage of in aphid control measures, if a more subtle means of deranging the symbiotes can be achieved than by applying broad spectrum antibiotics to food plants (Ehrhardt et al., 1966). One may speculate that the success of aphids as a group depends to a large extent on the biosynthetic capabilities of the symbiotic microorganisms they harbour. Aphids can therefore thrive on inadequate or imbalanced dietaries such as they must normally encounter when ingesting the sap of their host plants.

Of comparable interest are the studies that have been made on systemic insecticides incorporated into artificial diets (Mittler and Pennell, 1964; Parry and Ford, 1971; Raccah and Tahori, 1971; Meier and Pantanetti, 1966). While changes in the insecticides that may occur in a plant must be taken into account, the use of the artificial feeding technique for evaluating the potency of insecticides and their mode of action holds further promise.

Hemipterous insects have long been known to inflict injury on their host plants by sapping the plants' nutrients, by causing mechanical damage as a result of their feeding activities, and because the saliva they inject into their host plants may contain certain components (amino acids, plant growth factors or their precursors) detrimental to the plants, and, as far as has been ascertained, occasionally beneficial to these plant parasites (Kennedy and Stroyan, 1959; Miles, 1968; Forrest, 1971). Sensitive counting and autoradiographic techniques (Kloft et al., 1968; Lamb et al., 1967) for determining radioactivity in aphid salivary secretions undoubtedly have great potential for studies on the translocation of these and other materials in the phloem sieve tubes of plants (Forrest and Noordink, 1971). Studies on the composition of, and effect of diet on salivary components, such as pectinases (Adams, 1967) and other enzymes, amino acids, plant growth regulating substances, etc., will no doubt be greatly facilitated by collecting these secretions behind membranes.

That aphids could take up via membranes small amounts of liquid at ambient pressure was, of course, demonstrated already in 1935 by Hamilton. For 25 years thereafter numerous attempts to feed and maintain aphids for longer periods on artificial diets did not meet with much success. In retrospect, the reasons for failure were numerous. However, one factor that attained prominence was the role that plant turgor-pressure may play in assisting aphids to ingest sap from their hosts. This view stemmed, of course, from the demonstration by Kennedy and Mittler (1953) that the plant sap ingested by *Tuberolachnus salignus*, an aphid that colonizes the stems of willow, was under pressure; a finding that led to the reasonable assumption that the pressure may assist or contribute to a large extent to the recorded rates of sap intake and honeydew excretion by this aphid and probably by aphids in general.

Uptake rates of nutritionally adequate artificial diets at ambient pressures, while possibly only half the rate of sap uptake from host plants (Mittler, 1970a), are nevertheless surprisingly high considering that the aphids have to rely entirely on their sucking efforts to ingest them. Efforts at providing aphids with simple solutions under as much as 12 atmospheres of pressure resulted in only slight improvements in the uptake of these solutions, or in the aphids' survival or larviposition (Wearing, 1968; van Hoof, 1958; van Emden, 1967). The inescapable conclusion is that aphids are able to suck fairly efficiently. There is also no doubt that aphids can regulate their rate of feeding in relation to diet composition (Mittler, 1970b; Srivastava and Auclair, 1971a; Leckstein and Llewellyn, 1973; Auclair, in this symposium) or ant attendance (Hertzig, 1937; Banks and Nixon, 1958; Kleinjan, 1969). However, it has not yet been established whether an increase in uptake over a period of say a day is due to an increase in the actual rate of ingestion or results from more frequent periods of ingestion, each at a constant and possibly maximal rate of ingestion.

Electrical recording techniques (McLean, 1971; Friend and Smith, 1972) should be of value in clarifying this aspect. These and other aspects of aphid feeding were reviewed at the insect/plant symposium in London in 1971.

Recent studies with Drs. H. F. van Emden and K. P. Lamb have shown that when *M. persicae* were subjected throughout their development to daily periods of nutritional inadequacies by giving them a poor diet for 8–10 h each day, the aphids were able to compensate and possibly even over-compensate for lengthy and repeated periods of nutritional deprivation by a faster intake and better utilization of an optimal nutrient diet when it was available to them. Clearly such a regulatory ability should allow aphids to cope with, or even benefit from, diurnal changes in the composition of the sap of their host plants.

The advantages of being able to maintain aphids on artificial diets may be broadly divided into two categories. On the one hand, artificial feeding enables one to study aphids divorced from their host plants. This is important when one wants to ascertain the effects of environmental condition on the aphid itself rather than on the plant as well. In this regard the influence of such factors as temperature, photoperiod and light quality on settling behaviour development and growth rates, or on polymorphism, may be studied to good advantage (Cartier, 1966; Auclair, 1967; Tsitsipis and Mittler, 1970; Sutherland and Mittler, 1969).

On the other hand, there are numerous facets of nutrition that can best and sometimes only be resolved by getting away from the uncertainties inherent when natural dietaries are considered. One such application is the study of the effects that different diets have on the expression and production of different morphs (Harrewijn, 1972, and present symposium; Leckstein and Llewellyn, 1973; and review by Mittler, 1973). While a start has been made with regard to alary dimorphism, other aspects of aphid polymorphism and related behavioural responses of ecological significance, will be amenable to critical study by these means.

An interesting new application of the artificial feeding technique is in the study of the effect aphid nutrition has on the growth and development of their hymenopterous parasites (Mackauer and Cloutier, personal communication).

In the area of phagostimulation, numerous studies attest to the value of artificial feeding techniques. Regarding chemicals associated with specific host-plants of special importance to monophagous or oligophagous aphids, the classical demonstration by Wensler (1962) that sinigrin applied to bean plants makes these acceptable to cabbage aphids, has led to studies with artificial diets that have elucidated the response to this and other compounds by several aphid species (Moon, 1967; Wearing, 1968; Nault and Styer, 1972; Klingauf, 1971; Klingauf et al., 1972; Chawla et al., 1974).

There can be no doubt that our understanding on the basis for the selection and susceptibility of host-plants by aphids will be enhanced by further such studies; particularly so, if the analytical skills of plant biochemists are combined with the ingenuity that aphid workers have shown for quantitatizing the responses of aphids to diets. Much can be done with the simplest of methods (Mittler and Dadd, 1964b, 1965; Mittler, 1970b) but recent electrical and radiolabelling techniques for recording probing, salivation, and liquid uptake by aphids (McLean, 1971; Arn and Cleere, 1971; Harrewijn and Noordink, 1971) will advance the field enormously.

Differences have been encountered in the survival and growth on diet of biotypes from one geographical area (Cartier and Auclair, 1965) or from different continents (Mittler, unpublished obs.). Differences between biotypes have been observed in relation to systemic insecticides (Parry and Ford, 1964). As was emphasized for *Drosophila* (Cooke and Sang, 1970), nutritional tests with different races of aphids will be of value for elucidating their nutritional specializations.

The difficulties encountered in raising certain biotypes, and the need for comparative work in laboratories across the world on races that have been exposed to different insecticidal pressures, or that differ in their virus transmission efficiencies or their polymorphism, etc. have necessitated sending aphids over long distances. That this can readily be accomplished using sachets of artificial diet rather than plants has not only facilitated a number of such shipments in the past few years, but has avoided plant quarantine problems.

The continuous maintenance of aphids on nutritionally adequate artificial diets has been achieved by a number of workers with over a dozen different

176

species (Krieger, 1971; Dadd and Mittler, 1966; Dadd and Krieger, 1967; Ehrhardt, 1968b; Akey and Beck, 1971; Srivastava and Auclair, 1971b). We were able to maintain *Aphis fabae* on artificial diet for more than 100 successive generations during a period of three years (Mittler and Kleinjan, unpublished obs.). Clearly there are advantages in being able to maintain an experimental animal on a standardizable artificial diet, whether this is synthetic and completely defined or not. This was realized in work on *Drosophila* many years ago, and applies to the mass-rearing efforts devoted to an ever increasing number of insects — particularly those of economic importance. We have not reached the stage where it is more convenient and cheaper to maintain aphid cultures on artificial diet than on plants. In this connection, I am pleased to report that at the time of this symposium at least 6 successive generations of *M. persicae* have been maintained on oligidic and meridic diets containing inexpensive protein hydrolyzates and yeast extracts (Mittler and Koski, in preparation).

We have also established that *M. persicae* can utilize a number of dietary dipeptides. This confirms the results of in vitro studies on gut homogenates of the pea aphid by Srivastava and Auclair (1963). These authors, in contrast to Bramstedt (1948), did not detect any in vitro proteolytic activity. However, recent experiments using unhydrolyzed dietary protein indicate that *M. persicae* may have the ability to digest proteins. If this is substantiated, it would mean that aphids could exploit proteinaceous sources in their host plants — a view recently expressed by Leckstein and Llewellyn (1974).

Feeding aphids on chemically defined diets not only permits one to assess their nutrient requirements (e.g. in terms of optimal developmental, growth and larviposition rates) but allows one to assess the relative and absolute efficiencies with which nutrients are used. Such considerations depend on quantitizing the amounts and composition of the excreta, as well as the food ingested and assimilated. Clearly, direct comparisons of the composition of the diet ingested and honeydew excreted are practical (Bragdon and Mittler, 1963). Further studies along these lines are called for, not only because they are of interest from the insect nutritional point of view but because they could be of great value to plant physiologists. These have begun to lean more and more in recent years on aphids to help them elucidate translocation mechanisms in the phloem elements of plants (Bornman and Botha, 1973).

The examples given illustrate the growing interest that has developed in the past decade for feeding aphids on artificial diets. Clearly, the technique has great potential for future research seeking further insights into the inter-relationships between aphids and their hostplants.

REFERENCES

ADAMS, J. B. (1967): A physiological difference in aphids (Homoptera) raised on excised leaves and on intact plants. *Can. J. Zool.* **45**, 588–590.
AKEY, D. H. and BECK, S. D. (1971): Continuous rearing of the pea aphid, *Acyrthosiphon pisum*, on a holidic diet. *Ann. ent. Soc. Am.* **64**, 353–356.
AKEY, D. H. and BECK, S. D. (1975): Programmable automated system for continuous rearing of the pea aphid, *Acyrthosiphon pisum* on holidic diets. *Ent. exp. & appl.* **18**, 1—16.

ARN, H. and CLEERE, J. S. (1971): A double-label choice-test for the simultaneous determination of diet preference and ingestion by the aphid *Amphorophora agathonica*. *Ent. exp. & appl.* **14**, 377–387.

AUCLAIR, J. L. (1967): Effects of light and sugars on rearing the cotton aphid, *Aphis gossypii*, on a germ-free and holidic diet. *J. Insect Physiol.* **13**, 1247–1268.

AUCLAIR, J. L., MALTAIS, J. B. and CARTIER, J. J. (1957): Factors in resistance of peas to the pea aphid, *Acyrthosiphon pisum* (Harr.) (Homoptera: Aphididae). II. Amino acids. *Can. Ent.* **89**, 457–464.

BANKS, C. J. and NIXON, H. L. (1958): Effects of the ant, *Lasius niger* L., on the feeding and excretion of the bean aphid, *Aphis fabae* Scop. *J. exp. Biol.* **35**, 703–711.

BORNMAN, C. H. and BOTHA, C. E. J. (1973): The role of aphids in phloem research. *Endeavour*, **32**, 129–133.

BRADLEY, R. H. E. (1956): Effects of depth of stylet penetration on aphid transmission of potato virus Y. *Can. J. Microbiol.* **2**, 539–547.

BRAGDON, F. G. and MITTLER, T. E. (1963): Differential utilization of amino acids by *Myzus persicae* (Sulzer) fed on artificial diets. *Nature, Lond.* **198**, 204–210.

BRAMSTEDT, F. (1948): Über die Verdauungsphysiologie der Aphiden. *Z. Naturf.* **39**, 14–24.

CARTIER, J. J. (1966): Aphid responses to colors in artificial rearings. *Bull. ent. Soc. Am.* **12**, 378–380.

CARTIER, J. J. and AUCLAIR, J. L. (1965): Effets des couleurs sur le comportement de diverses races du puceron du pois, *Acyrthosiphon pisum* (Harris), en élevage sur un régime nutritif de composition chimique connue. *Proc. 12th Int. Congr. Ent., London* 1964, 414.

CHAWLA, S. S., PERRON, J. M. and CLOUTIER, M. (1974): Effects of different growth factors on the potato aphid, *Macrosiphum euphorbiae* (Aphididae: Homoptera), fed on an artificial diet. *Can. Ent.* **106**, 273–280.

COOKE, J. and SANG, J. H. (1970): Utilization of sterols by larvae of *Drosophila melanogaster*. *J. Insect Physiol.* **16**, 801–812.

DADD, R. H. (1973): Insect nutrition: current developments and metabolic implications. *Ann. Rev. Ent.* **18**, 381–420.

DADD, R. H. and KRIEGER, D. L. (1967): Continuous rearing of the *Aphis fabae* complex on sterile synthetic diet. *J. econ. Ent.* **60**, 1512–1514.

DADD, R. H., KRIEGER, D. L. and MITTLER, T. E. (1967): Studies on the artificial feeding of the aphid *Myzus persicae* (Sulzer). IV. Requirements for water soluble vitamins and ascorbic acid. *J. Insect Physiol.* **13**, 249–272.

DADD, R. H. and MITTLER, T. E. (1966): Permanent culture of an aphid on a totally synthetic diet. *Experientia*, **22**, 832.

EHRHARDT, P. (1966): Entwicklung und Symbionten geflügelter und ungeflügelter Virgines von *Aphis fabae* Scop. unter dem Einfluß künstlicher Ernährung. *Z. Morph. Ökol. Tiere*, **50**, 295–319.

EHRHARDT, P. (1968a): Nachweis einer durch symbiotische Microorganismen bewirkten Sterinsynthese in künstlich ernährten Aphiden (Homoptera, Rhynchota, Insecta). *Experientia*, **24**, 82.

EHRHARDT, P. (1968b): Die Wirkung verschiedener Spurenelemente auf Wachstum, Reproduktion und Symbionten von *Neomyzus circumflexus* Buckt. (Aphidae, Homoptera, Insecta) bei künstlicher Ernährung. *Z. vergl. Physiol.* **58**, 47–75.

EHRHARDT, P., JAYARAJ, S. and SCHMUTTERER, H. (1966): Die Wirkung verschiedener, über die Pflanze zugeführter Antibiotica auf Entwicklung und Fertilität der schwarzen Bohnenblattlaus *Aphis fabae*. *Ent. exp. & appl.* **9**, 332–342.

EHRHARDT, P. and SCHMUTTERER, H. (1966): Die Wirkung verschiedener Antibiotica auf Entwicklung und Symbioten künstlich ernähter Bohnenblattläuse (*Aphis fabae* Scop.). *Z. Morph. Ökol. Tiere*, **56**, 1–20.

FORREST, J. M. S. (1971): The growth of *Aphis fabae* as an indicator of the nutritional advantage of galling to the apple aphid *Dysaphis devecta*. *Ent. exp. & appl.* **14**, 477–483.

FORREST, J. M. S. and NOORDINK, J. P. W. (1971): Translocation and subsequent uptake by aphids of ^{32}P introduced into plants by radioactive aphids. *Ent. exp. & appl.* **14**, 133–134.

FRIEND, W. G. and SMITH, J. J. B. (1972): Feeding stimuli and techniques for studying the feeding of haematophagous arthropods under artificial conditions, with special reference to *Rhodnius prolixus*. In: *Insect and Mite Nutrition*. North-Holland Publ. Co., Amsterdam.

178

HAMILTON, M. A. (1935): Further experiments on the artificial feeding of *Myzus persicae* (Sulz.). *Ann. appl. Biol.* **22**, 243–258.

HARREWIJN, P. (1972): Wing production by the aphid *Myzus persicae* related to nutritional factors in potato plants and artificial diets. In: *Insect and Mite Nutrition*. North-Holland Publ. Co., Amsterdam.

HARREWIJN, P. (1973): Functional significance of indole alkylamines linked to nutritional factors in wing development of the aphid *Myzus persicae*. *Ent. exp. & appl.* **16**, 499–513.

HARREWIJN, P. and NOORDINK, J. P. W. (1971): Taste perception of *Myzus persicae* in relation to food uptake and developmental processes. *Ent. exp. & appl.* **14**, 413–419.

HERTZIG, J. (1937): Ameisen und Blattläuse. *Z. ang. Ent.* **24**, 367–435.

KENNEDY, J. S. and MITTLER, T. E. (1953): A method of obtaining phloem sap via the mouth-parts of aphids. *Nature, Lond.* **171**, 528.

KENNEDY, J. S. and STROYAN, H. L. G. (1959): Biology of aphids. *Ann. Rev. Ent.* **4**, 139–160.

KLEINJAN, J. E. (1969): The influence of ants (*Formica* spp.) on the growth and form of the black bean aphid, *Aphis fabae* Scop., reared on artificial diet. *M. Sc. Thesis in Entomology*, Univ. of California, Berkeley, 39.

KLINGAUF, F. (1971): Die Wirkung des Glucosids Phlorizin auf das Wirtswahlverhalten von *Rhopalosiphum insertum* (Walk.) und *Aphis pomi* De Geer (Homoptera: Aphididae). *Z. ang. Ent.* **68**, 41–55.

KLINGAUF, F. SENGONCA, C. and BENNEWITZ, H. (1972): Einfluß von Sinigrin auf die Nahrungsaufnahme polyphager und oligophager Blattlausarten (Aphididae). *Oecologia*, **9**, 53–57.

KLOFT, W., EHRHARDT, P. and KUNKEL, H. (1968): Radioisotopes in the investigation of interrelationships between aphids and host plants. In: *Isotopes and radiation in entomology*, IAEA, 23–30.

KRIEGER, D. L. (1971): Rearing several aphid species on synthetic diet. *Ann. ent. Soc. Am.* **64**, 1176–1177.

LAMB, K. P., EHRHARDT, P. and MOERICKE, V. (1967): Labelling of aphid saliva with rubidium-86. *Nature, Lond.* **214**, 602–605.

LECKSTEIN, P. M. and LLEWELLYN, M. (1973): Effect of dietary amino acids on the size and alary polymorphism of *Aphis fabae*. *J. Insect Physiol.* **19**, 973–980.

LECKSTEIN, P. M. and LLEWELLYN, M. (1974): The role of amino acids in diet intake and selection and the utilization of dipeptides by *Aphis fabae*. *J. Insect Physiol.* **20**, 877–885.

MALTAIS, J. B. (1952): A simple apparatus for feeding aphids aseptically on chemically defined diets. *Can. Ent.* **84**, 291–294.

McLEAN, D. L. (1971): Probing behaviour of the pea aphid, *Acyrthosiphon pisum*. V. Comparison of *Vicia faba*, *Pisum sativum*, and a chemically defined diet as food sources. *Ann. ent. Soc. Am.* **64**, 499–503.

VON MEIER, W. and PANTANETTI, P. (1966): Über Bioteste mit Blattläusen zum Nachweis geringer Insektizidmengen in Zuckerlösungen. *Schweiz. Landw. Forsch.* **5**, 469–480.

MILES, P. W. (1968): Insect secretions in plants. *Ann. Rev. Phytopath.* **6**, 137–164.

MITTLER, T. E. (1970a): Uptake rates of plant sap and synthetic diet by the aphid *Myzus persicae*. *Ann. ent. Soc. Am.* **63**, 1701–1705.

MITTLER, T. E. (1970b): Effects of dietary amino acids on the feeding rate of the aphid *Myzus persicae*. *Ent. exp. & appl.* **13**, 432–437.

MITTLER, T. E. (1971): Dietary amino acid requirements of the aphid *Myzus persicae* affected by antibiotic uptake. *J. Nutrition*, **101**, 1023–1028.

MITTLER, T. E. (1972): Interactions between dietary components. In: *Insect and Mite Nutrition*. North Holland Publ. Co., Amsterdam.

MITTLER, T. E. (1973): Aphid polymorphism as affected by diet. In: *Perspectives in Aphid Biology*. Bull. 2, Ent. Soc. New Zealand, 65–75.

MITTLER, T. E. and DADD, R. H. (1964a): An improved method for feeding aphids on artificial diets. *Ann. ent. Soc. Am.* **57**, 139–140.

MITTLER, T. E. and DADD, R. H. (1964b): Gustatory discrimination between liquids by the aphid *Myzus persicae* (Sulzer). *Ent. exp. & appl.* **7**, 315–328.

MITTLER, T. E. and DADD, R. H. (1965): Differences in the probing responses of *Myzus persicae* (Sulzer) elicited by different feeding solutions behind a parafilm membrane. *Ent. exp. & appl.* **8**, 107–122.

MITTLER, T. E. and PENNELL, J. T. (1964): Simple screening test for systemic aphicides. *J. econ. Ent.* **57**, 302–303.

MOON, M. S. (1967): Phagostimulation of a monophagous aphid. *Oikos*, **18**, 96–101.

NAULT, L. R. and STYER, W. E. (1972): Effects of sinigrin on host selection by aphids. *Ent. exp. & appl.* **15**, 423–432.

PARRY, W. H. and FORD, J. B. (1964): The artificial feeding of phosphamidon to *Myzus persicae:* I. Intraspecific differences exhibited by this aphid on feeding through a parafilm membrane. *Ent. exp. & appl.* **10**, 437–452.

PARRY, W. H. and FORD, J. B. (1971): Effects of phosphamidon on the longevity, fecundity and liquid uptake. *Ent. exp. & appl.* **14**, 389–398.

RACCAH, B. and TAHORI, A. S. (1971): Wing dimorphism influencing resistance or toxicity tests and food uptake in *Myzus persicae. Ent. exp. & appl.* **14**, 310–314.

RACCAH, B., TAHORI, A. S. and APPLEBAUM, S. W. (1971): Effect of nutritional factors in synthetic diet on increase of alate forms in *Myzus persicae. J. Insect Physiol.* **17**, 1385–1390.

SRIVASTAVA, P. N. and AUCLAIR, J. L. (1963): Characteristics and nature of proteases from the alimentary canal of the pea aphid, *Acyrthosiphon pisum* (Harr.) (Homoptera, Aphididae). *J. Insect Physiol.* **9**, 469–474.

SRIVASTAVA, P. N. and AUCLAIR, J. L. (1971a): Influence of sucrose concentration on diet uptake and performance by the pea aphid, *Acyrthosiphon pisum. Ann. ent. Soc. Am.* **64**, 739–743.

SRIVASTAVA, P. N. and AUCLAIR, J. L. (1971b): An improved chemically defined diet for the pea aphid, *Acyrthosiphon pisum. Ann. ent. Soc. Am.* **64**, 474–478.

SUTHERLAND, O. R. W. and MITTLER, T. E. (1969): Sexual forms of the pea aphid, *Acyrthosiphon pisum*, produced on an artificial diet. *Ent. exp. & appl.* **12**, 240–241.

TSITSIPIS, J. A. and MITTLER, T. E. (1970): Convenient lighting system for inducing the production of sexual forms of aphids feeding on artificial diets. *Ann. ent. Soc. Am.* **63**, 1665–1667.

VAN HOOF, H. A. (1958): Onderzoekingen over de biologische overdracht van een non-persistent virus. *Meded. Inst. plziektenk. Onderz.* **161**, 1–96.

VAN EMDEN, H. F. (1967): An increase in the longevity of adult *Aphis fabae* fed artificially through parafilm membranes on liquids under pressure. *Ent. exp. & appl.* **10**, 166–170.

VAN EMDEN, H. F. and BASHFORD, M. A. (1971): The performance of *Brevicoryne brassicae* and *Myzus persicae* in relation to plant age and leaf amino acids. *Ent. exp. & appl.* **14**, 349–360.

WATSON, M. A. and PLUMB, M. A. (1972): Transmission of plant-pathogenic viruses by aphids. *Ann. Rev. Ent.* **17**, 425–452.

WEARING, C. H. (1968): Responses of aphids to pressure applied to liquid diet behind parafilm membrane. Longevity and larviposition of *Myzus persicae* (Sulz.) and *Brevicoryne brassicae* (L.) (Homoptera: Aphididae) feeding on sucrose and sinigrin solutions. *N. Z. J. Sci.* **11**, 105–121.

WENSLER, R. J. D. (1962): Mode of host selection by an aphid. *Nature, Lond.* **195**, 830–831.

Symp. Biol. Hung. 16, pp. 181–185 (1976)

DISTRIBUTION OF ADULT INSECTS IN RELATION TO THE HOST-PLANTS

by

J.-P. MOREAU

STATION CENTRALE DE ZOOLOGIE, C.N.R.A., 78000 VERSAILLES, FRANCE

With the help of several methods: trapping, sweeping, hand picking, visual assessment, etc., one can study the distribution of adult insects in relation to the species or variety of plant submitted to competition, directly or indirectly for a source of nutritional needs.

The insects studied were on cereals: *Oscinella frit* and *O. pusilla* (Dipt., Chloropidae); on potato: *Leptinotarsa decemlineata* (Col., Chrysomelidae); and *Coccinella septempunctata* (Col., Coccinellidae).

In the case of *O. frit* and *C. septempunctata*, the distribution is not in relation with the possibility of development of progeny, while the adults of *O. pusilla* and *L. decemlineata* make the provision of nutrition for the successive generation as related to their distribution.

INTRODUCTION

Among the insects utilizing plants as direct or indirect source of food, there are species of which larvae have different feeding regime than that of their adults while in others, this regime is stationary throughout their entire development.

In the first category, the egglaying behaviour of the female differs in time and space from that of the feeding behaviour. Consequently, the adults are localized not only according to their specific nutrients but also in the function of their search for suitable egglaying site. The stimulus (physical and/or chemical) which attracts more or less strongly the females and occasionally the males, towards a plant or plant-part determines the survival of the further progeny. This happens because, in the majority of cases, the young larvae are incapable to disperse, to find more favourable food than that found by their parents. Moreover, mortality of the offspring will be heavier, if the eggs are laid on a plant unfavourable for larval growth, as in the case of *Hydrellia griseola* Fall. (Stavrakis, 1964). It may also occur that eggs are not laid on the host-plant, but the newly hatched larvae are able to migrate: in the case of *Cnephasia pumicana* Zell., the young larva produces a silk thread through which it is transported to the food plants by wind (Chambon, 1969).

In the second category, on the contrary, the feeding and reproduction are often closely associated and can rapidly follow each other. In fact, larvae feeding on the same plant-part as their parents often show dispersal (correlated with qualitative and/or quantitative deterioration of food material), or escape in time (embryonal or larval diapause). It has also been noted that the egglaying site generally differs from the feeding site;

moreover, certain *Homoptera* (Jassids, Cercopids, etc.) insert their hibernating eggs in the more lignified tissues than those of feeding sites.

MATERIAL AND METHODS

To illustrate these facts, we studied four insects in the field. In the first category, attack by *Oscinella frit* L. and *O. pusilla* Meig. (Dipt., Chloropidae) on oat, barley and wheat plants was observed (Chevin et al., 1971). The second category was studied in three varieties of potato (Ackersegen, Bintje and Kennebec) infested by *Leptinotarsa decemlineata* Say (col., Chrysomelidae) and by *Coccinella septempunctata* L. (Col., Coccinellidae). With the help of coloured traps — Moericke type — the distribution of frit flies could be studied (Moreau, 1963). The number of insects trapped does not represent the population in the field, especially in small samples, moreover, sex ratio is likewise different (Moreau and Durand, 1974). Sweepnet was used to assess the size of the population. After counting, the insects were released in the same plot. In the case of Colorado beetles and coccinellids, the sampling by direct observations has been preferred, although this method does not allow counting males and females separately.

RESULTS

Oscinella frit and *O. pusilla*

The larvae of *O. frit* consume the young stems or unripened grains, whereas *O. pusilla* feeds only on young stems. The adults feed the pollen and honeydew. They need contact with the host-plant for mating and normal egg laying (Moreau, 1967). However, in absence of plants, egg laying (though reduced) may, nevertheless, take place (Le Berre, 1959).

In autumn and spring, the eggs of *O. frit* are not only laid on the young plants, but also on the soil surface nearby (Ibbotson, 1961), and on young ears of cereals in summer; though larval development is not observed on wheat ears. On the contrary, *O. pusilla* lays eggs only on the young plants of wheat and barley (Chevin et al., 1971).

Damage on the young plants and ears in the month of July can be avoided by well timing the sowing. The sweep-net samples show the preference of *O. frit* for the ears, and that of *O. pusilla* for young plants (Table 1). A high number of males and females of *O. frit* meet on the bare soil and adjunct culture of non host-plants, while *O. pusilla* mostly settles exclusively on the young plants and at the base of the aged plants of barley and wheat.

TABLE 1

Net samples of adult frit flies on cereals
(Versailles, 17–19 and 28–30 July 1963)

	On young plants		On ears	
	♂	♀	♂	♀
O. frit	193	297	619	872
O. pusilla	37	77	4	6

Leptinotarsa decemlineata

Both adults and larvae feed on the potato plants. The variety "Kennebec" appears to be less favourable for larval development and for the reproduction of Colorado beetles than "Bintje" and "Ackersegen" (Moreau, 1971; Derridj, 1973). This is true for young plants. On the contrary, when the foliage becomes aged in August "Bintje" becomes the most unfavourable among the three varieties, whereas "Ackersegen" remains the best (Table 2).

The egg-laying and larval development are in relationship with the distributions. This has been observed in the first generation of *L. decemlineata* in 1972 (Table 3). In the second generation, larvae were found only on "Ackersegen".

TABLE 2

Distribution of Colorado potato beetles (%)
in 3 varieties of potato
(Versailles, 1970–71–72)

Date	Ackersegen	Bintje	Kennebec
August 1970 (1)	66	9	25
August 1971 (1)	92	0	8
June 1972 (2)	41	33	26
August 1972 (1)	96	1	3

(1) summer adults.
(2) spring adults.

TABLE 3

Number of fourth instar larvae
of L. decemlineata *noted on 48 plants*
of each of 3 varieties of potato

Date	Ackersegen	Bintje	Kennebec
27 June 72	217	115	74
3 July 72	392	266	130

Coccinella septempunctata

Both the pre-adult stages and adults of this predatory insect feed on aphids, such as *Macrosiphum euphorbiae* Kalt. and *Myzus persicae* Sulz. Thus, potato plants constitute an indirect source of nutrition.

In the experimental field "Ackersegen", "Bintje" and "Kennebec" were alternately planted. The flight of summer adults of coccinnellids started on the 21st of July in 1971. Their distribution indicated a preference for "Kennebec", although the colonies of aphids, present in large numbers were distributed equally on the host varieties. After a few days, the distribution of coccinellids has become more homogeneous, however, with a majority on "Kennebec" (Table 4).

The physiological state of the plant affects the feeding of aphids and ultimately influences the diapause of these insects (Rolley et al., 1974).

TABLE 4

Distribution of Coccinellid adults on 3 varieties of potato
(Versailles, 1971)

Date	No. of adults	Ackersegen	Bintje	Kennebec
21 July 71	total	37	19	184
	mean/plant	0.77	0.47	3.83
24 July 71	total	81	43	145
	mean/plant	1.69	0.93	3.02

It is possible that the choice of potato variety by *C. septempunctata* is an important factor in the further evolution of its population, however, we do not know whether the plant is beneficial or not for the succeeding generations.

DISCUSSION

To study the distribution pattern in adult insects in relation to their host-plants, four species (three phytophagous and one predator) have been investigated.

In phytophagous insects, a close relation with their host-plants is often found through a frequent coincidence by the adults and possibilities of larval development. This fact is observed in *Oscinella pusilla* (belonging to the first category) and *Leptinotarsa decemlineata* (included in the second category). On the contrary, in *Oscinella frit*, this relationship is distant, i.e. adults are of more wandering type and are less attracted by the host-plants of the larvae.

The host-plants of the aphids interfere with the feeding habit of the predatory species, *Coccinella septempunctata*, and this relation can elicit a change in the distribution of adults and in the possibilities of development of their descendants.

Moreover, the relations of adults with their host-plants are not specially influenced by the fact that the feeding material of the adults and of larval stages is different or the same, but rather by the degree of polyphagous habit of the species. In fact, *Oscinella pusilla* and *Leptinotarsa decemlineata*, both being oligophagous, are more associated with the plants than the more polyphagous species: *O. frit* and *Coccinella septempunctata*.

REFERENCES

CHAMBON, J. P. (1969): Extension d'un foyer et dispersion des populations d'une tordeuse (*Cnephasia pumicana* Zeller, Lépidoptère, Tortricidae). *Ann. Zool. Ecol. anim.* **1**, 433–444.

CHEVIN, H., DURAND, Y., LE BERRE, J. R. and MOREAU, J.-P. (1971): Etude de quelques facteurs intervenant dans la prolifération de deux espèces d'Oscinies, *O. frit* L. et *O. pusilla* Mg. *Ann. Zool. Ecol. anim.* **3**, 347–359.

DERRIDJ, S. (1973): Influence de la pomme de terre (*Solanum tuberosum*) et en particulier de la variété Kennebec sur la biologie de *Leptinotarsa decemlineata* Say (Col., Chrysomelidae) dans différentes conditions. *Thèse 3ème cycle* Paris VI, 78.

IBBOTSON, A. (1961): Host selection by frit fly in Britain. *Ann. Epiphyties*, **12**, 445–452.

LE BERRE, J. R. (1959): Etudes entreprises sur les Oscinies dans le cadre des recherches relatives aux immunités végétales à l'égard des insectes. *Meded. Landbouw. Opzoekingstns, Gent*, **24,** 593–610.

MOREAU, J. P. (1963): Contribution à l'étude éthologique de deux espèces d'Oscinies *Oscinella frit* L. et *O. pusilla* Mg. (Dipt., Chloropidae). *C. R. Acad. Sci. Paris* **256,** 1831–1833.

MOREAU, J. P. (1967): Analyse expérimentale de quelques causes de résistance des céréales aux insectes (l'exemple des Oscinies). *Meded. Rijks. Landbouw. Wetensch.* **32,** 318–327.

MOREAU, J. P. (1971): Das Verhalten des Kartoffelkäfers (*Leptinotarsa decemlineata* Say) gegenüber drei Kartoffelsorten. *Acta Phytopath. Acad. Sci. Hung.* **6,** 165–168.

MOREAU, J. P. and DURAND, Y. (1974): Captures d'adultes d'oscinies (*O. frit* L. et *O. pusilla* Mg.) au moyen de pièges colorés. (In print)

ROLLEY, F., HODEK, I. and IPERTI, G. (1974): Influence de la nourriture aphidienne (selon l'âge de la plante-hôte à partir de laquelle les pucerons se multiplient) sur l'induction de la dormance chez *Semiadalia undecimnotata* Schn. (Col., *Coccinellidae*). *Ann. Zool. Ecol. anim.* **6,** 53–60.

STAVRAKIS, G. (1964): Rapports entre la morphologie des feuilles et le comportement de ponte d'*Hydrellia griseola* Fall. (Dipt., Ephydridae). *Ann. Epiphyties,* **15,** 326.

Symp. Biol. Hung. 16, pp. 187–190 (1976)

HOSTS AND NON-HOSTS IN SUBSPECIES OF *AULACORTHUM SOLANI* (KALTENBACH) AND INTRASPECIFIC HYBRIDIZATIONS (HOMOPTERA: APHIDIDAE)

by

F. P. MÜLLER

FORSCHUNGSGRUPPE PHYTO-ENTOMOLOGIE, SEKTION BIOLOGIE,
UNIVERSITÄT ROSTOCK, 25 ROSTOCK, DDR

Many phytophagous insect species are split off into biotypes or bionomical races distinguished only by their host-plant selection. Even in the case of authorized subspecies the food plants may concern the most outstanding feature. The knowledge of such biotypes warrants attention both in practical and in theoretical regard. Practical aspects relate to the facts that host races may be of very different economic importance as pests or as virus vectors, and that their existence renders the breeding of resistant cultivars more difficult. As to the theoretical aspect the study of host races serves for analysing the mechanisms of host acceptability and host avoidance. Further on, such research appears essential for the understanding of evolutionary processes.

Particularly the superfamily Aphidoidea exhibits many examples of species or form complexes being composed of taxa differing by their host-plants. In most cases intraspecific units within aphid species still possess the potential hybridization power. However, in nature they are reproductively isolated owing to the choice of different plants. For example, it had been possible to achieve a hybrid between the black cherry aphid, *Myzus cerasi* (F.), and its subspecies *veronicae* Walker which performs no host alternation and settles *Galium* and *Veronica*.

Very interesting host-plant relations have been revealed in the complex of the pea aphid, *Acyrthosiphon pisum* (Harris). In our open air insectary at Rostock we succeeded in obtaining hybrids between the green subspecies *destructor* Johnson, which lives on pea but refuses *Trifolium pratense*, and a red biotype the main host-plant of which is red clover and which cannot develop on pea. The hybrids comprised green as well as red clones for the red biotype is heterozygous with respect to the colour whereas red was pointed out to be dominant over green. The red clones developed and propagated on pea as heavily as the green parent. In nature, however, on pea red aphids are extremely rare. This fact, that pea is infested only by green *A. pisum*, is expressed in the German plant protection literature in which *A. pisum* living on pea is called "Grüne Erbsenblattlaus". The red *Trifolium* strain and the green pea strain occur in nature side by side. But their behaviour directed to seek for the right host-plant represents a highly important isolating mechanism and effectively prevents interbreeding.

Similar experiments were extended on the complex of the foxglove aphid, *Aulacorthum solani* (Kaltenbach). In all hybridization experiments a typical polyphagous form was used as one parent. This strain had been collected

187

near Rostock from the ground ivy, *Glechoma hederacea* L., on which the fundatrices are to be found very frequently. It settles readily the foxglove *Digitalis purpurea* L., potato and many other plant species. The typical *A. solani* is called in the German language "Grünfleckige Kartoffelblattlaus". This means green-spotted potato aphid because the specimens have a green spot around the cornicle base.

Males of the polyphagous strain (no. 1925) were mated in October 1972 with hitherto unpaired females of strain no. 1138. The latter strain had been collected from lungwort, *Pulmonaria officinalis* L., on which it also was reared. It belongs to a form originally described as a separate species, *Aulacorthum solani langei* Börner. Subsequent authors have treated it partly as a synonym, partly as a subspecies of *A. solani*. This aphid which is to be classified as a subspecies lives monophagously on *Pulmonaria officinalis* and produces on this plant a specific leaf roll. On the contrary, the polyphagous mating partner is not able to subsist on the lungwort. The siphuncular spots characteristic in the typical *A. solani* are missing in the *Pulmonaria* infesting ssp. *langei*. When brought together normal copulations were observed. The females laid on the lungwort some eggs out of which three young fundatrices hatched in March 1973. They were transferred on a best growing *Pulmonaria officinalis* plant. Only one reached the adult stage and yielded normal colonies. On the lungwort they caused distortions similar to those produced by the female partner.

At the end of June transference experiments were carried out with each of the parents and with the hybrid. Twenty 3rd instar larvae were put on plants of potato ('Ora'), foxglove and lungwort. In each case 10 replications were started. After 21 days the aphids per plant were counted. All experiments including egg overwintering and all rearings were performed in an open air insectary which approximated natural conditions. The results of the transference experiments are given in Table 1.

The polyphagous parent attained its highest multiplication on potato and increased also heavily but a little less on foxglove. On lungwort, however, only a few aphids survived. On the contrary, the subspecies *langei* showed its best reproduction on *Pulmonaria* and did not increase on potato. The propagation on *Digitalis* proved to be moderate and being significantly different from that on the two other plants. As evident from the table, the propagation of the hybrid on the three plants represented itself intermediate between the parents. A very notable reproduction occurred on potato, but

TABLE 1

Number of aphids Aulacorthum solani *(Kaltenbach) after 21 days on three plant species each infested initially with 20 3rd instar larvae*
(Means of 10 replicates)

Strain	Potato	*Digitalis purpurea*	*Pulmonaria officinalis*
A. solani sensu stricto no. 1925	1321 Aa*	718 ABb	8 Ff
Hybrid 1925 ♂ × 1138 ♀	567 Bb	113 Dd	38 DEe
A. solani langei Börn. no. 1138	26 EFe	98 Dd	264 Cc

* Means not followed by the same capital or small letter are significantly different at the 1% or at the 5% level, respectively (*t*-test).

the number of aphids on potato proved significantly smaller in the hybrid in comparison with the polyphagous parent.

As indicated above, the two hybridized forms differ somewhat in their colour. The polyphagous male partner is yellowish green coloured with green siphuncular spots, whereas the *Pulmonaria* inhabiting subspecies differs in being evenly light green without any trace of siphuncular spots. With respect to the colour the hybrid appeared intermediate between its parents. The apterous viviparous females were green to light green with a tinge of yellow. Green siphuncular spots were missing or indistinctly evolved.

It must be noted that besides the colour differences there exist also a small morphological distinguishing character between the two parents. Namely, the apterous viviparous females in the *Pulmonaria* inhabiting subspecies bear as a rule no rhinaria on the 3rd antennal segment. The facts are demonstrated by Table 2. About 96% of the antennae of the apterous viviparae of *Aulacorthum solani* sensu stricto have one or two rhinaria on the 3rd antennal segment. In the monophagous subspecies, however, about 98% of the corresponding antennal segments are without rhinaria. In this morphological character too the hybrid is intermediate between its parents as approximately half of the examined antennae possess no rhinaria.

TABLE 2

Percentage of the antennae with 0, 1, 2, and 3 rhinaria on the 3rd segment in Aulacorthum solani *(Kalt.) sensu stricto in comparison with the subspecies* langei *Börner and the hybrid* (n = number of examined antennae)

Strain	Rhinaria on ant. segm. 3				
	0	1	2	3	n
A. *solani* sensu stricto no. 1925	1.6	67.4	28.7	2.3	129
Hybrid 1925 ♂×1138 ♀	52.0	46.4	1.6	0	304
A. *solani langei* Börn. no. 1138	98.2	1.8	0	0	525

In October 1973 a further hybridization experiment was started and the male partner of the same polyphagous strain was used as in the previous year. The males were brought together with oviparous females of the subspecies *aegopodii* Börner which lives monophagously on *Aegopodium podagraria* L. called ash-weed, dog-elder or in North American English, masterwort. The *Aegopodium* inhabiting subspecies is characterized by a whitish colour and by the presence of yellow siphuncular spots. It settles the mature leaves of its host-plant and causes there large yellow leaf spots within which the leaf tissue close to the veins may be stained purple red. When aphids of the polyphagous male partner were transferred to the dog-elder they settled exclusively the youngest shooting leaves and the yellowing senescent ones without producing any discolouration. In contrast to the previous experiment pairing happened only hesitatingly. But the oviparous females deposited their eggs quite normally. Out of the eggs only one fundatrix was obtained which reached the adult stage on dog-elder also producing on this plant some progeny.

Certainly the features of host-plant selection and host-plant acceptability in phytophagous insects have emerged by gene mutations. It is easy to understand that the appearance of such heritable variations within an originally uniform taxon or species contains the first steps in evolution. By their behaviour in preferring or selecting decided host plants the males of the one and the females of the other biotype never or only exceptionally encounter each other. Thus, ethological factors prevent effectively the exchange of heritable characters between the groups, although they are still equipped with the potential hybridization power. Therefore the host-plant selection implies an important factor of isolation. After a long period of such sympatric ecological and ethological isolation small morphological divergences or colour variations may appear. This has obviously happened within *Acyrthosiphon pisum* and *Aulacorthum solani*. Regarding the fact that the artificial hybrids are not found in nature, it seems highly probable that race formation and further on speciation may be sympatric by sequence of mutative changes in host-plant selection. The evolution of a new form is possible without the action of geographical isolation and without occupying a niche by competition pressure which doubtlessly does not exist in the present examples.

Symp. Biol. Hung. 16, pp. 191–195 (1976)

HOST SELECTION OF THE EUROPEAN CORN BORER (*OSTRINIA NUBILALIS* HBN.) POPULATIONS IN HUNGARY

by

B. NAGY

RESEARCH INSTITUTE FOR PLANT PROTECTION, H–1525 BUDAPEST, PF. 102, HUNGARY

It can be supposed that wild hop (*Humulus lupulus*) and wild hemp (*Cannabis sativa* var. *spontanea*) were the original host plants of the European corn borer (*Ostrinia nubilalis* Hbn.) in Hungary since these species are regularly infested nowadays too. After the extension of agricultural production in Hungary during the last 100–300 years, maize, sorghum, cultivated hemp and cultivated hop became the "main" host plants of the corn borer, maize being the most important one. *Artemisia vulgaris* is practically not infested in Hungary.

In oviposition preference tests carried out with a corn borer strain originating from maize, there was no significant difference between maize, hemp, *Artemisia* and the non-host plant *Galinsoga parviflora* and *Amaranthus retroflexus*. A multivoltine strain reared from hemp preferred this plant significantly for oviposition as compared to sorghum, *Panicum miliaceum*, *Artemisia* and maize.

First instar larvae of the same strain, however, showed significant preference for hemp and *Artemisia* against the other plants in the leaf-disc test. That preference was altered only slightly when newly hatched larvae fed on maize leaves before the test. Thus, the host selection behaviour of larvae seems to have preserved some ancient features.

INTRODUCTION

The more or less rapid changes of strains of the European corn borer (*Ostrinia nubilalis* Hbn.) as related to the different aspects of its biology, following the introduction and dispersion of the species in North America is a well-known fact (Vance, 1939; Neiswander, 1947).

In the Old World, where the corn borer is native, no such conspicuous changes were known; however, during the last two decades several observations indicated changes in the biology of Central European corn borer populations. For example, the now common, huge second flight at midsummer — practically without starting a second generation — could not be observed in Hungary before the 1950s (Nagy, 1961). The area occupied by the corn borer in Europe has extended to the N-NW and the population in SW Switzerland should be considered as an advancing fringe population (Chiang, 1961; K. Murbach, personal communication).

A heavy infestation on hemp (*Cannabis sativa*) in 1945–55 in Hungary (Nagy and Csehi, 1955) drew again attention to the problem of geographical strains. The scarcity of pertinent data from Central Europe makes it desirable to summarize our preliminary experiments and observations.

EXPERIMENTS AND OBSERVATIONS

Oviposition preference tests were carried out with a univoltine strain, originating from maize (*Zea mays*), in field cages, on 80–100 cm high plants and there was no significant difference in the moths' responses to maize, cultivated hemp or common mugwort (*Artemisia vulgaris*). However, on dwarf specimens of *Galinsoga parviflora* and *Amaranthus retroflexus* grown scattered among the above plants, there were also plenty of egg-clusters. Many egg-clusters were found also on the lath-frame of the cage.

In another experimental series, adults of the multivoltine strain reared from hemp stems, collected in the field, preferred this plant significantly for oviposition, against common sorghum (*Sorghum bicolor*), millet (*Panicum miliaceum*), mugwort, and maize in different types of cages, fitted with cut plants (Table 1).

TABLE 1

Oviposition preference of a Cannabis-*strain of* O. nubilalis *expressed in percent of the egg-clusters laid on different plants*

	A	B
Cannabis	52	66
Zea	10	17
Sorghum	29	—
Panicum	6	—
Artemisia	3	3
Wall of the container	0	14
	100	100
No. of egg-clusters	31	183

A — wire cage, 20 l (in glass-house).
B — glass cylinder, 2 l.

Host-plant selection tests were made with newly hatched corn borer larvae, using leaf discs of maize, hemp, mugwort, sorghum and millet. The discs were placed on wet filter paper in petri dishes and the number of larvae gathering under these discs served as the criterion of evaluation.

Larvae of the multivoltine hemp strain showed significant preference for hemp and mugwort against the other plants mentioned here. The preference was altered slightly when the newly hatched larvae were conditioned by feeding on maize leaves for a day before the test (Table 2, C). A dominant preference for hemp was observed only when filter paper discs were soaked in leaf saps of the above plants (Table 2, D).

An additional preference test was carried out with newly hatched larvae of a maize strain reared for two generations on our semiartificial diet containing wheat germ and alfalfa meal as the only plant components (Nagy, 1970). The results were in accordance with former findings, and a distinct preference for maize leaf discs was observed (Table 2, E).

Preference for hemp, hop (*Humulus lupulus*) and maize was about the same, but only half as high as for the mugwort. Alfalfa was not attractive in spite of the possibility of conditioning for this plant via the diet.

192

TABLE 2

Host-plant selection by newly hatched O. nubilalis *larvae*
expressed in percent of larvae gathered below the leaf discs
of different plants

	Ac	Bc	Cc	Dc	Ez
Artemisia	56	69	41	11	40
Cannabis	22	26	35	72	17
Humulus	—	—	—	—	18
Zea	0	0	15	11	23
Sorghum	22	5	9	0	—
Panicum	0	—	—	6	—
Medicago	—	—	—	—	2
	100	100	100	100	100
No. of larvae	28	19	39	18	220

A — no feeding before experiment.
B — feeding on maize for 3 hours before experiment.
C — feeding on maize for a day before experiment.
D — filter paper discs with leaf sap; no feeding before experiment.
E — larvae of a laboratory strain reared on semiartificial diet for two generations; no feeding before experiment.
c — *Cannabis* strain.
z — *Zea* strain.

Observations made on cultivated plants indicated a slight trend toward extending the spectrum of the host plants. Before all the mass attack on hemp, mostly by the multivoltine strain, observed in 1954–55, then the appearance of the pest on *Vigna sinensis* and on different varieties of paprika (*Capsicum annuum*). These occasional attacks became noticeable in the 1950s when and where the multivoltine hemp strain had become significant.

DISCUSSION AND CONCLUSIONS

From the above finding the following conclusions can be drawn:
1. The European corn borer, like several polyphagous insect pests, has an interesting network of phenomena related to host finding and selection. Habitats of the corn borer in Hungary include both cultivated and natural areas. Adults occur also in many non-host stands, e.g. flowering alfalfa fields, and egg-clusters can be found occasionally on non-host plants, e.g. on *Galinsoga parviflora* which do not suffer even an initial attack by the newly hatched larvae.
Under these circumstances one should suppose that selection would result in broadening of the host range, however, this does not seem to be the case in Hungary, since maize, hop, partially hemp and sorghum even today are the most generally infested host plants among cultivated crops. The same situation was found many decades ago. Thus, the spectrum of the main host-plants practically does not seem to have changed during several decades.
2. It is a very interesting phenomenon that first instar larvae retained strong preference for the mugwort which was supposedly one of the primary

13

host-plants and which represents in France* and Ukraine a very common host-plant (Roubaud, 1928; Scsegolev, 1951). This character of the larvae did not change in the Hungarian populations despite the fact that this very common plant is practically never attacked in this country.

3. The above-mentioned experiments showed, on the other hand, that besides the constancy of basic host preference characters some small changes may occur. This could be proved in experiments regarding oviposition preference (e.g. the preference of hemp strain females for hemp), and in food choice of first instar larvae (e.g. slight increase of preference for maize in first instar larvae conditioned for maize at the beginning of their development). Decrease in larval preference for maize was shown by Rathore and Guthrie (1973) in the 87th generation of a continuous rearing on a meridic diet.

Also the above-mentioned observations concerning the appearance of the multivoltine strain on cultivated plants not formerly attacked, indicate such slight changes and remind one strongly of the broader host-plant complex of the North-American multivoltine strain (Arbuthnot, 1944).

4. Comparing the conservatism of the main host spectrum to the slight changes observed in host selection experiments and in newly appearing host-plants among the cultivated plants, the conclusion can be drawn that basic changes in host selection are rare in nature.

This is the more interesting from the evolutionary point of view, since the presence of quite extended masses of various cultivated plant species, among which there are several potential hosts, represent a continuous selection pressure for broadening host spectrum. Thus, in the case of corn borer populations, native in Hungary, the characters determining host selection behaviour seem to be very stable in the evolutionary process.

Another proof for the conservatism of host selection behaviour is the above-mentioned strong preference of newly hatched first instar larvae for mugwort. Thus, the Hungarian corn borer strain(s) retained this very ancient feature without having reinforced it by living at least in a part of the population on this plant.

The low larval preference for maize is also an interesting phenomenon, because maize represents today the most important host plant of the corn borer in Hungary, covering a quarter of the arable land. This huge area of maize arose during the last one or two centuries, thus, the preference for hemp may represent also an ancient feature of the host selection behaviour.

Wild hemp and wild hop (both in the family *Cannabinaceae*) may be considered the most important native host-plants of the European corn borer in Hungary, and have served as food plants for many thousands of years.

5. The conservatism of host selection behaviour found in relation to the corn borer populations of Hungary does not mean at all that differences among populations of different geographical regions are lacking. On the contrary, if one compares our findings with those of Soviet or American entomologists, very important differences appear.

---

* This earlier statement may need recent confirmation (J. P. Moreau and P. Robert, personal communication).

The International Project on *Ostrinia nubilalis*, at present involving 12 countries, carries out large scale comparisons on the corn borer resistance of corn varieties from many countries (Dolinka et al., 1973). It would be useful to complete this worldwide screening program by detailed investigations of the host selection behaviour of the corn borer populations present in different geographical regions, since such investigations could provide very valuable data for a better planned selection of resistant corn varieties.

ACKNOWLEDGEMENT

The author is grateful to S.D. Beck (Madison) and T. Jermy (Budapest) for their very helpful suggestions and criticisms of the manuscript.

REFERENCES

ARBUTHNOT, K. D. (1944): Strains of the European corn borer in the United States. *U.S. Dep. Agric. Washington, D.C. Techn. Bull.* **869.**

CHIANG, H. C. (1961): Fringe populations of the European corn borer, *Pyrausta nubilalis*, their characteristics and problems. *Ann. ent. Soc. Amer.* **54**, 378–387.

DOLINKA, B., CHIANG, H. C. and HADZISTEVIC, D. (1973): *Report of the International Project on* Ostrinia nubilalis. Phase I. Results 1969 and 1970. Martonvásár.

NAGY, B. (1961): Neuere Beobachtungen über die Flugzeit des Maiszünslers (*Ostrinia nubilalis* Hb.; Lepidopt.) in Ungarn. *Annales Inst. Prot. Plant. Hung.* **8**, 215–230.

NAGY, B. (1970): Rearing of the European corn borer (*Ostrinia nubilalis* Hb.) on a simplified artificial diet. *Acta Phytopath. Acad. Sci. Hung.* **5**, 73–79.

NAGY, B. and CSEHI, É. (1955): Néhány megfigyelés a kenderen károsító kukorica-molyról. *Magyar Tud. Akad. Agrártud. Oszt. Közl.* **8**, 106–108.

NEISWANDER, C. R. (1947): Variations in the seasonal history of the European corn borer in Ohio. *J. econ. Ent.* **40**, 407–412.

RATHORE, Y. S. and GUTHRIE, W. D. (1973): European corn borer (*Ostrinia nubilalis:* Lep., Pyralidae): leaf feeding damage on field corn (*Zea mays*) by larvae reared for 87 generations on a meridic diet. *J. econ. Ent.* **66**, 1195–1196.

ROUBAUD, E. (1928): Biological researches on *Pyrausta nubilalis* Hb. In: Ellinger, T. (ed.): *Internat. Corn Borer Invest. Scient. Rep.* Chicago.

SCSEGOLEV, V. N. (1951): *Mezőgazdasági rovartan* (Agricultural entomology). Budapest, Akadémiai Kiadó.

VANCE, A. M. (1939): Occurrence and responses of a partial second generation of the European corn borer in the Lake States. *J. econ. Ent.* **32**, 83–90.

Symp. Biol. Hung. 16, pp. 197–201 (1976)

PHYSICO-CHEMICAL ASPECTS OF THE EFFECTS OF CERTAIN PHYTOCHEMICALS ON INSECT GUSTATION*

by

D. M. NORRIS

UNIVERSITY OF WISCONSIN, MADISON 53706, USA

The chemical bases for host tree acceptance and non-host-tree rejection by the smaller European elm bark beetle, *Scolytus multistriatus*, were studied. Chemical messengers were shown to interact with chemosensory neurons in the insect's antennae through formation of an energy-transfer complex. Oxidation-reduction interactions also may occur. Hydroquinone is the most potent known stimulant for *S. multistriatus* feeding, and p-benzoquinone is inhibitory to feeding. Thus, the difference between stimulation and inhibition of feeding is the loss or gain of one or more electrons. Relative feeding inhibition among a group of variously substituted 1,4-naphthoquinones was positively correlated with the combined redox potential and intra-molecular hydrogen bonding capabilities of each molecule.

This group of 1,4-naphthoquinones showed the same relative order of feeding inhibition in *Periplaneta americana* as in the elm bark beetle. A receptor protein for quinone feeding inhibitors was isolated from nerve membranes in *P. americana* antennae. This receptor protein changes shape in response to messengers. Its change in shape is attributable to the making or breaking of disulfide bridges. Stimulatory quinols promote the reduction of disulfides; whereas the inhibitory quinones oxidize sulfhydryls of cysteines and promote the formation of disulfides. This folding and unfolding of the membrane protein in response to stimuli regulate the cationic permeabilities of the dendritic membranes of the sensory neuron. Our total findings allow us to present a functional energy-transfer mechanism for this chemoreception.

Our research (Norris, 1970) has reduced the physical chemical distinction between certain naturally occurring gustatory stimulants and inhibitors to donation versus acceptance of a proton and electron in *Scolytus multistriatus*. Quinones showed the same relative order of inhibiton in *S. multistriatus* (Norris, 1969; Norris et al., 1970a) and the very different *Periplaneta americana* (Norris et al., 1970b; Norris et al., 1971). Inhibitory activity was strongly correlated with the combined redox potential and intra-molecular hydrogen bonding capabilities of the messenger (Norris et al., 1970a). Intra-molecular hydrogen bonding capabilities unfailingly made a quinone more inhibitory than its relative redox potential would predict.

At least a few segments of one antenna were necessary for gustatory response to chemical messenger (Norris et al., 1971). Thus, major chemoreceptors are on the antennae, and include sensilla basiconica and sensilla trichodea types (Borg and Norris, 1971a; Norris and Chu, 1974a). Radiolabelled stimulants (Borg and Norris, 1971b) and inhibitors (Ferkovich and Norris, 1972) were shown *in situ* to adsorb on the cuticular surface of che-

* This research was supported by the Director of the Research Division, C.A.L.S., University of Wisconsin, Madison; and in part by research funds from grant No. GB-41868 from U.S. National Science Foundation. Thanks also are extended to my many associates who have worked on aspects of this research.

mosensitive sensilla, penetrate cuticular pores, move through pore tubules and gain access to dendritic membranes of sensory neurons innervating a sensillum. Whether exposed *in situ* or *in vitro* to receptor nerve membrane, the relative order of quinone binding matched the order of feeding inhibition (Norris et al., 1971); reduction of stimulant-induced depolarization (mV) in electroantennograms (Norris and Chu, 1974b); inhibition of endogenous nerve activity in ventral nerve cord, excluding the relatively non-lipid soluble quinone 2-hydroxy-1,4-naphthoquinone (Baker and Norris, 1971); and inhibition of ouabain-sensitive ATPase activity in nerve cord homogenates (Baker and Norris, 1971).

Our most studied quinone, 2-methyl-1,4-naphthoquinone (menadione), showed no significant (0.05 level) inhibition of the nerve ATPase activity (Baker and Norris, 1971); nevertheless, it is a very significant (0.01 level) feeding inhibitor (Norris et al., 1971). Its unique property among the studied quinones has proved highly useful in identifying the primary, readily reversible, receptor macromolecule in the nerve membrane (Ferkovich and Norris, 1972). In electroantennogram studies, increased exposure of the antenna to 2-methyl-1,4-naphthoquinone resulted only in significant (0.01 level) increase in the highly reversible effects on the sensory nerves (Norris and Chu, 1974b). In important contrast, increased exposure of the antenna to the most potent tested quinone inhibitor of gustation, 5-hydroxy-1,4-naphthoquinone (Norris et al., 1971) resulted in marked increases on both the highly reversible and relatively irreversible effects on the sensory nerves (Norris and Chu, 1974b). Several unique properties of the inhibitory 2-methyl-1,4-naphthoquinone thus lead us to conclude that it selectively interacts with the readily reversible primary receptor macromolecule (Norris and Chu, 1974b). Whether live insects are exposed to menadione in their environment, and then their antennae are washed, removed, homogenized, membrane-bound protein separated and protein-bound radiolabelled messenger quantitated by scintillation counting (Ferkovich and Norris, 1972); or insect antennae are washed, removed, homogenized in the presence of labelled menadione, membrane-bound protein separated and protein-bound radiolabelled messenger counted (Singer et al., 1975), the same protein is very selectively labelled. Thus, whether presented *in situ* or *in vitro* to antennal proteins, the gustatory inhibitor menadione interacts with the same protein: the primary reversible receptor (Norris and Chu, 1974b).

While the primary receptor protein was being further isolated, and purified, the interactions of the messenger quinones with various model candidate receptor molecules at physiological pH were analyzed. When mixed at a range of concentrations with numerous amino acids, immediate ultraviolet-difference spectra were obtained only between messenger quinones and the sulfhydryl-containing amino acid cysteine (Ferkovich and Norris, 1971–72). The sulfhydryl-containing peptide, reduced glutathione, also gave such spectra with quinones (Ferkovich and Norris, 1971–72). In such experiments, classical sulfhydryl reagents (e.g. N-ethyl maleimide, NEM) competed with the natural messenger quinones in forming immediate U. V. difference spectra with the receptor (Norris et al., 1970b). Quinones also reportedly (Webb, 1966; Morton, 1965) react with amines, but this type reaction is only slightly involved (Rothstein, 1970) in biological membranes at physiological pH. Reactions with sulfhydryls are the important means of such messenger-receptor interactions in controlling the cation permeability of plasma membranes (Rothstein, 1970; Farah et al., 1969; Scott et al., 1970). Our total research effort on this aspect over many years strongly supports the sulfhydryl as the site of quinone, and other electron-acceptor, messenger reactions with the primary chemoreceptor macromolecule. Very recent electroantennogram findings in D. Schneider's laboratory using silk moths also strongly support the major role of sulfhydryls in moth sensory nerve reactions with pheromone (Villet, 1974). This work (Villet, 1974) tried to make a scientific case for the major involvement of amino groups in nerve interactions with pheromone, but failed convincingly. Thus, these interesting results of Villet (1974) on pheromone reactions with sensory nerves in moth dramatically fit within the defined scope of Norris' (1971) hypothesized unifying mechanism for neurofunction. We have been perfectly happy just having experimental evidence of such sulfhydryl involvement in our studied species, but having the unquestionably glamorous pheromone messengers fit our basic SH vs. S–S equilibrium scheme for the energy transduction in the receptor does mildly please us.

Using affinity chromatography, our group more recently has highly purified the chemoreceptor protein from $P.$ $americana$ antennae demonstrated first by Ferkovich and Norris (1972). Incorporation of ^{35}S from cystine into the receptor in $situ$ also was studied and the proteinaceous receptor selectively incorporated ^{35}S within 1 hour after injection of (^{35}S)-cystine into the abdomen of adults. In a series of recent in $vitro$ competition-type experiments, the quinone messengers, juglone and menadione, and the classical sulfhydryl reagents, NEM and iodoacetic acid (IAA), all only reacted significantly with the receptor protein which selectively incorporated the ^{35}S. Pre-incubation of the receptor with ouabain did not affect the protein's binding of messenger quinone; thus, this primary, relatively reversible, energy-transduction mechanism in $P.$ $americana$ chemoreception is not dependent upon ouabain-sensitive ATPases. Our total studies however do indicate that the secondary, relatively irreversible, interactions of the more potent messenger quinones with sensory nerves are ouabain sensitive. This apparent sensitivity is associated with the "ion pumps" which serve to repolarize, or hyperpolarize, nerve membranes by "pumping" especially cations through the membrane. Such pumps require the energy of phosphorylation, and such energy is effectively uncoupled by the more effective quinone messengers.

The polarographic analyses (Rozental and Norris, 1973a, b) of messenger reactions with the primary $P.$ $americana$ receptor macromolecule prove conclusively that the $P.$ $americana$ receptor is predictably folded (aggregated) by electron-acceptor messengers, and predictably unfolded by electron-donor messengers. Though our understandings of the structures of biological membranes still are largely hypothetical (Avery, 1972), our tentative interpretation of the workings of the $P.$ $americana$ primary receptor protein within the nerve membrane is that this macromolecule operates as an ionophore. Electron-donor messengers react selectively with sulfides in disulfide bridges in the protein, unfold its three-dimensional structure and make it more leaky especially to cations. This variously allows the influx of such ions and allows depolarization of the membrane. Electron-acceptor messengers react selectively with sulfhydryls in the $P.$ $americana$ receptor, tighten (fold) its three-dimensional structure and make the ionophore less leaky especially to cations. This action tends to inhibit depolarization of the membrane, or even result in hyperpolarization. We always have felt that in some chemoreceptor energy-transduction mechanisms, electron-acceptor messengers would serve to allow depolarization of the membrane, and electron donors to inhibit depolarization. Our continuing basic enthusiasm in the area of such energy-transducing mechanisms thus is the now obvious fundamental importance of reversible interactions between the two, at least biphasic, systems of (a) electron donor-electron acceptor messengers and (b) disulfide-sulfhydryl equilibrium in the receptor.

REFERENCES

AVERY, J. (1972): *Membrane Structure and Mechanisms of Biological Energy Transduction.* Plenum Press, London.
BAKER, J. E. and NORRIS, D. M. (1971): Neurophysiological and biochemical effects of naphthoquinones on the central nervous system of *Periplaneta. J. Insect Physiol.* **17,** 2383–2394.
BORG, T. K. and NORRIS, D. M. (1971a): Ultrastructure of sensory receptors on the antennae of *Scolytus multistriatus. Z. Zellforsch.* **113,** 13–28.
BORG, T. K. and NORRIS, D. M. (1971b): Penetration of ³H-catechol, a feeding stimulant, into chemoreceptor sensilla of *Scolytus multistriatus. Ann. ent. Soc. Am.* **64,** 544–547.
FARAH, A., YAMODIS, N. D. and PESSAH, N. (1969): The relation of changes in sodium transport to protein-bound disulfide and sulfhydryl groups in the toad bladder epithelium. *J. Pharmacol. Exptl. Therap.* **170,** 132–144.
FERKOVICH, S. M. and NORRIS, D. M. (1971–72): Naphthoquinone inhibitors of *Periplaneta americana* and *Scolytus multistriatus* feeding: Ultraviolet difference spectra of reactions of juglone, menadione and 1,4-naphthoquinone with amino acids and the indicated mechanism of feeding inhibition. *Chem.-Biol. Interactions,* **4,** 23–30.
FERKOVICH, S. M. and NORRIS, D. M. (1972): Antennal proteins involved in the neutral mechanism of quinone inhibition of insect feeding. *Experientia,* **28,** 978–979.
MORTON, R. A. (1965): *Biochemistry of Quinones.* Academic Press, New York.
NORRIS, D. M. (1969): Transduction mechanism in olfaction and gustation. *Nature, Lond.* **222,** 1263–1264.
NORRIS, D. M. (1970): Quinol stimulation and quinone deterrency of gustation by *Scolytus multistriatus. Ann. ent. Soc. Am.* **63,** 476–478.
NORRIS, D. M. (1971): A hypothesized unifying mechanism in neural function. *Experientia,* **27,** 531–532.
NORRIS, D. M., BAKER, J. E., BORG, T. K., FERKOVICH, S. M. and ROZENTAL, J. M. (1970a): An energy-transduction mechanism in chemoreception by the bark beetle, *Scolytus multistriatus. Contrib. Boyce Thompson Inst.* **24,** 263–274.
NORRIS, D. M. and CHU, H. M. (1974a): Morphology and ultrastructure of the antenna of male *Periplaneta americana* as related to chemoreception (Blatt., Blattidae). *Cell. Tissue Res.* **150**(1), 1–9.
NORRIS, D. M. and CHU, H. M. (1974b): Chemosensory mechanism in *Periplaneta americana:* Electroantennogram comparisons of certain quinone feeding inhibitors. *J. Insect Physiol.* **20,** 1687–1697.
NORRIS, D. M., FERKOVICH, S. M., ROZENTAL, J. M., BAKER, J. E. and BORG, T. K. (1970b): Energy transduction: Inhibition of cockroach feeding by naphthoquinone. *Science,* **170,** 754–755.
NORRIS, D. M., FERKOVICH, S. M., BAKER, J. E., ROZENTAL, J. M. and BORG, T. K. (1971): Energy transduction in quinone inhibition of insect feeding. *J. Insect Physiol.* **17,** 85–97.
ROTHSTEIN, A. (1970): In: *Current Topics in Membranes and Transport.* Academic Press, New York.
ROZENTAL, J. M. and NORRIS, D. M. (1973a): Interaction of a proteinaceous receptor and messenger naphthoquinones. *Trans. Am. Soc. Neurochem.* **4,** 116.

Rozental, J. M. and Norris, D. M. (1973b): Chemosensory mechanism in American cockroach olfaction and gustation. *Nature, Lond.* **244**, 370–371.

Scott, K. M., Knight, V. A., Settlemire, C. T. and Brierley, G. P. (1970): Differential effects of mercurial reagents on membrane thiols and on the permeability of the heart mitochondrion. *Biochemistry*, **9**, 714–724.

Singer, G., Rozental, J. M. and Norris, D. M. (1975): [35]Sulfur incorporation in neural membrane protein involved in *Periplaneta americana* chemoreception. *Experientia* (in press).

Villet, R. H. (1974): Involvement of amino and sulfhydryl groups in olfactory transduction in silk moths. *Nature*, **248**, 707–709.

Webb, J. L. (1966): *Enzyme and Metabolic Inhibitors III*. Academic Press, New York.

Symp. Biol. Hung. 16, pp. 203–208 (1976)

ETHOLOGY OF *DASYNEURA BRASSICAE* WINN. (DIPT., CECIDOMYIDAE). I. LABORATORY STUDIES OF OLFACTORY REACTIONS TO THE HOST-PLANT

by

J. Pettersson

DEPT. OF PLANT PATHOLOGY AND ENTOMOLOGY, AGR. COLLEGE OF SWEDEN, UPPSALA 7, SWEDEN

Studies on the flight habits of *D. brassicae* indicates that an olfactory stimulus is involved in the migrations. In the present paper the occurrence of olfactory response to rape plants among mated females is demonstrated. The males do not show the same capacity. The odour seems to be perceived mainly by receptors on the antennae. A substance with considerable arresting effect on the females is sinigrin. This substance occurs in the undamaged plant, which indicates that sinigrin is part of a long distance orientation factor facilitating the finding of a suitable host-plant, rather than finding suitable oviposition sites on the plant.

INTRODUCTION

The brassica pod midge *Dasyneura brassicae* Winn. is a serious pest on rape in Sweden. It causes important losses and the control is a difficult problem due to the lack of suitable and efficient methods, chemical as well as others (Sylvén, 1970). For a successful oviposition the midge is dependent on damage done to the rape pod by other insects of which the by far most important one is *Ceutorrhynchus assimilis*, the rape weevil (Sylvén, 1949). The economic importance of the weevil itself is usually not too large, but as suitable control methods are lacking also for this insect, the combination of these two, the rape pod gall midge and the rape weevil, is a plant protection problem that must be given high priority.

Sylvén (1970) studying the flight habits of *D. brassicae* found among other things that the midges flew rather passively with the wind until they came leeward of host-plants when they started moving upwind usually at a low level in the vegetation cover. In a rape field the midges in the vegetation cover move upwind while in the air above the field the midges move with the wind. This is the behaviour that can be expected when an olfactorial factor is involved.

Olfactory reactions of *D. brassicae* as an aid in finding the host-plant is put forward as a probable mechanism by earlier workers.

Görnitz (1953, 1956) working with different extracts of rape and traps under field conditions got indications of such a relationship. The irregular behaviour of the midges and the lack of a working laboratory technique, however, made it difficult to come to a definitive conclusion. Görnitz, however, obtained most convincing results for several *Phyllotreta* species and for *Ceutorrhynchus assimilis* and *C. quadridens*. Coutin (1964) by observing the behaviour of females on the plant came to the conclusion that in the acceptance of a pod as a suitable oviposition site were involved an olfactorial and a tactile stimulus. He gave, however, no experimental evidence for his observations.

MATERIAL AND METHODS

Insect material. Gall midges were caught with exhaustors in a field of spring rape close to the department buildings. As there are great problems with an exact identification and the males are rare in the field, pods attacked by the midge were collected and stored for 48 hours on a net allowing the larvae to jump down in a vessel containing moist sand and peat mixed in equal parts. Jars with this content were stored in an insectarium until the midges hatched when they were collected and used in the experiments. No difference in activity and reactions between insects hatched in these jars and insects collected outdoors was found in comparative experiments.

Olfactometer technique. The olfactometer used in the laboratory studies has been described in connection with olfactorial investigations on aphids (Pettersson, 1970, 1973) and only a brief explanation will be given here. The construction of the apparatus is shown in Fig. 1. Five midges were placed in the observation chamber and observation of their positions at

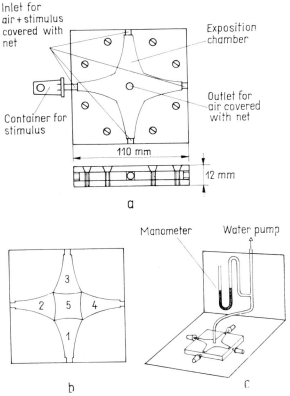

Fig. 1. Olfactometer used in the experiments. *a.* Dimensions and constructions of the working unit seen from above and from one side. *b.* Division of the chamber into five zones of which zones 1–4 each are overstreamed only by one stimulus. Zone 5 is regarded as indifferent and a visit in this zone cannot definitely be said to be dependent on the action of any of the applied stimuli. *c.* Olfactometer ready for work, lamp and surrounding paper screen omitted

intervals long enough to allow the midges to move from one place to any other place in the chamber. Ten such observations were plotted on one map and the total number of visits in the arm zones were used as a measurement of the attractivity of the stimulus applied to that specific arm. The experiments were carried out in series of ten in a suit from which the average was taken and in the tables the average and the standard deviation (S. D.) are given.

RESULTS

Reactions to the rape plant. The reactions of females and males to leaves and pods of rape when tested with the olfactometer are shown in Tables 1 and 2. Obviously the females are strongly attracted by the rape plant. Due to the experimental arrangement it is not possible to accept any other explanation than that there is an olfactory reaction. In the side tubes that have been called "empty" the air was allowed to pass moist blotting paper to avoid differences between the armzones regarding relative humidity.

The males seem to be indifferent about the host plant odour. This is obviously coupled to their biological function, being only mating partners.

No comparison was made between damaged and undamaged pods. In the experiments both leaves and pods were macerated to obtain maximum deliberation of attracting substances. Such a comparison will, however, be of interest in the further studies as it is of importance for the midge to find pods which are damaged in order to maximize the conditions for a successful oviposition.

Receptors for host-plant odour. Attempts were made to find the sites for perception of the host-plant odour. Two places were suspected as possible receptor carriers from observations of midges in the olfactometer. It was commonly seen how the midges stopped and raised the front tibiae in an indicating manner as well as antennal movements could be observed. Excising the front legs is a procedure that very much hampers the movements of the midge. No midge with excised front tibiae was able to move in a normal way. However, females on which the antennae were cut off did not seem to be affected concerning ability to walk and fly. The results of experiments with such females in the olfactometer are shown in Table 3. Obviously the capability to react to the olfactorial host-plant stimulus is lost when the antennae are cut. This must be taken as an indication that the main receptors are situated on the antennae.

TABLE 1

Averages of visiting frequencies in armzones in experiments with five mated females

(Ten repetitions)

Plant part	Applied stimulus							
	Plant material		Empty		Empty		Empty	
	$\bar{x}$	S.D.	$\bar{x}$	S.D.	$\bar{x}$	S.D.!	$\bar{x}$	S.D.
Pods	19.1	6.78	6.9	2.96	5.9	3.35	9.4	4.88
Leaves	19.3	7.20	8.9	4.38	8.2	3.49	9.0	4.12

TABLE 2

Averages of visiting frequencies in armzones in experiments with males

(The experiment is repeated ten times)

Plant part	Applied stimulus							
	Plant material		Empty		Empty		Empty	
	$\bar{x}$	S.D.	$\bar{x}$	S.D.	$\bar{x}$	S.D.	$\bar{x}$	S.D.
Pods	12.0	4.90	10.2	4.37	9.2	4.14	10.4	4.43
Leaves	10.1	4.33	9.7	1.94	12.1	3.76	13.4	2.67

TABLE 3

*Averages of visiting frequencies in armzones
in experiments with mated females
of* D. brassicae *with antennae excised*

(The experiment is repeated ten times)

Applied stimulus							
Rape pods		Empty		Empty		Empty	
$\bar{x}$	S.D.	$\bar{x}$	S.D.	$\bar{x}$	S.D.	$\bar{x}$	S.D.
9.6	3.1	7.9	2.47	8.7	2.21	8.8	3.05

TABLE 4

*Averages of visiting frequencies in armzones in experiments with mated females
of* D. brassicae

(The experiment is repeated ten times)

Concentration of sinigrin, %	Applied stimulus							
	Sinigrin		Empty		Empty		Empty	
	$\bar{x}$	S.D.	$\bar{x}$	S.D.	$\bar{x}$	S.D.	$\bar{x}$	S.D.
0.15	13.0	5.75	12.2	4.19	9.5	5.31	8.9	3.10
0.30	21.1	7.62	7.3	4.56	7.8	3.03	8.3	3.83

Active substances. The most characteristic substances of *Cruciferae* are undoubtedly the glucosinolates known for their goitrogenic effect and extensively studied by many authors. The general formula for this type of substances is shown in Fig. 2.

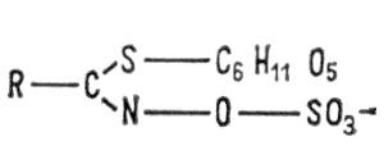

$$R - C \overset{\displaystyle S - C_6 H_{11} O_5}{\underset{\displaystyle N - O - SO_3^-}{\big\langle}}$$

Fig. 2. General formula for the glucosinolates. In sinigrin R corresponds to $CH_2 = CHCH_2-$. In the present investigations the potassium salt in water solution was used

About thirty different glucosinolates are known to occur in different species of *Crucifereae*. Some of the substances occur in many species while others are confined to one or two species. According to the present opinion

206

(Ettlinger and Kjaer, 1968; Josefsson, 1970) these substances exist in the undamaged plant. When the plant is mechanically wounded an enzyme, myrosinase, is liberated from specialized cells and decomposes the glucosinolates usually forming an isothiocyanate (Ettlinger and Kjaer, 1968).

One of the most well-known of these glucosinolates is sinigrin. The results with this compound are shown in Table 4.

The higher of the two used concentrations (0.30%) is the amount that occurs in pods of several *Cruciferae*.

It is a little difficult to understand how this substance can act as an odour. It is a solid substance occurring as an ion in water solutions. Obviously it is dependent upon water for being capable of function as an attracting odour. This means that the true stimulus the insect is perceiving is extremely small water particles containing sinigrin. No experimental work has been done on this subject due to the lack of a suitable model system. Perhaps the reactions of this insect and of the cabbage aphid *Brevicoryne brassicae* (L.) which also gives an olfactory response to sinigrin (Pettersson, 1973) can be used for this purpose.

According to the results sinigrin in water solution has an arresting effect on females of *D. brassicae*. The reaction is strong enough to explain the attractivity of rape plant parts. However, further substances with an arresting effect must be taken into consideration and will be tested in the further work.

SUMMARY

The results of the present investigation can be summarized in the following way:

1. Females of *D. brassicae* are capable of discovering a presumptive host-plant by odour.

2. Males do not react to the host-plant odour.

3. The main perceiving organs for the host-plant odour are situated on the antennae.

4. Sinigrin, which occurs in the undamaged rape plant has an arresting effect when it is applied in the same concentration as it occurs in the plant sap of macerated plant parts.

In the introduction two behavioural phenomena were mentioned in which odouriferous stimuli could be involved, viz. migration (Sylvén, 1970) and oviposition (Coutin, 1964). Chemically the stimuli affecting these processes can be identical and the different effect depends upon a concentration gradient effect. However, the possibility that different substances are involved must be kept open until further investigations are performed.

REFERENCES

COUTIN, R. (1964): Le comportement de ponte chez plusieurs Cécidomyies en relation avec l'état de développement chez la plante hôte des organes recherchés pour l'oviposition. *Rev. Zool. Agric. Appl.* **63**, 45–55.

ETTLINGER, M. G. and KJAER, A. (1968): Sulfur compounds in plants. In: Mabry, T. J., Alston, R. E. and Runeckles, V. C. (eds.): *Recent advances in phytochemistry*. North-Holland Publ. Co., Amsterdam.

GÖRNITZ, K. (1953): Untersuchungen über in Cruciferen enthaltene Insekten-Attrak-tivstoffe. *NachrBl. Dtsch. PflSchutzd.* **7**, 81–94.
GÖRNITZ, K. (1956): Weitere Untersuchungen über Insekten-Attraktivstoffe aus Cruciferen. *Ibid.* **10**, 137–146.
JOSEFSSON, E. (1970): *Pattern, content and biosynthesis of glucosinolates in some culti-vated Cruciferae.* Thesis, Lund.
PETTERSSON, J. (1970): An aphid sex attractant. I. Biological studies. *Ent. Scand.* **1**, 63–73.
PETTERSSON, J. (1973): Olfactory reactions of *Brevicoryne brassicae* (L.) (Hom.: Aph.). *Swed. J. agric. Res.* **3**, 95–103.
SYLVÉN, E. (1949): Skidgallmyggan, *Dasyneura brassicae* Winn. *Medd. Växtskyddsanst.* **54**, 1–120.
SYLVÉN, E. (1970): Field movement of radioactively labelled adults of *Dasyneura brassicae* Winn. (Dipt., Cecidomyiidae). *Ent. Scand.* **1** (3), 161–187.

Symp. Biol. Hung. 16, pp. 209–214 (1976)

HOST DETECTION BY WILD
AND LAB-CULTURED OLIVE FLIES

by

R. J. PROKOPY and E. G. HANIOTAKIS

GROUP OF ENTOMOLOGY, "DEMOCRITOS" NUCLEAR RESEARCH CENTER,
AGHIA PARASKEVI, ATTIKIS, GREECE

We studied the processes by which female olive flies, *Dacus oleae* (Gmelin), detect olive fruit after arriving on natural or caged olive trees. We found (1) that some chemical or physical character of the olive leaves or branches acts to hold females longer than on trees of other species, (2) that females are attracted to wooden models of olive fruit, with no enhancement of attraction in the presence of an olive fruit odour stream, (3) that females locate individual fruits not by odour but solely by visual characters, in particular, fruit shape, colour, and size, and (4) that after arriving on to a fruit, some chemo- or physico-tactile cue enables them to distinguish real olives from wax and wooden olive models as potential oviposition sites.

In some instances, we compared the responses of lab type females cultured artificially for ca. 85 generations with the responses of the wild type. The lab type proved just as adept as the wild type in distinguishing the location of olives by shape and at distinguishing a real olive from a wooden olive model after arrival. However, the lab type was slightly more attracted to red olive models and slightly less attracted to yellow olive models than the wild type, and was more prone to attempt oviposition into ceresin wax olive models (= the oviposition substrate used in artificial culturing) than the wild type.

INTRODUCTION

The chemical and physical stimuli eliciting the arrival of olive flies, *Dacus oleae* (Gmelin), on the leaves of their only known host-plant, olive trees, have been investigated and reported elsewhere (Fiestas ros de Ursinos et al., 1972; Prokopy and Haniotakis, 1974). No one, however, has studied how, after such arrival, olive fly females select olive fruit for oviposition.

Here, we report on the processes of female arrestment on olive trees, fruit-searching activity, location of individual fruits and initiation of oviposition. In some instances we compare the responses of different types of olive flies: a natural-population wild type; a wild type obtained from infested olives collected in nature and released when mature (= wild-released type); and a lab type cultured artificially for ca. 85 generations (Tsitsipis, 1974) and released when mature (= lab-released type).

GENERAL METHODOLOGY

Experiments with wild-population flies were conducted in unsprayed olive trees in groves at Spata (about 10 km from "Democritos"). Except in the first test, experiments with released flies were conducted in saranscreen cages (2.5 m diam × 2.5 m tall) located outdoors or in greenhouses, into which small fruitless potted olive trees were placed.

14

The process for marking the released flies for identification (Neon Red dye for lab-released flies and Arc Yellow dye for wild-released flies), the method of maintaining the flies prior to release, the loading of flies into release containers and the release procedure itself are described elsewhere (Prokopy and Economopoulos, 1974; Prokopy and Haniotakis, 1975).

Where fly responses to wooden or ceresin wax olive-shaped models were tested, the models were 31 mm tall × 20 mm diam at the center. The wooden models were coloured with the same paints described by Prokopy et al. (1974): Chropilux black, red or blue; Pentalux white; Chromgelb yellow; or mixtures of 95% Chromgelb yellow with 5% blue or red to produce green or orange. In some experiments a sticky coating of Bird Tanglefoot was applied to the models to capture arriving flies. All olives utilized during the experiments in the cages were an unknown variety, large (ca. 20 mm diameter), mature (black) but very firm, without evidence of insect injury and picked within 8 hours of utilization.

RESULTS AND DISCUSSION

Arrestment by olive trees

In order to test whether some character of olive leaves or branches acts to hold olive flies longer on olive trees than on trees of other species, we released simultaneously at Spata 200 females into each of 6 olive and 6 apricot trees, neither type having fruit or harbouring wild-population flies. The females were mature individuals of a 10-generation lab type whose first-generation parentage consisted of wild males and females cultured artificially for ca. 75 generations. Just prior to the release we sprayed the leaves of each tree with 1 liter of aqueous solution of 10% protein hydrolysate and 1% sucrose as food for the flies and hung 6 sticky-coated rectangles in each tree to capture flies and thereby gain an estimate of the population density. We repeated this experiment three times each time removing the flies from the rectangles three days after the release.

In all, we captured 466 of the females released on olive trees but only 130 of those released on apricot trees. Three days after release we observed several released females on the olive trees but none on the apricot trees. Therefore, we consider the capture data as indicative that a larger number of released females remained on the olive trees than on the apricot trees. Because all trees were fruitless, approximately the same size and provided with large amount of identical food, we suggest that some chemical or physical character of the olive leaves or branches acts to arrest the females on olive trees longer than on apricot trees.

This suggestion was supported by observations gained from another experiment. Here we observed that a few hours after release, accumulation of immature flies released equidistant from a pear tree and a similar-sized olive tree was greater on the pear tree. However, two days later, when most flies were mature, only about one-fourth as many were observed on the pear tree as on the olive tree.

To test whether olive fruit odour influences fruit-searching activity, we released ca. 120 female wild type flies into a cage in an isolated greenhouse room, hung 15 wooden models of olive fruit on the olive trees in the cage and counted the number of flies arriving on the wooden models during consecutive 10 or 15-minute periods in which, successively, no fruit, 2–6 whole apples, 2–6 crushed apples, 1/2–1½ kg of whole olives and 1/2–1½ kg of crushed olives were brought into the cage from another isolated room. The fruit was placed 30–50 cm below the wooden models. A gentle breeze from a fan blew over the fruit-location site toward the flies on the leaves and twigs of the olive trees. There were 9 replicates, with at least a two-hour interval.

The total number of flights by females onto the wooden models under the respective treatment conditions was as follows: no fruit = 61; whole apples = 52; crushed apples = 75; whole olives = 64; and crushed olives = 62. We conclude that females actively search for olive fruit in the absence of olive fruit odour and that the presence of such odour (at least in the case of the particular olives utilized) did not affect this activity.

Location of individual fruits

A. By Odour. We conducted three types of experiments to determine if females locate individual olive fruits on the basis of fruit odour.

In the first experiment conducted in Spata, we hung 10 wooden models of olives of the same size, shape, and approximate colour (green) and 10 real olives growing in the trees. We ringed the middle of both the wooden models and the real olives with a band of Bird Tanglefoot to capture some of the arriving flies. In all, 10 wild-population females were captured on the wooden models, 8 on the real olives.

In the second experiment we compared the responses of wild-released and lab-released flies to 5 real olives, 5 ceresin wax olive models and 5 wooden olive models, all of the same size, shape and colour (black) and hung ca. 30 cm apart in olive trees within the greenhouse cage. There were 3 replicates, each lasting ca. 1 hour. The total number of female flights to these 3 types of olives was as follows. For the wild-type flies: real olives = 125; ceresin wax models = 145; and wooden models = 131. For the lab-type flies: real olives = 103; ceresin wax models = 108; and wooden models = = 114.

In the third experiment we assessed the responses of wild-type flies in the greenhouse cage to 3 types of wooden olive models hung ca. 30 cm apart in the olive trees: 5 untreated models; 5 models rubbed with fresh apple juice every 10 min; and 5 models rubbed with fresh olive juice every 10 min. There were 2 replicates, each lasting ca. 30 minutes. The total number of female flights to the models of each type was as follows: untreated = = 25; rubbed with apple juice = 19; rubbed with olive juice = 24.

We conclude from these 3 experiments that both wild and lab-type olive flies locate individual olive fruits solely by the eye.

B. By Colour. To assess the importance of olive fruit colour in the flies' olive-detection mechanism, we compared the responses of wild-population,

TABLE 1

*Capture of wild-population (WP), wild-released (WR), and lab-released (LR)
type olive fly females on different colours of sticky-coated wooden olive models
hung in olive trees*
(Six replicates/treatment, positions rotated 8 times/replicate)

Fly type	No. captured/colour as % of total no. captured in experiment								total no. captured	total no. released
	black	red	yellow	green	orange	blue	white	alum. foil		
WP	18.3	18.3	16.7	14.8	11.7	10.1	4.7	5.5	257	—
WR	17.8	16.9	20.1	15.2	15.8	7.4	2.9	4.0	349	450
LR	18.8	21.3	13.4	14.8	12.6	9.0	5.8	4.3	277	450

wild-released and lab-released flies to 8 different colours of sticky-coated
wooden olive models hung ca. 30 cm apart in olive trees in Spata or in
outdoor and greenhouse cages.

We found (Table 1) that females of both wild types were about equally
attracted to black, red, and yellow models, slightly less attracted to green
and orange models and least attracted to blue, white and aluminium foil-
covered models. Females of the lab-type were slightly more attracted to red
models and slightly less to yellow models than females of either wild type,
with degrees of attraction to the other colours being essentially the same as
that of the wild types. Earlier (Prokopy et al., 1974) the lab-type flies were
found to be more attracted to red and less attracted to yellow 15×20 cm
rectangles than were the wild types.

Ripening olives gradually change colour from yellow-green or yellow-
orange to reddish and, eventually, black. Olives in all of these colour stages
have been found suitable for fly oviposition (Sacantanis, 1953); although
green ones are preferred over black ones (Cirio, 1972).

We conclude from our colour response experiments that the colour, *per se*,
of olives and/or the degree of contrast against the background is a cue
utilized by the flies in locating olives and that the flies are not very attracted
toward olives of unnatural colour (blue, white, and aluminium foil — all of
which reflect more light below 500 nm than any of the other 5 colours).

C. By Shape. To assess the importance of olive fruit shape in the flies'
olive detection mechanism we compared the responses of wild-population,
wild-released and lab-released flies to sticky-coated wooden models of
6 different shapes, each having a surface area of 19.6 cm^2 and painted black.
The models were hung ca. 30 cm apart in olive trees in Spata or in outdoor
and greenhouse cages.

We found (Table 2) that females of all 3 fly types were most attracted
to olive-shaped and spherical models, considerably less attracted to cubical
models and very little attracted to cylindrical and rectangular models.
The degrees of attraction to olive and spherical shape was the same for all
3 fly types. We conclude that the shape of olives is a highly important cue
utilized by the flies in locating olives.

D. By Sizes. To assess the importance of olive fruit size in the flies' olive
detection mechanism, we tested the responses of wild-released flies to 4 different sizes of black, spherical, sticky-coated wooden models hung together
in an outdoor cage.

TABLE 2

Capture of wild-population (WP), wild-released (WR), and lab-released (LR) type olive fly females on different shapes (each having surface area 19.6 cm²) of sticky-coated black wooden models hung in olive trees
(Six replicates/treatment, with treatment positions rotated 6 times/replicate)

Fly type	No. capture/shape as % of total number captured in experiment							
	olive shape	sphere	cube	horizontal cylinder	vertical cylinder	rect-angle	total no. captured	total no. released
WP	31.0	31.0	12.4	4.8	9.0	11.7	145	—
WR	36.4	32.9	17.1	7.9	2.9	2.9	140	300
LR	31.6	34.6	12.5	8.1	8.1	5.1	136	300

The total number of females captured on 6 models of each size was as follows: 16 mm diam = 14; 25 mm diam = 33; 37 mm diam = 69; and 75 mm diam = 114. We conclude that these flies can more readily locate large olives than small ones and that a sphere 75 mm in diam is a super-normal olive to the flies. Based on previous work with another tephritid, *Rhagoletis pomonella* (Walsh) (Prokopy, 1968) we suspect that the size of an "olive" cannot be increased indefinitely before it is no longer a super-normal olive to the flies but instead becomes a neutral or repulsive object.

Initiation of oviposition

In those preceding experiments when the real or artificial fruit was not sticky-coated, we did not terminate the experiment at this point of fly arrival on the fruit but instead continued to watch the flies for as long as they remained on the fruit. We were particularly interested in whether or not the females attempted to bore (i.e. oviposit) into the fruit.

Of the wild-type females arriving on wooden olive models in the "Fruit-Searching Activity" experiment, the following percentages attempted to bore into the models under the respective treatment conditions: no fruit = 23%; whole apples = 13%; crushed apples = 11%; whole olives = 8% and crushed olives = 22%. Of the females arriving on the real olives and the ceresin wax and wooden olive models in the second experiment of "A. By Odour", the following percentages attempted to bore. Among the wild-type flies: real olives = 56%; ceresin wax models = 25%; and wooden models = 16%. Among the lab-type flies: real olives = 59%; ceresin wax models = 42%; and wooden models = 22%. Finally, of the wild-type females arriving on the wooden olive models in the third experiment of "A. Odour", the following percentages attempted to bore: untreated wooden models = 25%; wooden models rubbed with apple juice = 0%; and wooden models rubbed with olive juice = 4%.

We conclude from this information that the odour of the nearby olive fruit does not influence the number of boring attempts into wooden olive models, that both wild and lab-type females have a strong preference for boring into real olives compared with wooden olive models. Nevertheless, the lab-type is more prone than the wild type to bore into ceresin-wax olive models (= the oviposition substrate used in artificial culturing) and the presence of apple or olive juice on the fruit surface somehow deters boring. The last-mentioned finding agrees with the finding by Cirio (1972) of an oviposition deterrent in olive juice.

Activity of males

Although both sexes were released in about equal numbers into the outdoor and greenhouse cages, very few males, compared with females, were observed flying onto the non-sticky real or artificial olives. Some males were observed feeding from olive juice exuding from the oviposition punctures and a few were seen attempting to mate with ovipositing females, but a higher proportion of mating attempts were observed on the leaves than on the fruit. Unlike the situation with *R. pomonella* (Prokopy et al., 1971) the host fruit does not appear to be a specific rendezvous site for mating in *D. oleae*. In contrast to the few males seen visiting non-sticky real and artificial fruit, male visits to sticky-coated real and artificial fruit were just as frequent as female visits. Observation suggested that the presence of captured females on the sticky-coated fruit was a strong positive stimulus to the males.

ACKNOWLEDGEMENTS

We thank the College of Agriculture of Athens for making their olive groves in Spata available to us, the "Democritos" staff for supplying us with pupae and Dr. G. L. Bush for loaning us the saran-screen cages. This study was made possible through the support of UNDP (SF Greece 69/525): "Research on the control of olive pests and diseases in Continental Greece, Crete and Corfu" and the Greek Atomic Energy Commission.

REFERENCES

Cirio, U. (1972): Reperti sul meccanismo stimolo-risposta nell ovideposizione del *Dacus oleae* Gmelin (Diptera: Trypetidae). *Redia* **52**, 577–600.

Fiestas ros de Ursinos, J. A., Constante, E. G., Duran, R. M. and Roncero, A. V. (1972): Etude d'un attractif naturel pour *Dacus oleae*. *Ann. Entomol. Soc. Fr.* **8**, 179–188.

Prokopy, R. J. (1968): Visual responses of apple maggot flies, *Rhagoletis pomonella* (Diptera: Tephritidae): Orchard studies. *Ent. exp. & appl.* **11**, 403–422.

Prokopy, R. J., Bennett, E. W. and Bush, G. L. (1971): Mating behavior in *Rhagoletis pomonella* (Diptera: Tephritidae). I. Site of assembly. *Can. Ent.* **103**, 1405–1409.

Prokopy, R. J. and Economopoulos, A. P. (1975): Attraction of laboratory-cultured and wild *Dacus oleae* flies to sticky-coated McPhail traps of different colors and odors. *Environ. Ent.* **4**, 187—192.

Prokopy, R. J. and Haniotakis, G. E. (1974): Responses of wild and laboratory-cultured *Dacus oleae* flies to host plant color. *Ann. Ent. Soc. Am.* **68**, 73–77.

Prokopy, R. J., Economopoulos, A. P. and McFadden, M. W. (1975): Attraction of wild and laboratory-cultured *Dacus oleae* flies to small rectangles of different hues, shades, and tints. *Ent. exp. & appl.* **18**, 141—152.

Sacantanis, K. (1953): Facteurs déterminant le comportement de *Dacus oleae* Gmel. vis-à-vis des variétés d'oliviers. *Rev. Path. Végét. Entomol. Agric. Fr.* **32**, 50–57.

Tsitsipis, J. A. (1974): Mass rearing of the olive fruit fly, *Dacus oleae* (Gmel.) at "Democritos". In: *Sterile Insect Method for Fruit Fly Suppression*. Proc. Symp. Vienna 12–16 Nov. 1973. IAEA, Vienna (in press).

Symp. Biol. Hung. 16, pp. 215-216 (1976)

THE ROLE OF THE HOST-PLANT IN THE REPRODUCTION OF *EURYGASTER MAURA* L. AND *E. AUSTRIACA* SCHRK.

by

V. RÁCZ

RESEARCH INSTITUTE FOR PLANT PROTECTION,
H-1525 BUDAPEST, PF. 102, HUNGARY

From the *Eurygaster* Lap. (Het., Scutelleridae) species only *E. maura* L. and *E. austriaca* Schrk. occur in the cultivated areas of Hungary, mostly as pests of winter cereals.

E. austriaca lives — except its annual migrations — nearly exclusively in cereal stands, while *E. maura* occurs also in other plant associations (meadows, pastures, forest clearings, embankments) on wild grasses, where it also reproduces.

E. integriceps Put. is one of the most dangerous pests of cereals, living in areas to the S and SE of Hungary. The study of relations between *E. integriceps* and its host-plants has been started in the 1960s by I. D. Shapiro and his coworkers. In their extensive study the authors established that against this species — related to the Hungarian ones — the susceptibility of different wheat varieties shows significant differences.

We began to study the effect of different wheat varieties in 1971. Since at the beginning no highly significant differences were expected between the influence of wheat varieties, only laboratory experiments were conducted. The observations comprised practically the mutual effect of the seed and that of the seedlings. The larvae and adults were reared on dry wheat seeds and on young wheat seedlings, germinated in little bundles of wet cotton.

In the first year three wheat varieties were tested (Bezostaya 1, Mironovskaya 808 and Fertődi 293) whereas in 1972 and 1973 four varieties (the former ones and Kiszombori 1) were studied.

Table 1 shows the egg production of the two *Eurygaster* species reared on different varieties.

TABLE 1

Fecundity of Eurygaster maura *and* E. austriaca

Wheat variety	E. maura			E. austriaca		
	1971	1972	1973	1971	1972	1973
Bezostaya 1	38.1	75.8	43.2	32.8	32.4	59.8
Mironovskaya 808	39.0	68.5	60.9	47.4	28.8	60.1
Fertődi 293	40.0	63.1	65.1	23.2	42.9	49.4
Kiszombori 1	—	26.8	20.2	—	42.9	19.2
LSD$_{5\%}$	*	36.03	32.40	*	*	*
LSD$_{10\%}$	*	*	26.97			

* Not significant.

In the case of *E. maura* the varieties Bezostaya 1, Mironovskaya 808 and Fertődi 293 formed a group with similar results while the number of eggs laid by females feeding on the variety Kiszombori 1 was considerably lower.

No significant differences were observed, however, between the varieties in the case of *E. austriaca*, which seems to possess a high degree of tolerance against variety differences. By studying the rate of development of *E. maura*, the periods observed between 50 per cent egg hatch and 50 per cent of last moult (transformation into adults) were compared, but no significant differences were observed during the three years.

Half of the young (first larval instar) larvae hatched first on the variety Bezostaya in each year: in 1971 by 3, in 1972 by 3–6, and in 1973 by 5–7 days earlier.

The influence of different varieties on the rate of development needs further studies.

We may conclude that the comparison of wheat varieties for suitability as host-plants for the two *Eurygaster* species gave with the applied methods unambiguous results. At present only trends of differences are known but it is hoped that further research will provide clearer information on this question which is important both from practical and theoretical points of view.

Symp. Biol. Hung. 16, pp. 217–221 (1976)

EFFECTS OF CERTAIN ALLELOCHEMICS ON THE GROWTH AND DEVELOPMENT OF THE BLACK CUTWORM

by

J. C. Reese and S. D. Beck

DEPARTMENT OF ENTOMOLOGY, UNIVERSITY OF WISCONSIN,
MADISON, WISCONSIN 53706, USA

The black cutworm, *Agrotis ipsilon*, is a highly polyphagous species and therefore the larvae are subject to a wide variety of allelochemics (secondary plant substances having effects on other organisms). We have undertaken the experimental testing of the hypothesis that some of these allelochemics may have in addition to immediate effects on survival or feeding behavior, subtle chronic effects, even at low concentrations, on rate of growth, utilization of food and pupation. These subtle effects are best seen when the insect chronically ingests the chemical, as it would in a monoculture situation. We have found, for example, that certain chemicals which do not significantly inhibit 10-day survival, or have a slight effect on this parameter, may markedly inhibit pupation at concentrations of 10^{-4} M or lower. The concentrations at which we have found biological activity are well within the range at which many allelochemics occur in plants. We have been particularly interested in the chronic effects of substituted phenols and their oxidized forms (quinones).

INTRODUCTION

Although a great deal of attention has been paid to the chemical basis of host selection and host-plant resistance, little work has been done on the chronic effects of various allelochemics. We therefore have undertaken the testing of the hypothesis that some allelochemics may have, in addition to immediate effects on survival and feeding behavior, subtle chronic effects, even at low concentrations, on rate of growth, utilization of food and pupation. Allelochemics may affect many other aspects of the plant–insect relationship, too. Though the compounds contained in a particular plant may not be acutely toxic to the insect feeding on it, the chronic effects of a lower developmental rate could increase through longer exposure to parasites, predators, pathogens and physical factors, such as weather. The life cycle could even be extended enough to put it out of synchrony with the seasonal changes in weather and in host-plant availability. In addition to indirect effects on mortality through alteration of the life cycle, allelochemics may increase mortality through a build-up in the insect tissues. Depending upon the biochemistry of the processes affected, these effects might show up much later in the life cycle of the insect, such as at the time of pupation, adult development, or reproduction.

Allelochemics might affect the longevity of the adult and this might cause a reduction in either the number or viability of the eggs. Allelochemics might also cause eggs to take longer to hatch, forcing longer exposure to predators, parasites and adverse weather. Considering the literature of synthetic chemosterilants it seems surprising that there is so little data on this last aspect, although Norris (personal communication) is now getting

217

some evidence for allelochemic effects on fecundity. The host-finding and oviposition behavior of the adult may also be affected by allelochemics.

Some of the above effects of allelochemics may have some very basic and as yet little understood effects on host-plant specificity and speciation of both the insect and the plant. For example, some of the chronic effects may not immediately save the plant from being partially consumed. However, if there are enough deleterious effects on the insect, there may not be large populations attacking that particular plant. This partial resistance is in fact a rule rather than an exception, since highly susceptible plants do not long survive. Even the insect's normal host-plant contains various deleterious phenols, flavonoids, alkaloids and glycosides; this accounts at least in part for our observation that our best laboratory methods for rearing certain insects promote faster growth, better survival and better adult longevity and reproduction than the insect's normal host. Those insects that do survive may be those individuals that happened to be slightly better, through a slightly different genotype, at detoxifying or deactivating the deleterious allelochemics, though becoming too successful might endanger their own food supply. The plant too, in surviving insects attacks, will tend toward some separation from parental lines. Thus, the plant–insect interactions and the allelochemics that modify them, have an impact on speciation. The chronic effects seem to us to be at least as important as the more readily observed acute effects.

Much excellent work has been done on the immediate effects of allelo-chemics. The short term effects, however, may be quite different from the chronic effects (Matsumoto, 1962). Also the chronic effects may be apparent at much lower concentrations and as discussed above, these effects may be somewhat subtle and in some cases more difficult to measure experimentally. A few investigators, including Beck (1960), Beck and Smissman (1960, 1961), Harley and Thorsteinson (1967), Maxwell et al. (1967) and Todd et al. (1971) have examined the effects of certain allelochemics on growth or pupation. Fraenkel (1969) has discussed the effects of different plant species on the utilisation of food and postulates that these effects could have been due to allelochemics. We have evidence to support this hypothesis for certain allelochemics.

In our investigations we are particularly interested in a number of common phenolics and quinones that are known to occur in plant tissue. The frequent occurrence of these compounds coupled with the highly polyphagous behavior of the black cutworm make these studies especially interesting to us, because it is hoped that results of our studies will be more likely to give leads for future research with other insects, than if we were working with an insect more restricted in its host range, or if we were working with relatively rare allelochemics. Also, the black cutworm is an important pest and often appears in large numbers with little warning, making the availability of resistant varieties especially desirable.

RESULTS AND DISCUSSION

Compounds were incorporated into an oligidic diet (Reese et al., 1972). Eighteen larvae were in each of five replicates per treatment. Statistical comparisons of experiments to control were only made within generations.

218

Using this design we have examined the effects of about nine substituted phenolics and quinones. When catechol was incorporated into the diet, no significant effects on 10-day survival could be observed. However a significant negative correlation between catechol molarity and 10-day larval weights was observed (Fig. 1; single asterisks indicate statistical significance at $P < .05$; double asterisks, $P < .01$). Thus the compound appeared to be inhibiting weight gain, even though acute toxicity was not observed. The correlation coefficient between catechol and 28-day pupation percent was also statistically significant (Fig. 2). In choice test experiments, however, there was no relationship between concentrations and feeding behavior. Thus it appeared that catechol did not act through a sensory modality.

TABLE 1

Effects of p-benzoquinone on nutritional indices and related parameters

	3.75×10^{-2} M p-benzoquinone	Control	Percent of control
Dry wt. eaten	0.4568 g	0.6404 g	71.3*
Dry wt. gain	0.0915 g	0.1279 g	71.5*
ECI	20.5%	20.0%	102.5
ECD	83.6%	59.8%	139.8*
AD	25.7%	34.4%	74.7*

* Statistically significant at $P < 0.01$.

p-Benzoquinone, a common quinone, similarly yielded no significant correlation coefficient between concentration and 10-day survival. The correlation coefficient for 10-day weight was significant (Fig. 3). Also, the correlation coefficient for pupation percent and concentration was significant (Fig. 5). When choice experiments were performed, the results clearly showed that at higher concentrations black cutworm larvae exhibit a strong non-preference for p-benzoquinone (Fig. 4).

Further understanding of the way p-benzoquinone was acting on black cutworm larvae was gained by calculating various nutritional indices and related parameters using the methods of Waldbauer (1968) as modified by Reese and Beck (in preparation). As shown in Table 1, the larvae ate less of the treated diet. This is quite reasonable in light of the choice test results. They also gained less weight, which is again to be expected. What was extremely interesting to us was that the approximate digestibility (AD) of the p-benzoquinone diet was much lower (i. e. much more of the diet simply passed through the gut tract without passing through the gut wall). To explain this, we hypothesize that the p-benzoquinone may, by its strong electron acceptor properties be changing the $-SH/S-S-$ equilibrium of important digestive enzymes or membrane proteins of the microvilli of the midgut, in a way not unlike the changes of chemosensory receptor molecules studied by Norris (1971) and Rosenthal and Norris (1973). Either of these possibilities could have the net effect of allowing less material to pass across the gut wall. The efficiency of conversion to tissues of that material which passed across the gut wall (ECD), however, was much higher in the insects exposed to p-benzoquinone. Apparently the smaller volume of material

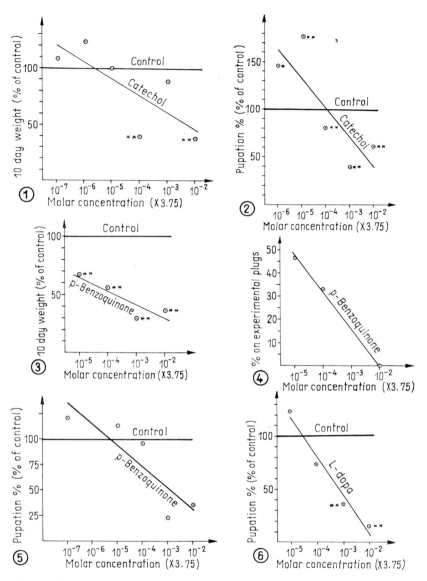

Fig. 1. Effects of catechol on 10-day weight of black cutworm larvae
Fig. 2. Effects of catechol on 28-day pupation percent of black cutworm larvae
Fig. 3. Effects of p-benzoquinones on 10-day weight of black cutworm larvae
Fig. 4. Effects of p-benzoquinones on choice of diet of black cutworm larvae
Fig. 5. Effects of p-benzoquinone on 28-day pupation percent of black cutworm larvae
Fig. 6. Effects of L-dopa on 28-day pupation percent of black cutworm larvae

passing across the gut wall permitted more efficiency by the anabolic machinery of the midgut epithelium and other tissues of the insect. Thus, the higher ECD compensated for the lower AD so that the efficiency of con-

220

version of ingested food (ECI) was essentially the same for the experimental insects as for the controls.

Partly as the result of an article by Rehr et al. (1973), we decided to test L-dopa. Of the compounds tested, L-dopa was the only one to give significant correlation coefficient for 10-day survival. However, even at the 3.75×10^{-2} M concentration, the 10-day survival was only reduced to 59.2% of the controls. The inhibition was more evident for 10-day weights. The correlation coefficient for 28-day pupation was highly significant (Fig. 6), and at the highest concentration the pupation percent was only 14.2% of the controls.

CONCLUSIONS

These experiments give preliminary evidence for the hypothesis that some allelochemics may exert subtle chronic effects, even at rather low concentrations, on larval rate of growth, utilization of food and pupation. The concentrations at which we have found biological activity are well within or below the range at which many allelochemics occur in plants.

REFERENCES

BECK, S. D. (1960): The European corn borer, *Pyrausta nubilalis* (Hubn.) and its principal host plant. VII. Larval feeding behavior and host plant resistance. *Ann. ent. Soc. Am.* **53**, 206–212.

BECK, S. D. and SMISSMAN, E. E. (1960): The European corn borer, *Pyrausta nubilalis* and its principal host plant. VIII. Laboratory evaluation of host resistance to larval growth and survival. *Ann. ent. Soc. Am.* **53**, 755–762.

BECK, S. D. and SMISSMAN, E. E. (1961): The European corn borer, *Pyrausta nubilalis* and its principal host plant. IX. Biological activity of chemical analogs of corn resistance A (6-methoxybenzoxazolinone). *Ann. ent. Soc. Am.* **54**, 53–61.

FRAENKEL, G. (1969): Evaluation of our thoughts on secondary plant substances. *Ent. exp. & appl.* **12**, 473–486.

HARLEY, K. L. S. and THORSTEINSON, A. J. (1967): The influence of plant chemicals on the feeding behavior, development and survival of the two-striped grasshopper, *Melanoplus bivittatus* (Say), Acrididae: Orthoptera. *Can. J. Zool.* **45**, 305–319.

MATSUMOTO, Y. (1962): A dual effect of coumarin, olfactory attraction and feeding inhibition, on the vegetable weevil adult, in relation to the uneatability of sweet clover leaves. Studies on host plant discrimination of the leaf-feeding insects. VI. *Jap. J. Appl. Ent. Zool.* **6**, 141–149.

MAXWELL, F. G., JENKINS, J. N. and PARROTT, W. L. (1967): Influence of constituents of the cotton plant on feeding, oviposition and development of the boll-weevil. *J. econ. Ent.* **60**, 1294–1297.

NORRIS, D. M. (1971): A hypothesized unifying mechanism in neural function. *Experientia* **27**, 531–532.

REESE, J. C. and BECK, S. D.: Effects of certain allelochemics on black cutworm larvae; I. The use of nutritional indices (in preparation).

REESE, J. C., ENGLISH, L. M., YONKE, T. R. and FAIRCHILD, M. L. (1972): A method for rearing black cutworms. *J. econ. Ent.* **65**, 1047–1050.

REHR, S. S., JANSEN, D. H. and FEENY, P. P. (1973): L-dopa in legume seeds: A chemical barrier to insect attack. *Science* (Washington) **181**, 81–82.

ROSENTAL, J. M. and NORRIS, D. M. (1973): Chemosensory mechanism in American cockroach olfaction and gustation. *Nature* **244**, 370.

TODD, G. W., GETAHUN, A. and CRESS, D. C. (1971): Resistance in barley to the greenbug, *Schizaphis graminum*. I. Toxicity of phenolic and flavonoid compounds and related substances. *Ann. ent. Soc. Am.* **64**, 718–722.

WALDBAUER, G. P. (1968): The consumption and utilisation of food by insects. *Adv. Insect Physiol.* **5**, 229–288.

Symp. Biol. Hung. 16, pp. 223–227 (1976)

INHIBITORY ACTION OF CHESTNUT-LEAF EXTRACTS (*CASTANEA SATIVA* MILL.) ON OVIPOSITION AND OOGENESIS OF THE SUGAR BEET MOTH (*SCROBIPALPA OCELLATELLA* BOYD.; LEPIDOPTERA, GELECHIIDAE)

by

P. Ch. Robert

STATION DE ZOOLOGIE, I.N.R.A., 68 COLMAR, FRANCE

The beet or an aqueous extract of its leaves 1. attracts the egg-laying females of *Scrobipalpa ocellatella*, 2. releases oviposition, and 3. stimulates oogenesis.

An aqueous leaf-extract of chestnut tree *Castanea sativa* Mill. 1. is repellent for egg-laying females of *Scrobipalpa ocellatella*, 2. inhibits oviposition, and 3. does not stimulate oogenesis.

An aqueous leaf-extract of chestnut tree, sprayed on a sugar-beet, masks the stimulating actions of the host-plant. A beet-plant having received such a treatment 1. is no more attractive for egg-laying females, 2. does not release oviposition, and 3. does not stimulate oogenesis. Thus, the sugar-beet is transformed into a non-host-plant.

Extracts of chestnut tree leaves only have these properties; fruit or wood extracts are of no effect.

INTRODUCTION

The beet moth is an oligophagous insect. The larvae feed on plants belonging to the genus *Beta* (Chenopodiaceae), especially the cultivated beets (*Beta vulgaris*). Females lay eggs on the same plant.

There is a close relationship of chemical nature between egglaying females and host-plant. The beet plant or an aqueous leaf-extract of this plant stimulates oogenesis, releases oviposition and determines the choice of the egg-laying site (Robert, 1965, 1970). However, the effect of the non-host-plants is not indifferent always. Many plant extracts are repellent to females when choosing a site for oviposition (Robert, 1971): this choice results in positive reactions towards the host-plant and in negative ones against non-host-plants. This behaviour leads the mated females with certainty towards the beet, a suitable food plant for the progeny.

Having determined the role of non-host-plants in the choice of egg-laying sites (Robert, 1971), we study here the influence of one of these non-host-plants on the oviposition, and on the oogenesis of the beet moth.

MATERIAL AND METHODS

The insects used in the present work were taken from natural populations, collected in the field in autumn prior to pupation.

Chestnut leaves were selected from various non-host-plants because their aqueous extract is strongly repellent in trials for egg-laying site selection. The leaves were collected before dropping in autumn. Water extract of leaves was utilized in concentration about that in fresh plants. Then it was sprayed on a neutral substrate (strips of chromatograph-paper) or on the beet plants (red beets).

The egg-laying activity was expressed by the fecundity in each mated female during 21 days.

INFLUENCE OF CHESNUT-LEAF EXTRACTS
ON OVIPOSITION BEHAVIOUR

The beet plant or an aqueous beet extract releases oviposition behaviour in all females. Fecundity is usually higher than 100 eggs per female (Robert, 1965, 1970).

In the absence of beet stimuli egg-laying is completely inhibited in many females: some lay only a few eggs and only few females deposit more than 100 eggs (Robert, 1965, 1970).

These results are confirmed by the following experiment in which the action of aqueous chestnut-leaf extracts is studied.

Experiment 1: Chestnut-leaf extracts on neutral substratum. Three groups of pairs were provided as follows:
— group 1 : paper impregnated by aqueous beet-leaf extract,
— group 2 : untreated paper,
— group 3 : paper impregnated by aqueous chestnut-leaf extract.

TABLE 1

Egg-laying behaviour of females in the presence of plant extracts on paper

Fecundity Egg classes	No. of egg-laying females (in %) for each egg-class		
	Group 1 beet-leaf extract	Group 2 control	Group 3 chestnut-leaf extract
0	0	22.4	87
1– 10	0	22.4	12.5
11– 50	9.7	20.4	0
51–100	29.0	26.5	0
>100	61.3	8.2	0
Average fecundity	108 eggs	38 eggs	0.7 egg
No. of females	32	49	32

The results are summarized in Table 1. The amount of eggs laid is ranged in classes. The beet extract releases oviposition in all the females of group 1; while in absence of plant extract (group 2), 22% of the females do not lay eggs, only 8% lay more than 100 eggs. The inhibition of egg-laying increases in group 3: 87% of the females do not lay any egg and not one lays more than 10 eggs. The same is observed in the average fecundity.

In the absence of beet stimuli, the females do not lay eggs or lay only a few. *The inhibition of egg-laying is considerably increased in the presence of chestnut-leaf extract.*

Expreriment 2: Aqueous chestnut-leaf extract on beet plants. Two groups of pairs were formed:
— group 1 : beet plant (control),
— group 2 : beet plant sprayed with aqueous chestnut-leaf extract.

As indicated in Table 2, all females lay eggs on the control beet plants (group 1). On the treated beet plants (group 2) egg-laying is inhibited in 58% of females, moreover, not one lays more than 100 eggs. This latter result can be seen in fecundity data also: on the control beet plants there were 134 eggs, on the treated ones only 7.5 eggs.

TABLE 2

*Egg-laying behaviour of females in the presence
of beet plants treated with chestnut extract*

Fecundity Egg classes	No. of egg-laying females (in %) for each egg-class	
	Group 1 beet plants	Group 2 beet plants + chestnut-leaf extract
0	0	58
1– 10	0	21
11– 50	3	14
51–100	23	7
>100	73	0
Average fecundity	134 eggs	7.5 eggs
No. of females	30	29

The behaviour of females in the presence of treated beet plant is similar to that of females in the presence of non-host-plants, such as garden pea. So chestnut-leaf extract sprayed on beet plant masks the stimulatory action of the host-plant and inhibits the egg-laying behaviour.

The concentration of chestnut extract is important. When the concentration is doubled, 90% of the females do not lay eggs and the average fecundity decreases to 1.5 eggs.

The inhibitory substances are not found in all parts of the chestnut tree: the aqueous solution of chestnut fruits does not influence the egg-laying behaviour of females.

Conclusion

An aqueous extract of the chestnut tree inhibits egg-laying not only in the absence of beet (experiment 1), but also in the presence of the host-plant (experiment 2). In the latter case, it seems that the specific stimulation of beet plant was masked by unknown substances.

In *Scrobipalpa ocellatella*, the ovogenesis is very intense at the beginning of adult life. Mature ovocytes are formed rapidly and are stored in case no egg-laying takes place, their number varies from 60 to 80 in 2–4-day-old females. The aged, mature ovocytes are resorbed (Robert, 1970).

When the host-plant is absent, the resorbed ovocytes are not renewed and the number of mature ovocytes decreases until reaching 20–30 in 7-day-old females. In the presence of beet plant, ovogenesis becomes stimulated and the resorbed ovocytes are replaced by new mature ones so the stock remains constant.

Experiments were carried out to see whether the aqueous chestnut-leaf extract, inhibiting egg-laying, acts also on ovogenesis.

Experiment 3: Five groups of virgin females were observed. The presence of beet does not release oviposition in virgin females so the study of the development of ovocytes in various situations was possible.

— group 1: virgin females on beet plant,
— group 2: virgin females without beet plant and without chestnut-leaf extract,
— group 3: virgin females on beet plant treated with chestnut-leaf extract at the concentration 1 (i. e. about that of the plant),
— group 4: virgin females on beet plant treated with chestnut-leaf extract at 1/16 concentration,
— group 5: virgin females on beet plant treated with chestnut-leaf extract at double concentration.

The females were observed until the 7th day of their life. The number of mature ovocytes was determined by dissection (Table 3).

The average number of mature ovocytes found in groups 1 and 2 was similar to those found in the preceding experiments, i.e. 73 and 28 ovocytes in the presence and absence of beet plant, respectively.

But in group 3 (in the presence of treated beet plants), there were only 38 mature ovocytes, significantly less than in the first group.

The concentration of chestnut-leaf extract plays a role: at 1/16 concentration (group 4) ovogenesis is not inhibited; at concentration 2 inhibition is complete (group 5); the number of stored ovocytes is not significantly

TABLE 3

Effect of different concentrations of chestnut leaf extract on ovogenesis

	No. of mature ovocytes up to 7th day				
	Group 1 beet	Group 2 control	Group 3 beet + extract conc. 1	Group 4 beet + extract conc. 1/16	Group 5 beet + extract conc. 2
Average number of ovocytes	73	28	38	70	29
No. of females	83	47	110	20	14

different from that of the control (without beet plant). *Thus, the chestnut extract is able to cover the stimulative effect of host-plant on ovogenesis.*

It has also been observed that the masking substances are not found in all parts of chestnut. Extracts of fruits sprayed on beet plant do not exhibit this activity.

DISCUSSION

Females of *S. ocellatella* usually do not lay eggs in the absence of beet plants. The presence of aqueous chestnut-leaf extract increases considerably the proportion of non-egg-laying females (Exp. 1; Table 1).

Oviposition is released in all females when beet is present. An aqueous chestnut leaf-extract sprayed on the beet inhibits oviposition completely in more than half of the females; in the remaining females oviposition activity is reduced (Exp. 2; Table 2).

An aqueous chestnut-leaf extract sprayed on a beet plant masks the stimulative effect of this host-plant on ovogenesis (Exp. 3; Table 3).

Consequently, there are chemical substances which are able to mask the stimulative effect of the host-plant on oviposition in an oligophagous plant feeding insect. The reproductive potential is strongly reduced: it was reduced to 17 times under experimental conditions. This decrease of the reproductive potential is the result of the slowness of ovogenesis, on the one hand, and of the inhibition of the egg-laying behaviour, on the other.

These results encourage us to continue our investigations: substances, having such a restraining effect on reproduction, may have a role in plant protection reducing pests below "economic injury level". The work ahead concerns the isolation of the active chemical substances and their use in field experimentation.

The chestnut extracts act on the egg-laying behaviour and at the same time on ovogenesis. Their action appears to be very different from that of sterilants or repellents which are used even now. These plant extracts are not toxic for beet moth and one can imagine that they are also unharmful for other living organisms.

Until now, we used only leaf extracts of chestnut but it can be expected that substances of other plants and of different chemical nature may be equally active. This is an important aspect to be considered in future studies.

REFERENCES

ROBERT, P. CH. (1965): Influence de la plante-hôte sur l'activité reproductrice de la Teigne de la Betterave *Scrobipalpa ocellatella* Boyd. (Lépidoptère, Plutellidae). *Proc. XIIth Int. Congr. Ent. London*, 552–553.

ROBERT, P. CH. (1970): Action stimulante de la plante-hôte sur l'activité reproductrice chez la Teigne de la Betterave *Scrobipalpa ocellatella* Boyd. (Lépidoptère, Gelechiidae). In: *L'influence des stimuli externes sur la gamétogenèse des insectes.* Colloque International du C.N.R.S. **189**, 147–172.

ROBERT, P. CH. (1971): Der Einfluß der Wirtspflanzen und der Nichtwirtspflanzen auf Eibildung und Eiablage der Rübenmotte *Scrobipalpa ocellatella* Boyd. (Lepidoptera, (Gelechiidae). *Acta Phytopath. Acad. Sci. Hung.* **6**, 235–241.

Symp. Biol. Hung. 16, pp. 229–235 (1976)

VARIOUS HOST—INSECT INTERRELATIONS IN HOST-FINDING AND COLONIZATION BEHAVIOR OF BARK BEETLES ON CONIFEROUS TREES

by

J. A. RUDINSKY

OREGON STATE UNIVERSITY, CORVALLIS, OREGON 97330, USA

Three aspects of the relationship beween the coniferous host tree and bark beetles of the family Scolytidae are analyzed, and recent results are summarized. 1) Host-plant-terpenes serve in the host-recognition mechanism by attracting or repelling certain species and by synergizing or enhancing insect pheromones. 2) Proposed biosynthetic pathways for bark beetle pheromones produced from host-plant substances include 3-methyl-2-cyclohexen-1-one from the monoterpene terpinolene through piperitenone followed by cleavage of the isopropylidene side chain, and the analog alcohol 3-methyl-2-cyclohexen-1-ol from reduction of the ketone. Also, *trans*-verbenol, verbenone, and myrtenol are believed to result from oxidation of alpha-pinene, and similarly pinocarvone by oxidation of beta-pinene. However, the suggestion that certain kinds of pheromone production occur only on the freshly attacked host-tree is not supported by evidence that newly emerged beetles already release the same pheromones. 3) The host-plant also serves as the dense medium or substrate for sonic communication by those bark beetles that produce sound signals – an aspect that has received no study. In various *Dendroctonus* species it is established that the acoustic and chemical signals function as both stimulus and response in host-finding and colonization behavior.

It is well established that a complex host-plant—insect relationship has evolved in many bark beetles of the family Scolytidae infesting coniferous trees in North America, including the major pests that annually cause great environmental damage with high direct and indirect costs. Of the various aspects in this complex relation, only three will be considered here, which are recent and somewhat unsettled or controversial questions we are now studying at Oregon State University.

HOST TREE SUBSTANCES AS ATTRACTANTS, REPELLENTS, AND PHEROMONE SYNERGISTS

Extensive and intensive research attention has been given to insect uses of volatile substances emanating from the host-plant. After demonstrations by several workers that certain host terpene hydrocarbons serve as attractants for bark bettles while other host substances are repellent or inactive, it is believed that many bark beetle species recognize their host tree primarily through olfactory mechanisms using tree volatiles. The work of von Rudloff (1972) demonstrating the reliability of volatile needle oil as a chemosystematic character of coniferous trees indirectly supports this belief.

Natural host tree volatiles are also utilized in the pheromone blend of each beetle species as synergists or additives that greatly increase response to the insect-produced components. This fact led Renwick and Vité (1970) to propose the hypothesis that decrease in the quantity of host resin exuded during attack by *Dendroctonus ponderosae* Hopk. is the critical factor in stopping attack on the tree. However, our studies at Oregon State University (Rudinsky et al., 1974a) contradict this hypothesis because *D. ponderosae* releases an antiaggregative pheromone which seems to mask or repress aggregation in the same way as the corresponding antiaggregative pheromones of *D. frontalis* Zimmerman, *D. brevicomis* Lec., and *D. pseudotsugae* Hopk. The abundance or shortage of host-tree resin probably affects aggregation, but resin may be no more important for *D. ponderosae* than for other pest *Dendroctonus*, all of which utilize host-tree-produced synergists of the aggregative pheromone.

PHEROMONE BIOSYNTHESIS FROM HOST SUBSTANCES

Host terpenes also serve as a source of material which is biosynthesized into the pheromones released by bark beetles. This biosynthesis may occur early, during larval or pupal development or maturation, so that the pheromone is present in the newly emerged adult beetle before flight or attack of the host-plant. For example, volatiles collected from newly emerged female *D. pseudotsugae* contained the antiaggregative pheromone 3-methyl-2-cyclohexen-1-one, called MCH (Rudinsky et al., 1973). We have suggested that it likely comes from the host monoterpene terpinolene through piperitenone followed by cleavage of the isopropylidene side chain. Terpinolene was found in Douglas fir in all its range tested by von Rudloff (1972) using needle oil and by Zavarin (1972) using cortical oleoresin. The analog alcohol 3-methyl-2-cyclohexen-1-ol, which is also produced by the female beetle (Kinzer et al., 1971; Vité et al., 1972) and is released as a potent attractant (Rudinsky et al., 1974b), could come from reduction of the ketone by the female.

Myrtenol and pinocarvone were independently identified from *D. frontalis* by two research groups (Renwick et al., 1973; Rudinsky et al., 1974a) and they are believed to be synthesized by oxidation of the host monoterpenes alpha- and beta-pinene respectively (Renwick et al., 1973). Similarly, with *D. brevicomis* myrtenol, pinocarvone, and pinocarveol were identified by both groups (Hughes, 1973; Libbey et al., 1974) and are believed to be also oxidation products of the pinenes.

The presence of these (and other) oxidation products in the hindguts of beetles exposed to large quantities of host oleoresin or individual terpenes led to the hypothesis by Vité et al. (1972) and Hughes (1973) that such pheromones represent a special class, i.e. "contact" pheromones originating when the beetle is in the "resinous environment" of the freshly attacked tree. However, this hypothesis is not supported by our own identification of these same substances from newly emerged beetles which were untreated and unexposed to fresh resin (Rudinsky et al., 1974a; Libbey et al., 1974). Figure 1 shows gas-liquid chromatograms of these substances in the volatiles collected from living *D. frontalis* and *D. ponderosae* pairs without prior exposure to fresh resin. Thus while attack of the host would appear to be a possible stimulus for biosynthesis of these substances, it cannot be a necessary stimulus.

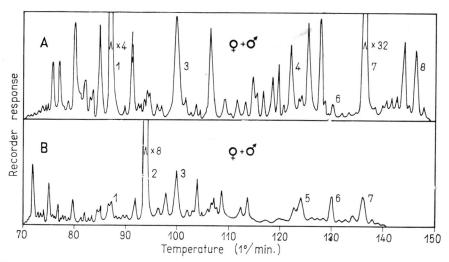

Fig. 1. Gas-liquid chromatograms of volatiles collected from unfed, newly emerged pairs of A. *Dendroctonus frontalis* and B. *D. ponderosae*. Numbered peaks are 1 = = frontalin, 2 = *exo*-brevicomin, 3 = *endo*-brevicomin, 4 = pinocarvone, 5 = MCH, 6 = *trans*-verbenol, 7 = verbenone, 8 = myrtenol. Identifications were confirmed by mass spectrometry

SONIC CUES OF PHEROMONE RELEASE ON THE HOST PLANT

Pheromone *production* has usually been equated to pheromone *release* in bark beetle studies, because it was believed that release occurred automatically through defecation as the beetle attacked and fed in the host tree. However, studies with *D. pseudotsugae* showed that the female releases its antiaggregative pheromone under apparent control of the central nervous system in response to a sonic signal of the male beetle without feeding or defecation (Rudinsky, 1968, 1969; Rudinsky and Michael, 1972). This phenomenon required development of new techniques to collect the volatiles released by living beetles on porous polymer traps for gas-liquid chromatographic/mass spectrometric analysis (Rudinsky et al., 1973, 1974a). With five *Dendroctonus* species we have demonstrated that inter-sex response stimulates pheromone release before feeding or mating can occur and without new host attack or resin "contact." Besides the previously mentioned behavior of *D. pseudotsugae*, female presence evokes release of the male antiaggregative pheromone in *D. frontalis*, *D. ponderosae*, and *D. brevicomis* (Rudinsky et al., 1974a; Libbey et al., 1974). Similar behavior occurs with *D. rufipennis* Kirby (Rudinsky et al., 1974c). *Intra*-sex behavior also evokes pheromone release. The male antiaggregative pheromones of the studied pine beetles were released when 2–3 males were placed together, and can thus be called "rivalry" pheromones; the possibility of such male-to-male pheromones in *D. frontalis* was first suggested by Yu and Tsao (1967). The conclusion that inter- and intra-sex response stimulates pheromone release is also supported by acoustic studies of the sonic signals of these species. For example, with *D. frontalis* the synthetic male pheromone evokes

231

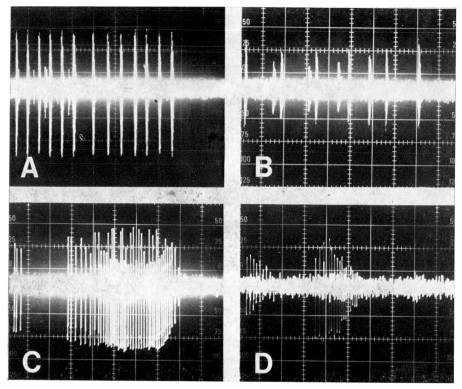

Fig. 2. Oscillograms of typical chirps evoked in male *Dendroctonus frontalis* by A. the natural rivalry pheromone of another male; B. synthetic rivalry pheromone including 0.01% verbenone; C. the natural attractant of the female; and D. synthetic attractant of the female including 0.0001% verbenone. 23 November 1972, 28 °C. Time scale on the abscissa was 125 msec/division in A–B and 12.5 msec/division in C–D

the "rivalry" chirp characteristic of male stridulation caused by the natural male pheromone and by male fighting (Rudinsky, 1973b; Rudinsky and Michael, 1974). Figure 2 A–B shows typical male chirps induced by natural and synthetic male pheromone. These chirps differ greatly from typical male chirps which are evoked by the female attractant (Fig. 2 C–D); differences were described and statistically analyzed by Rudinsky and Michael (1974).

Much of the flexibility and complexity in bark beetle pheromone release comes from the interaction of sonic cues with *multifunctional* pheromones which are released in either low or high concentrations at different stimuli. For example, with *D. frontalis* verbenone is produced by both sexes, i. e. in small quantity by the female and in large quantity by the male (Pitman et al., 1969), and was found to be differentially released by each sex (Rudinsky et al., 1974a). It has corresponding effects in bioassay, with the small quantity typical of the female evoking the male attractant chirp and the larger quantity typical of the male evoking the male "rivalry" chirp (Rudinsky, 1973b; Rudinsky et al., 1974a). This is shown in Fig. 2 B–D,

232

where distinct chirps were evoked by synthetic pheromones that differ only in the quantity of verbenone present. Myrtenol also appears to be a multifunctional pheromone; used in small quantity (as released by the female) it synergized the other attractants, but used in larger quantity (as released by the male) it repressed male attraction (Rudinsky et al., 1974a). With *D. pseudotsugae* the female releases different quantities of 3-methyl-2-cyclohexen-1-one with different effects; a low concentration in the feces is synergistically attractive with the aggregative pheromone, but a high concentration of the same substance released in the presence of a male masks the aggregative pheromone and prevents further attraction. Field and laboratory tests showing the multifunctionality of this pheromone were reported by Kinzer et al. (1971), Rudinsky et al. (1972), Furniss et al. (1972), Rudinsky (1973a), and Rudinsky et al. (1974b).

These results indicate the necessity to distinguish clearly between production and release of bark beetle pheromones, and emphasize the need for further study of this complex communication system of interacting olfactory and auditory signals.

HOST-PLANT AS TRANSMITTING MEDIUM FOR ACOUSTIC COMMUNICATION

The biosynthetic possibilities of both the host tree material and insect physiological processes must be utilized by a very large number of under-bark dwelling species, and there is already considerable evidence that the same host attractants and pheromones are shared by different insect families. These limitations probably increase selective pressure toward development of alternative communication such as acoustic signals. Also, the favorable sound-transmitting qualities of wood as well as the "noisy" chitinous covering of the beetle body probably make acoustic communication an inevitable evolutionary development for bark beetles. Sexual dimorphism of acoustic apparatus in Scolytidae is so well known as to have provoked the generalization that only one sex stridulates in each species (Barr, 1969; Schönherr, 1970). That belief was invalidated by electronic recordings of stridulation in species or sexes previously believed to be silent, e.g. female *Dendroctonus* (Rudinsky and Michael, 1973) and various *Ips* (unpublished data). However, sexual dimorphism remains a striking source of acoustic variety in scolytids. It is probably significant that the gallery-establishing sex of each species has "quiet" sonic signals that are well suited for dense-medium or substrate transmission, while the other sex of each species, which often begins stridulating on the bark and outside the gallery, has "louder" stridulation more capable of airborne transmission.

Untouched in any study so far is the relation of the dynamic and mechanical properties of different host tree genera as sound-transmitting media to the development of acoustic communication by the insect species infesting each kind of host tree. Considering the well known differences in sound-producing qualities of various woods used in construction of musical instruments, for example, this aspect of host–insect interaction could be an interesting study.

ACKNOWLEDGEMENT

The studies at Oregon State University were partially supported by NSF Grant No. GB–36892.

REFERENCES

BARR, B. A. (1969): Sound production in Scolytidae (Coleoptera) with emphasis on the genus *Ips. Can. Ent.* **101**, 636–672.

FURNISS, M. M., KLINE, L. N., SCHMITZ, R. F. and RUDINSKY, J. A. (1972): Tests of three pheromones to induce or disrupt aggregation of Douglas-fir beetles (Coleoptera: Scolytidae) on live trees. *Ann. ent. Soc. Am.* **65**, 1227–1232.

HUGHES, P. R. (1973): *Dendroctonus:* Production of pheromones and associated compounds in response to host monoterpenes. *Ph.D. Dissertation,* University of California, Davis.

KINZER, G. W., FENTIMAN, A. F., Jr., FOLTZ, R. L. and RUDINSKY, J. A. (1971): Bark beetle attractants: 3-methyl-2-cyclohexen-1-one isolated from *Dendroctonus pseudotsugae. J. econ. Ent.* **64**, 970–971.

LIBBEY, L. M., MORGAN, M. E., PUTNAM, T. B. and RUDINSKY, J. A. (1974): Pheromones released during inter- and intra-sex response of the scolytid beetle *Dendroctonus brevicomis. J. Insect Physiol.* **20**, 1667–1671.

PITMAN, G. B., VITÉ, J. P., KINZER, G. W. and FENTIMAN, A. F., Jr., (1969): Specificity of population-aggregating pheromones in *Dendroctonus. J. Insect Physiol.* **15**, 363–366.

RENWICK, J. A. A. and VITÉ, J. P. (1970): Systems of chemical communication in *Dendroctonus. Contrib. Boyce Thompson Inst. Plant Res.* **24**, 283–292.

RENWICK, J. A. A., HUGHES, P. R., TY TANLETIN DE J. (1973): Oxidation products of pinene in the bark beetle, *Dendroctonus frontalis. J. Insect Physiol.* **19**, 1735–40.

RUDINSKY, J. A. (1968): Pheromone-mask by the female *Dendroctonus pseudotsugae* Hopk., an attraction regulator. *Pan-Pac. Ent.* **44**, 248–250.

RUDINSKY, J. A. (1969): Masking of the aggregation pheromone in *Dendroctonus pseudotsugae* Hopk. *Science,* **166**, 884–885.

RUDINSKY, J. A. (1973a): Multiple functions of the Douglas-fir beetle pheromone 3-methyl-2-cyclohexen-1-one. *Environ. Ent.* **2**, 579–585.

RUDINSKY, J. A. (1973b): Multiple functions of the southern pine beetle pheromone verbenone. *Ibid.* **2**, 511–514.

RUDINSKY, J. A. and MICHAEL, R. R. (1972): Sound production in Scolytidae: Chemostimulus of sonic signal by the Douglas-fir beetle. *Science* **175**, 1386–1390.

RUDINSKY, J. A. and MICHAEL, R. R. (1973): Sound production in Scolytidae: Stridulation by female *Dendroctonus* beetles. *J. Insect Physiol.* **19**, 689–705.

RUDINSKY, J. A. and MICHAEL, R. R. (1974): Sound production in Scolytidae: "Rivalry" behavior of male *Dendroctonus* beetles. *J. Insect Physiol.* **20**, 1219–1230.

RUDINSKY, J. A., FURNISS, M. M., KLINE, L. N. and SCHMITZ, R. F. (1972): Attraction and repression of *Dendroctonus pseudotsugae* (Coleoptera: Scolytidae) by three synthetic pheromones in traps in Oregon and Idaho. *Can. Ent.* **104**, 815–822.

RUDINSKY, J. A., MORGAN, M. E., LIBBEY, L. M. and MICHAEL, R. R. (1973): Sound production of Scolytidae: 3-methyl-2-cyclohexen-1-one released by female Douglas-fir beetle in response to male sonic signal. *Environ. Ent.* **2**, 505–509.

RUDINSKY, J. A., MORGAN, M. E., LIBBEY, L. M. and PUTNAM, T. B. (1974a): Anti-aggregative-rivalry pheromone of the mountain pine beetle, and a new arrestant of the southern pine beetle. *Environ. Ent.* **3**, 90–98.

RUDINSKY, J. A., MORGAN, M. E., LIBBEY, L. M. and PUTNAM, T. B. (1974b): Additional components of the Douglas-fir bettle (Col., Scolytidae) aggregative pheromone and their possible utility in pest control. *Z. ang. Ent.* **76**, 65—77.

RUDINSKY, J. A., SARTWELL, Jr. C., GRAVES, T. M. and MORGAN, M. E. (1974c): Granular formulation of methylcyclohexenone: an antiaggregative pheromone of the Douglas-fir and spruce bark beetles (Col., Scolytidae). *Z. ang. Ent.* **75**, 254–263·

VON RUDLOFF, E. (1972): Chemosystematic studies in the genus *Pseudotsuga.* I. Leaf oil analysis of the coastal and Rocky Mountain varieties of the Douglas-fir. *Can. J. Bot.* **50**, 1025–1040.

234

SCHÖNHERR, J. (1970): Stridulation einheimischer Borkenkäfer. *Z. ang. Ent.* **65**, 309–312.

VITÉ, J. P., BAKKE, A and RENWICK, J A. A. (1972): Pheromones in *Ips* (Coleoptera: Scolytidae): Occurrence and production. *Can. Ent.* **104**, 1967–1975.

YU, C. C. and TSAO, C. H. (1967): Gallery construction and sexual behavior in the southern pine beetle, *Dendroctonus frontalis* Zimm. *J. Ga. Ent. Soc.* **2**, 95–98.

ZAVARIN, E. (1972): Chemical races of Douglas-fir. *Symposium on Chemistry in Evolution and Systematics*, Strasbourg, France, 3–8 July 1972.

Symp. Biol. Hung. 16, pp. 237–240 (1976)

INVESTIGATIONS ON THE INFLUENCE OF FRUIT ODOUR ON THE ORIENTATION OF CODLING MOTH (*LASPEYRESIA POMONELLA* L.)

by

K. Russ

BUNDESANSTALT FÜR PFLANZENSCHUTZ,
WIEN II., TRUNNER-STRASSE 5, AUSTRIA

Investigations of Sutherland (1972), Sutherland and Hutchins (1973) as well as Wearing and Hutchins (1973) have shown significant orientation of neonate larvae of the codling moth as well as females to components of fruit odour, especially to α-farnesene.

Own investigations have shown that neonate larvae of the codling moth significantly prefer apple fruit extract as well as apple fruit as compared to pear extract or pear-fruit. It is supposed that pears contain less α-farnesene and are therefore less attractive for neonate larvae.

INTRODUCTION

In the course of a project on the genetic control of codling moth in connection with an integrated control program, investigations on the biology of this pest led to observations on its orientation during the egg-laying period on its main host-plants.

It seemed to be important to clarify, if there were differences in the oviposition behaviour or in the differentiation of neonate larvae between apple and pear plants.

Investigations of Sutherland (1972), Sutherland and Hutchins (1973), as well as Wearing and Hutchins (1973) showed that neonate larvae as well as the female of the codling moth respond and orientate themselves to α-farnesene, an odour compound which is present in various pomaceous fruits. However, according to these authors the sphere of action of these odours at larvae and adults is very small, in fact, it is only within a few centimetres.

Because our experience showed that the heaviness of attack on apple and pear fruits by the codling moth is different, it was expected that caterpillars and adults are attracted to their host fruits differently depending on the various intensity of the odour, emitted by these two host plants. Sutherland (1972) showed clearly that the orientation of the neonate larvae depends on the distance from the source of odour and, consequently, on the gradient of the odour itself. This means that caterpillars and adults are able to perceive differences in the concentration of this fruit odour.

Assuming that possibly pear fruits and leaves or other parts of this host plant are less attractive for the codling moth, while these parts of the apple tree are especially attractive, we initiated a test series in order to confirm this assumption.

EXPERIMENTS

a) *Y-tube test with an odour-extract of apple and pear fruits*

Apple and pear fruit odour extract originating from a fruit juice factory were used in a quantity of 0.5 ml in 100 ml H_2O. This solution was dropped on cotton wool in glass vials and fixed on the two ends of a Y-tube. Neonate larvae were placed at the main end of the tube. A weak air-stream passed the tubes in direction from the odour sources to the larvae. In all experiments only neonate larvae were used. The tests were carried out in darkness and checked after 24 hours.

b) *Petri-dish tests with pear and apple fruit extract*

The same odour-extract solution as described in a) was used on filter-paper strips in petri dishes (diameter 95 mm). Experiments were carried out only by using neonate larvae, not older than 12 hours. The neonate larvae were cooled in a refrigerator until the beginning of the test to enable us to place the immobilized larvae in the center of the petri dish between the apple- and pear-odour sources. The tests were terminated and checked after 2 hours. All experiments were carried out in complete darkness.

c) *Tests using ripe apple and pear fruits in petri dishes*

In these experiments pieces of the surface of ripe apple and pear fruits, stored until the experiments in CO_2 were used. The pieces were fixed *vis-à-vis* in a petri dish. The test-larvae were cooled in a refrigerator and then placed in the center of the petri dish. The tests were terminated and checked after 2 hours.

d) *Investigations on the attraction of neonate larvae of codling moth to apple and pear fruits at different stages of fruit development*

Beginning with the fall of the petals, tests using petals and developing fruits were carried out. The test materials (petals, fruits and leaves) were placed *vis-à-vis* in petri dishes and then the cooled and immobilized larvae were placed in the center of the petri dish.

These tests were carried out in darkness and terminated and checked after 2 hours.

e) *Experiments in a field-cage*

An apple and a pear tree were caged together in a field cage (length: 1.5 m; width: 1.5 m; height: 2.8 m) to investigate the responses of the adults, especially the females, to these two host-plants.

Beginning with the 27th of May 1974 daily releases of newly emerged males and females (from diapausing larvae collected in the field) into this cage and a daily check of the number of eggs laid on the leaves, fruits or branches were made.

(The test has not yet been completed !)

238

RESULTS

1) *Results of tests a)—c)*

The results of the tests a—c are given in Table 1. As Table 1 shows, the neonate larvae of the codling moth significantly favour the apple-juice extract and the ripe apple fruit itself. Whereas it seems to be very clear that the extract of pears as well as the pear fruits had a much less attracting effect on the larvae.

The test series a—c shows also that the neonate larvae of the codling moth significantly favoured apple fruits and less so the pear fruits. A substantially lower percentage of larvae remained indifferent.

TABLE 1

Orientation of neonate codling moth larvae to the odour of ripe apples and pears

Test	Orientation of larvae to		Larvae indifferent, %	No. of larvae used per test
	apples, %	pears, %		
a) Y-tube test with odour-extract of fruits	54.02	8.05	37.93	84
b) Petri-dish test with odour-extract of fruits on filter paper	58.33	13.11	28.56	87
c) Petri-dish test with pieces of apples and pears	51.66	24.47	23.78	161

2) *Attraction of neonate larvae to apple and pear fruits at different stages of fruit development*

The results of this test series are summarized in Table 2.

In this experiment it was possible to show a distinct preference of the neonate larvae to certain parts of the apple tree. Sometimes, however, a part of the pear tree was more attractive, especially during the development of fruits.

It seems that the different stages of fruits have different amounts of attractive odour. Further investigations are needed to clarify this problem.

It was also very interesting to note that parts of the walnut tree compared to the apple tree, were not attractive to the neonate codling moth larvae.

3) *Experiments in the field cage*

Our investigations in the field cage are not yet finished nevertheless they clearly show that the codling moth favours apple trees much more than pear trees for oviposition. This assumption is clearly reflected from the data of Table 3 showing that the number of eggs laid on pears was much less than those on apples.

This preliminary experiment seems to indicate why pear fruits are less heavily infested than apple fruits. It shows also that an orientation of females during oviposition to fruit odour is probable and that the females are able to distinguish between pears and apples.

TABLE 2

Preference of neonate codling moth larvae for the different parts of apple and pear tree

Date	No. of replicates	Plant part and diam. of fruit (cm)	No. of larvae	apple		pear		indifferent	
				No.	%	No.	%	No.	%
19 Apr.	1	petals	8	3	37.5	3	37.5	2	25.0
	1	leaves	8	0	0.0	3	37.5	5	62.5
19 Apr.	3	syncarp (1/2)	23	10	43.5	9	39.1	4	17.4
26 Apr.	6	apples = 0.64 pears = 1.02	33	17	51.5	8	24.3	8	24.2
2 May	3	apples = 0.92 pears = 1.19	30	14	46.6	10	33.4	6	20.0
10 May	4	apples = 1.33 pears = 1.41	39	9	23.1	16	41.1	14	35.8
17 May	3	apple and pear fruit pieces = 1.6*	21	5	23.8	8	38.1	8	38.1
24 May	8	ditto**	96	44	45.8	33	34.3	19	19.9
7 June	5	ditto***	41	9	22.0	24	58.5	8	19.5

* diam. of apples = 1.43 cm.   ** diam. of apples = 1.96 cm.   *** diam. of apples = 3.37 cm.
diam. of pears = 1.79 cm.   diam. of pears = 1.95 cm.   diam. of pears = 2.50 cm.

TABLE 3

Oviposition on apple and pear trees under field conditions in a field cage

Date	Apples			Total No. of eggs	Pears		Total No. of eggs
28 May	No. of eggs laid on	fruits:	23	174	fruits: 12		54
		leaves:	149		leaves: 31		
		branches:	2		branches: 1		
	No. of	fruits:	108		fruits: 26		
		leaves:	1222		leaves: 968		
7 June 1974	No. of eggs laid on	fruits:	38	268	fruits: 0		14
		leaves:	213		leaves: 14		
		branches:	17		branches: 0		
	No. of	fruits:	58		fruits: 17		
		leaves:	1148		leaves: 942		

REFERENCES

SUTHERLAND, O. R. W. (1972): The attraction of newly hatched codling moth (*Laspeyresia pomonella*) larvae to apple. *Ent. exp. & appl.* **15**, 481–487.
SUTHERLAND, O. R. W. and HUTCHINS, R. F. N. (1973): Attraction of newly hatched codling moth larvae (*Laspeyresia pomonella*) to synthetic stereo-isomers of farnesene. *J. Insect Physiol.* **19**, 723–727.
WEARING, C. H. and HUTCHINS, R. F. N. (1973): α-Farnesene, a naturally occurring oviposition stimulant for the codling moth (*Laspeyresia pomonella*). *J. Insect. Physiol.* **19**, 1251–1256.

Symp. Biol. Hung. 16, pp. 241–245 (1976)

OVIPOSITION BEHAVIOUR OF *CEUTORRHYNCHUS MACULAALBA* HERBST. (COL.: CURCULIONIDAE)

by

Gy. Sáringer

LABORATORY OF THE RESEARCH INSTITUTE FOR PLANT PROTECTION, H-8360 KESZTHELY, HUNGARY

The shape of the flowers of the garden poppy (*Papaver somniferum* L.) does not visually stimulate the females in the choice of the oviposition site. Olfactometer tests showed that the scent of flowers attracts the females and not the scent of the leaves. Antennectomy showed that the choice of egg-laying site is regulated by receptors situated on the scape. In experiments carried out with pieces of capsule-wall it was found that the convexity of the capsule was the decisive factor in eliciting oviposition. Visual stimuli do not influence the choice of the egg-laying site. In order to inhibit egg laying, 15 compounds had been examined. No egg laying took place on capsules treated with the following compounds: 4'-(dimethyltriazeno)-acetanilide, 3,5-dichlorophenoxyacetic acid, 3,5-dichlorophenoxyacetamide, 3,5-dichlorophenoxy-N-diethylacetamide, 2,4,6-trichlorophenoxyethanol, and 2,4,6-tribromophenoxyethanol.

The adults of *C. maculaalba* in Hungary overwinter in the soil; they begin to emerge in the middle of May and continue until the middle of June. Females lay the eggs after the petal-fall when the young capsules become accessible (Fig. 1). First they peel the outer wall of the capsule then they gnaw a hole in the peeled surface and lay their eggs into the capsule. Generally the female lays one egg, rarely two eggs on the septum walls inside the capsule. Oviposition is not performed in capsules older than three days. The larvae develop in the capsules feeding on seed buds (Szelényi, 1939; Zsoár, 1950; Sáringer, 1964, 1970a, b).

In order to find out whether the shape of the flower plays a role in the choice of the egg-laying site, seventy-three artificial, filter-paper flowers, resembling poppy flowers, were placed on a poppy field at the end of the flowering period (second half of June), they were scattered over the whole plot, fastened on stalks after removing the capsules and were left in the field for five days. Observations made for two hours on each of these five days showed that adults were never found in the artificial flowers. At the same time a great number of beetles occurred in the nine natural poppy flowers. This observation suggests that *the form of the flower does not play a role in choosing the egg-laying site.*

The effect of the odour of flowers was investigated by the vertical olfactometer, constructed by T. Jermy (unpublished). The round wire screen arena of the olfactometer was divided into four sectors. The first sector was supplied with scent of freshly cut flowers, the second with odour of freshly cut upper leaves; the third and fourth parts were odourless, with clean air continuously flowing. The insects (seven females and eight males) were placed into the olfactometer after a twenty-four hours' period of starvation. Observations were made for four hours and twenty minutes; during this period,

Fig. 1. A one-day-old poppy capsule with *Ceutorrhynchus maculaalba*
(Photograph by Zsoár)

based on forty-four observations, the displacement and distribution of
the insects in the four sectors were recorded. Statistical analysis showed
that the odour of the flower had a significant attractive effect on the adults
in contrast to the odour of the leaf. The number of adults found in the leaf-
odour sectors and in the control sectors did not show significant differences.
According to these results the odour of the flower plays the most important
role in host selection.

In the next series of experiments *the stimuli of the shape* of the egg-laying
site, i.e. *the convexity of the capsule* was studied. In one of the experiments
pieces of 1.5 cm² were cut out from young capsule walls and these were
fastened by fine needles to cork pieces, to form flat surfaces; these pieces were
then presented to the females. In another experiment the females were
placed singly into glass tubes (4 mm in diam.), which were fastened with their
mouth on young capsules so that the covered part formed practically a flat
surface. In both series feeding was intensive but females did not bore any
holes nor did they lay any eggs. In the control, however, egg-laying was
observed. We may conclude therefore that the hole-boring reflex prior to
egg-laying is elicited by the stimuli of the capsule. Presumably proprioreceptors in the legs are responsible and the female perceives the position of
its straddled legs in relation to the body.

Observations on the egg-laying behaviour revealed that just before the
laying of eggs the females move their antennae intensively and many times
pass them downward over the capsule wall. On the basis of this observation
*we suggest that a decisive role is played by the antennae in choosing the egg-
laying site.* In other words, contact receptors on the antennae of females
may be responsible for choosing a place suitable for egg-laying.

Antennae of *C. maculaalba* consist of 10 segments (1 basal segment known
as the scape, 6 segments of the pedicel and 3 segments of the flagellum or

TABLE 1

Egg production of C. maculaalba *influenced by various treatments of the antennae*
(Sáringer, 1970b)

Treatment	No. of eggs per females after 24 hours		No. of females	Feeding
	in the capsule	on the surface of the capsule		
I. Flagellum segments removed	4	1	4	+++
Flagellum segments stained	6	3	5	++
Control	7	—	5	+++
II. Flagellum segments + 6 pedicel segments removed	3	—	5	++
Flagellum segments + 6 pedicel segments stained	4	1	6	++
Control	9	—	7	+++
III. Entire antenna removed	—	—	6	++
Entire antenna stained	—	1	6	+++
Control	11	1	6	+++

+++ = intensive feeding.
++ = medium feeding.

clavola, geniculate-clavate). The question arises whether these segments are equivalent when choosing the egg-laying site. This question was answered by applying two methods. In one of them, the antennal segments (scape, pedicel, flagellum) were gradually removed, while in the other, they were covered with thick black Indian ink. The latter method was at the same time the control of the former one, since it excluded the strong traumatic effect of amputation.

In the course of a preliminary study of mere informatory character, segments were gradually removed and covered respectively, first the three segments of the flagellum, then those of the pedicel and finally the whole antenna. These adults (4–7 specimens per treatment) were placed onto 1–2 days old poppy capsules. From the same population untreated adults of the same age were simultaneously kept under observation on similarly old poppy capsules. The treatments were evaluated after 24 hours. The results are presented in Table 1.

According to the data of Table 1, females are able to choose the egg-laying site even with only the scape left. This means that the contact receptors present on the scape ensure this behaviour. Further investigations are required to clarify the role of the mouth parts.

The role of vision in egg-laying was also studied in our experiments. Egg-laying females were kept in two parallel series on young poppy heads, one series in complete darkness and the other under natural light conditions. the experiment was repeated twice. Females kept in the darkness laid only about one-third of the eggs compared to those kept under natural light. The fact that egg-laying occurred also in complete darkness indicated that vision may not play a decisive role in selecting the site for oviposition.

These results show that the stimuli directing the egg-laying of *C. maculaalba* are as follows:

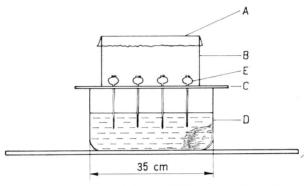

Fig. 2. Setup for screening oviposition inhibitors on *Ceutorrhynchus maculaalba*.
A = cellophane cover; B = glass cylinder; C = cardboard sheet with holes; D =
= dish with water; E = poppy capsules; F = lamp

1. the odour of poppy flowers through the olfaction attracts the females;
2. the substances of the poppy heads (capsules) act as chemostimuli, attracting the females to feed on the capsules;
3. both the contact chemostimuli and the convexity of the capsule (the latter as shape stimulus) provoke hole boring activity and egg-laying on the capsule.

A further question arises whether the effect of stimuli eliciting egg-laying can be inhibited by certain chemicals. Here fifteen compounds were tested; in each case four young poppy heads were immersed in the 1% solution of a given compound, to which 3% Sandovit was added as surfactant. The four control capsules were immersed into the pure solvent plus 3% Sandovit. The capsules (4 treated and 4 untreated ones) were then inserted in holes punctured in a circle (20 cm diam.) in a cardboard sheet which was placed on a large dish filled with water. The stalks immersed in water prevented the wilting of capsules. The cardboard sheet was covered with a glass cylinder (30 cm diam. and 10 cm high), closed on its upper end by cellophane. Into this setup 10 males and 10 females were placed. A lamp placed over the setup provided even illumination (Fig. 2). Results were evaluated after 48 hours by counting the eggs in each capsule. From the 15 compounds listed in Table 2, six compounds completely prevented egg-laying.

244

TABLE 2

Antioviposition effect of different compounds
(After Jermy and Matolcsy, 1967; Matolcsy et al., 1968)

Compound	Antioviposition effect on *C. maculaalba* (adults) on poppy capsules
1. S-Carboxymethyl-N, N-dimethyl dithiocarbamate	+ +
2. 4'-(Dimethyltriazeno)-acetanilide	+ + +
3. Triphenyltinacetate	+
4. Triphenyltinhydroxide	∅
5. 4-chlorophenoxyethanol	+
6. 4-methoxyphenoxyethanol	+ +
7. 3,5-dichlorophenoxyacetic acid	+ + +
8. 3,5-dichlorophenoxyacetamide	+ + +
9. 3,5-dichlorophenoxy-N-diethylacetamide	+ + +
10. 2,4,6-trichlorophenylmethyl-ether	+ +
11. 2,4,6-trichlorophenylethyl-ether	∅
12. 2,4,6-trichlorophenoxyethanol	+ + +
13. 2,4,6-trichlorophenoxyacetic acid	+
14. α-(2,4,6-trichlorophenoxy)-propionic acid	+ +
15. 2,4,6-tribromophenoxyethanol	+ + +

+ + + = no oviposition (strongest inhibition).
+ + = few eggs.
+ = more eggs.
∅ = many eggs (no inhibition).

REFERENCES

JERMY, T. and MATOLCSY, G. (1967): Antifeeding effect of some systemic compounds on chewing phytophagous insects. *Acta Phytopath. Acad. Sci. Hung.* **2**, 219–224.
MATOLCSY, G., SÁRINGER, GY., GÁBORJÁNYI, R. and JERMY, T. (1968): Antifeeding effect of some substituted phenoxy compounds on chewing and sucking phytophagous insects. *Acta Phytopath. Acad. Sci. Hung.* **3**, 275–277.
SÁRINGER, GY. (1964): Der Wintermohn und die Mohnschädlinge. *Ann. Inst. Prot. Plant. Hung.* **9**, 185–194.
SÁRINGER, GY. (1970a): The life-history of *Ceuthorrhynchus maculaalba* Herbst. (Col., Curculionidae) in Hungary I. Effect of environmental conditions on the emergence of hibernating adults. *Acta Phytopath. Acad. Sci. Hung.* **5**, 375–387.
SÁRINGER, GY. (1970b): Role played by the contact receptors of antennae in the egglaying process of *Ceuthorrhynchus maculaalba* Herbst. (Coleoptera: Curculionidae). *Acta Agron. Acad. Sci. Hung.* **19**, 393–394.
SZELÉNYI, G. VON (1939): Die Schädlinge des Ölmohns in Ungarn. *Vehr. 7. Intern. Kongr. Entom.*, Berlin **4**, 2625–2639.
ZSOÁR, K. (1950): Data on the biology of poppy-weevil (*Ceuthorrhynchus maculaalba* Herbst.). *Agrtud. Egyet. Mezőgazd. Kar Évkönyve*, Gödöllő, **1**, 130–136.

Symp. Biol. Hung. 16, pp. 247–253 (1976)

THE ROLE OF GRAMINACEOUS HOST-PLANTS IN THE INDUCTION OF AESTIVATION-DIAPAUSE IN THE LARVAE OF *CHILO ZONELLUS* SWINHOE AND *CHILO ARGYROLEPIA* HAMPS.

by

P. SCHELTES

INTERNATIONAL CENTRE OF INSECT PHYSIOLOGY AND ECOLOGY,
P. O. BOX 30772, NAIROBI, KENYA

Stemborers in the temperate and subtropical regions have a diapause during the cold season, induced by short daylength or low temperature. In the tropical regions, however, the host-plant seems to be the factor which triggers the aestivation-diapause.

A new criterion for the determination of aestivation-diapause in field collected stemborer larvae is suggested. Preliminary experiments in the field and laboratory indicate that decreasing water, protein, and carbohydrate content of the stem may be involved in the induction of aestivation in *Chilo zonellus* and *Chilo argyrolepia*. High temperatures inside the stem may also have an impact.

Most plant feeding insects face problems of survival during certain periods of the year due to lack of food and/or extreme climatic conditions. In the temperate and subtropical regions the unfavourable conditions prevail during winter; in the tropics, it is the hot, dry season, which may be detrimental to insect life. Many insects are known to overcome unfavourable periods by entering a state of dormancy. Stemborers (order Lepidoptera) are no exception in this respect. They enter diapause as mature larvae at the end of the growing season of the host-plant and remain inside the old stalks until the return of more favourable conditions after which they pupate and develop further. This type of dormancy has been observed among stemborers of the temperate region, subtropics and tropics. Although the phenomenon of diapause can be found in stalk borers all over the world the environmental token responsible for diapause induction and termination may differ considerably from one species to another. In the temperate and subtropical regions we often find that daylight and temperature are the main diapause-inducing factors. Diapause in the European cornborer, *Pyrausta nubilalis*, is induced by short photoperiods but the rate of induction is decreased at high temperatures (Beck and Hanec, 1960). Fukaya (1967) also could induce diapause in the rice stemborer, *Chilo suppressalis*, by exposing the larvae to short daylengths. The Southwestern cornborer, *Diatraea grandiosella*, is only found below about 37.5 °N. In this species temperature appears to be the critical token for the onset of diapause. Larvae reared at temperatures lower than 25 °C entered diapause; photoperiod was less important (Chippendale and Reddy, 1972, 1973). The dormancy of stemborers in the tropics (whence most of them originated) has been studied much less. No author mentions photoperiod as a factor which has an impact on the aestivation diapause of stalkborers in the tropics. A rise in temperature as the possible inducing factor has only been mentioned recently by Hummelen (1974) for *Rupela albinella* in Surinam.

Diapause in stemborers is often thought to be induced by the physiological conditions of the stalk, independently of temperature. Swaine (1957) working with the maize stemborer, *Busseola fusca*, in Tanzania mentions the water content of the maize stem as the diapause inducing factor. Usua (1973) working with the same species in Nigeria found that the diapause was largely induced by the state of maturity and composition of food taken by the larvae. He proposes that the carbohydrate, protein and water content of the maize plant are important factors to induce diapause. Hynes (1942) also suggests changes in the chemical nature of the food as critical factors in the onset of dormancy in *Diatraea lineolata* in Trinidad. In Indonesia Van der Goot (1925) reports a high incidence of aestivating larvae of *Scircophaga innotata* only in mature rice plants, even if mature plants are under wet conditions. He therefore does not believe that the water content of the stem is important but other chemical changes in the rice plant might very well be responsible.

My own experimental work has focused on *Chilo zonellus* Swinhoe and *Chilo argyrolepia* Hamps. Literature data on *Chilo zonellus*, however, are rather contradictory: Ingram (1958) reports that in Uganda the insect has no resting stage although development is probably slowed down in the dry season. Nye (1960) found larvae in a resting phase in dry maize stems in Tanzania; larvae in stalks of sorghum growing nearby however, were fully active. He concluded therefore that a larva only undergoes a resting stage when its host-plant dries out completely. Such a situation is not likely to occur in Uganda. On the other hand, resting larvae have been found in irrigated fields in Sudan (Schmutterer, 1969), and in Kenya among larval populations bred on 7–8 weeks old maize stems in an outdoor insectary (Mathez, 1972).

In our study of *Chilo zonellus* and *Chilo argyrolepia* we have considered factors in the host-plant as well as climatic factors. Populations of both species were followed closely in the experimental fields at Kikambala in the Coast Province of Kenya. Samples of larvae were regularly sent to the laboratory where the percentage of aestivating larvae was determined. As the criterion for diapause, we used the failure of a larva to pupate under non-diapausing conditions within the period required normally by an active larva of that particular stage and sex. This criterion has been used by many other workers dealing with larval diapause in Lepidoptera (Beck and Hanec, 1960; Katiyar and Long, 1961; Usua, 1973). They bred diapausing larvae on fresh segments of maize stalk or on diet. When aestivating *Chilo* larvae are provided with such food, the aestivation may be rapidly broken; however, when fed on dry old stems, the larvae remain in aestivation. On the contrary, active larvae invariably die or have accelerated pupation when put on dry food, even if the climatic conditions are suitable for normal development. Therefore, for our tropical species we propose a new criterion for larval aestivation: the capability of a larva to survive on dry food for a significantly longer period than the period within which active larvae die or pupate on this food. It should be emphasized that the method where field collected larvae are bred on diet in order to determine the percentage of aestivation (the method we used initially) still has a relative value: the percentages obtained may be rather low, but the variations in those percentages are most likely the same as when we use the dry food method.

THE HOST-PLANT AS AN AESTIVATION-INDUCING FACTOR

The impact of the host-plant on the aestivation of the larvae is not very clear. A few remarks, however, can be made. The host-plant does not have to be completely mature before aestivating larvae can be found in it. As can be seen in Fig. 1, aestivating larvae are found in host-plants which had fully developed cobs (1A), in plants where the cobs had only developed a little (1B) and even in plants where drought had stopped development at the formation of the first tassels (1C). In most cases a low water per-

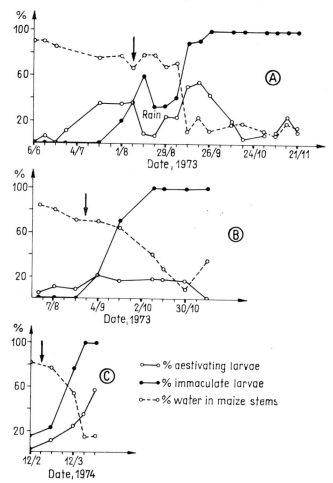

Fig. 1. Relationship between percentage aestivating larvae of *Chilo zonellus* and *Chilo argyrolepia*, percentage immaculate larvae and the maize plant in three experimental fields at Kikambala, Coast Province, Kenya. A, B: % aestivation was determined by breeding the larvae on diet after collection from the field. C:% aestivation was determined by breeding the larvae on dry old maize stems after collection from the field. Arrow indicates maximal development of the maize plant. At that time the plants in field A were mature and had well developed cobs; the cobs in field B, however, never developed fully and remained stunted due to drought. The growth of plants in field C stopped already at the beginning of the tasseling stage, also due to drought

centage of the stem is correlated with a relatively high incidence of aestivation; on one occasion, however, we had a fairly high percentage of aestivating larvae even in maize with 70% water content; but those larvae had the spotted colour pattern characteristic of active larvae [aestivating larvae are usually unspotted (Jepson, 1954)].

The inverse relationship between the percentage of immaculate larvae in the population and the water content of the maize is striking (Fig. 1).

Since many authors claim that a chemical change in the composition of the maize stalk induces aestivation, it is important to know how the maize stalk induces aestivation and how it changes at various stages of its growth: during the growing season crude fiber, fat, and nitrogen-free extracts remain fairly constant; only crude protein and water content decrease continuously. After maturity, however, also the fat and nitrogen-free extracts decrease rapidly (Jones and Huston, 1914; Usua, 1973). To investigate whether a decreased protein carbohydrate or water content could induce aestivation, we bred *Chilo zonellus* larvae on artificial diet (Chatterji et al., 1968) where these factors were reduced. It should be noted that all the diets used contained 5.6% proteins and 9.9% carbohydrates (% of dry weight) originating from the wheat germs, the quantity of which remained constant in all the experimental diets. Particularly in the absence of casein, we were able to increase the larval life-span, but also the elimination of glucose retarded pupation. Surprisingly enough a lower water content did not increase the length of larval life (Table 1).

TABLE 1

Time (days) required for pupation of 1st instar
Chilo zonellus *larvae on diets containing decreased amounts of casein, glucose and/or water* ($\pm$S.D.)

	Females	Males
Normal diet (control)	26.6$\pm$ 4.1	23.4$\pm$5.0
Without casein	47.6$\pm$ 9.0	35.5$\pm$9.4
Without glucose	34.7$\pm$10.6	28.0$\pm$1.8
Reduced water content (75%)	28.7$\pm$ 2.3	26.5$\pm$3.1
Without casein, glucose, reduced water content (75%)	54.3$\pm$ 7.2	34.5$\pm$9.7

CLIMATE AS AESTIVATION-INDUCING FACTOR

No evidence could be found that photoperiod, air temperature or relative humidity are involved in the aestivation induction. The first aestivating larvae in the field could be found at maximum/minimum air temperatures of 32 °C/24 °C but also of 28 °C/21 °C. Relative humidity could be 80% as well as 65%. The factor which, however, has an effect is rainfall: it seems to prevent and terminate the diapause. A factor complementary to rainfall is hours of bright sunshine. Laboratory experiments are yet to be carried out.

THE MICROCLIMATE IN THE HOST-PLANT AS
AESTIVATION-INDUCING FACTOR

Even with the same outside temperature, it is very likely that temperatures inside the different types of maize stems are quite different, due to differences in water content (thus evaporation). The temperature in the microclimate of the maize stem increases as the plant dries out and in this way temperature may have an impact on the induction of aestivation of the larva in this stem. To test our hypothesis, we placed 1st, 3rd and 5th instar larvae to different temperature regimes: 25 °C continuously, 35/25 °C, 37/25 °C and 41/25 °C (6 hours day and 18 hours night temperature). The

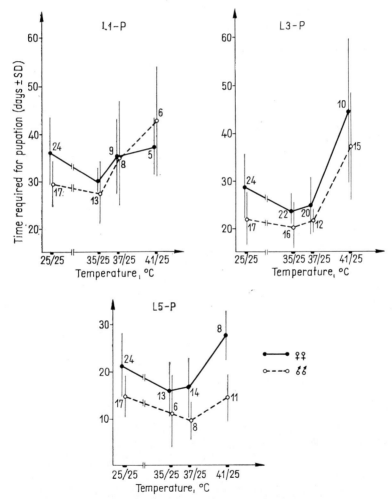

Fig. 2. Influence of temperature on the time required for pupation of 1st, 3rd and 5th instar larvae. Larvae were at 30 °C before exposure to the experimental temperatures. Temperatures have period of 6 h "high" temperature and 18 h "low" temperature. Numbers refer to numbers of larvae which completed development

3rd and 5th instar larvae were initially bred at 30 °C (neutral conditions) (Fig. 2). Initially increase in temperature reduced the time until pupation as one would expect in the development of a poikilotherm; further increase, however, resulted in a longer larval life. A similar result was found with continuous temperatures of 25°, 30°, 33.5° and 37 °C; the most rapid development occurred at 30 °C. In the above mentioned experiments the larvae finally did pupate, perhaps due to the absence of dry conditions, necessary for retaining the state of aestivation.

In a small scale experiment where 5th instar larvae (initially bred at 30 °C on diet) were placed on old (beyond maturity) maize stems at 30°, 33.5° and 37 °C, all 60 individuals at 30° died or pupated within a month, but 4/60 larvae at 33.5 °C and 6/60 larvae at 37 °C assumed the unspotted coloration characteristic of aestivating larvae. The 5th instar larvae used were, however, rather old and we have some evidence that larvae of that age are not particularly sensitive to high temperatures.

DISCUSSION

An extension of the larval life by exposing the larvae to different food or climate conditions is in itself not sufficient proof that those conditions are aestivation inducing. It is for example very well possible that the slower growth rate obtained when larvae were bred on diet without casein and/or glucose, should be explained in terms of deficiencies in the metabolism rather than in terms of diapause. This is quite normal for many insects.

Also the retarding effect of high temperatures on the pupation of larvae cannot immediately be related to aestivation although the occurrence of immaculate larvae in old maize at high temperatures is more meaningful.

In the present research we are trying to see if larvae exposed to the presumed diapause inducing factors are comparable in their morphology, behaviour (particularly their capability to survive under dry conditions) and physiology to the field-collected aestivating larvae.

REFERENCES

BECK, S. D. and HANEC, W. (1960): Diapause in the European corn borer, *Pyrausta nubilalis* (Hübn.). *J. Insect Physiol.* **4**, 304–318.

CHATTERJI, S. M., SIDDIQUI, K. H., PANWAR, V. P. S., SHARMA, G. C. and YOUNG, W. R. (1968): Rearing of the maize stem borer *Chilo zonellus* Swinhoe on artificial diet. *Indian J. Ent.* **30** (1), 8–12.

CHIPPENDALE, G. M. and REDDY, A. S. (1972): Diapause of the Southwestern corn borer, *Diatraea grandiosella*: transition from spotted to immaculate mature larvae. *Ann. ent. Soc. Am.* **65** (4), 882–887.

CHIPPENDALE, G. M. and REDDY, A. S. (1973): Temperature and photoperiodic regulation of diapause of the Southwestern corn borer *Diatraea grandiosella*. *J. Insect. Physiol.* **19**, 1397–1408.

FUKAYA, M. (1967): Physiology of rice stem borers, including hibernation and diapause. In: *The major insect pests of the rice plant*. Proceedings of a Symposium at the International Rice Institute, Sept. 1964, Chapter 8, 213–227. John Hopkins Press.

GOOT, P. VAN DER (1925): Levenswijze en bestrijding van den witten rijstboorder op Java. *Ph. D. Thesis:* Meded. Inst. Plziekten (Buitenzorg) **66**, 308 pp.

HUMMELEN, P. J. (1974): Relations between two rice borers in Surinam, *Rupela albinella* (Cr.) and *Diatraea saccharalis* (F.), and their Hymenopterous larval parasites. *Ph. D. Thesis.* Meded. Landbouwhogeschool, Wageningen **74–1**, 89 pp.

HYNES, H. B. N. (1942): Lepidopterous pests of maize in Trinidad. *Trop. Agriculture, Trin.* **19** (10), 194–202.

INGRAM, W. R. (1958): The lepidopterous stalk borers associated with Gramineae in Uganda. *Bull. ent. Res.* **49**, 367–383.

JEPSON, W. F. (1954): *A critical review of the world literature on the Lepidopterous stalk borers of tropical graminaceous crops.* London, Commonw. Inst. Ent.

JONES, W. J. and HUSTON, H. A. (1914): Composition of maize at various stages of its growth. *Bull. Agr. Exp. Station,* Purdue University **175** (17), 601–629.

KATIYAR, K. P. and LONG, W. H. (1961): Diapause in the sugar cane borer, *Diatraea saccharalis. J. econ. Ent.* **54**, 285–287.

MATHEZ, F. C. (1972): *Chilo partellus* Swinh., *C. orichalcociella* Strand (Lep., Crambidae) and *Sesamia calamistis* HMPS (Lep., Noctuidae) on maize in the Coast Province, Kenya. *Mitt. Schweiz. Ent. Ges.* **45** (4), 267–289.

NYE, I. W. B. (1960): The insect pests of Graminaceous crops in East Africa. *Colon. Res. Stud.* **31**, 48 pp.

SCHMUTTERER, H. (1969): *Pests of crops in Northeast and Central Africa.* Gustav Fischer Verlag, Stuttgart.

SWAINE, G. (1957): The maise or sorghum stalk-borer, *Busseola fusca* (Fuller), in peasant agriculture in Tanganyika Territory. *Bull. ent. Res.* **48**, 711–722.

USUA, E. J. (1973): Induction of diapause in the maise stemborer *Busseola fusca. Ent. exp. & appl.* **16**, 322–328.

Symp. Biol. Hung. 16, pp. 255–259 (1976)

THE INFLUENCE OF PHYTOHORMONES AND GROWTH REGULATING SUBSTANCES ON INSECT DEVELOPMENT PROCESSES

by

S. Scheurer

VVB AGROCHEMIE UND ZWISCHENPRODUKTE,
7101 CUNNERSDORF, DDR

The development of chemical substances to regulate plant bioprocesses has taken a tremendous swing. The use of such products enables man to influence shoot lengths, inhibit or stimulate bud formation, control the process of ageing and leaf abscission and to regulate carbohydrate, nitrogen and protein metabolisms within the plants. It [is plausible that such interference in the bioprocesses of the plant with the aim of achieving a specific result will have consequences upon the relationship of plant and insect. Such changes will come to bear on biocoenoses either for a short term or for a complete vegetation period. The author describes the immediate effects of chemical preparations such as abscissic acid, gibberellin, maleic acid anhydride, phosphone, alar and CCC on insects. It is reported how the changes induced by these substances in the plant affect the bioprocesses of selected insects. This aspect is of special importance as both acceleration and inhibition of development could be observed over and above insecticidal effects.

There are trends which indicate alterations in the active flight behaviour, shortening of multiplication periods and changes in the densities of infestation. The responses shown by insects to "artificially process-controlled" plant development shed new light on the close interrelation between host-plant, development and multiplication of phytophagous insects.

Research work carried out during the last two decades concerning the bio-processes of plants and animals has produced results which enable us to understand biological phenomena more fully. One part of this wide-scanned range of subjects is the work on phytohormones with the aim of gaining insight into the biosynthesis and the modes of action of these substances. It was found that among the native substances predominantly cytokinines, gibberellines (e.g. GA_3), auxins (e.g. IAA), abscissic acid (ABA) and ethylene participate decisively in complex plant bioprocesses. Germination, cell division, longitudinal growth, sprouting, time of flowering, ageing, leaf shedding, production of side shoots and the changes entailed in the carbohydrate and nitrogen metabolisms of plant tissues and the transport vessels are controlled or, at least influenced, by this endogenous system of regulation. The phytohormones are in their perpetual interaction in many instances the activating and initiating parts of this system. — The development of certain chemical preparations, the so-called plant growth regulators, has made it possible to interfere with this control system of plant bioprocesses or to direct it towards certain ends which is indeed being practiced by regulating the above processes. It is obvious that such purpose-oriented intervention into phenomena of crop plants should involve decisive consequences for the relations existing between plant and insect. The host–parasite relationship may be transformed to provide either better or even reduced developmental conditions for the parasite.

On the other hand, the phytohormones may act upon the insects directly. It has been established (Eidt and Little, 1970) that ABA, with a structure similar to farnesic acid, when injected into the chrysalis of *Tenebrio molitor* slows down its transformation into the fully developed adult. In the manner of ecdyson-like substances, GA_3 shortens the process of maturation of larvae of *Locusta migratoria migratorioides* and *Schistocerca gregaria*. Moreover, if GA_3 is added it is possible to maintain *Aphis* populations on an artificial diet. In the case of *Drosophila* larvae which are injected with GA_3 a buffering effect may be observed in the polytenous chromosomes of the salivary glands as well as retarded larval development; the number of adults emerged from the puparia was significantly reduced. It may be supposed that GA_3 has a specific effect upon the activity of the genome and in this way exerts its influence on the normal development of *Drosophila* (Alonso, 1971).

There are many data on direct effects of synthetic growth regulating substances on insects. They show that 2,4,5-T increases the hatching rate of thrips eggs in petri dishes while nothing is known yet on the way the preparation acts. — In larvae of *Spodoptera littoralis*, CCC, when applied in the final larval stage causes the production of sterile males and a reduction in the number of eggs laid by the females. Likewise, the growth regulating substances Alar 85 and phosphon cause sterility in both sexes of *Spodoptera littoralis;* the number of eggs is significantly smaller, their development totally cut off (El-Ibrashy, 1972).

The question, to what extent the changes caused by phytohormones and growth regulating substances in plants may turn out to be of significance for the bioprocesses of certain insects is still more important for ecologists and physiologists. It suggests that plant sucking, eating and mining insects which according to their habits are intimately related to the plant as a biological system should react very sensitively to any irregularity in the metabolism and the duration of the various processes. Taking into special consideration the physiology of nutrition of phloem, xylem and parenchyma feeders, interferences with the biological system "plant" may have drastic consequences on their population change all the more as acute changes in the stream of plant sap are produced.

ABA reduces the synthesis of protein in the leaves thus increasing the proportion of amino acids which, in turn, provides favourable conditions for the feeding of aphids as the content of nutritive sugars is rather high. This appearance is of special significance in autumn before leaves are shed. These results are reflected very well in our tests carried out with *Aphis fabae* Scop.

In our tests parts of three-week-old *Vicia faba* plants consisting of the basal part of the stem and the first primary leaf were used. These plant portions were kept in the solution of the preparation (10^{-2}, 10^{-3}, 10^{-4} per cent by weight; hereinafter the concentration is characterised simply as 10^{-2}, 10^{-3}, 10^{-4}) for three days. After transferring the explantates into a nutrient solution *Aphis fabae* virgins of uniform age were settled (the vapour phase effect was thus excluded). These virgins immediately started to produce larvae on those parts of the plant which had been treated with the preparation. Thus we were enabled to observe their development and multiplication under long-day conditions (16/8) over a period of three weeks. In addition, the mortality rate of the parent insects was established

256

after 24 and 48 hours; afterwards these and their descendants were killed off.

If, however, ABA was used, a marked increase in the size of the V_1 could be observed besides the fact that maturity was reached 3 to 5 days earlier than in the control. Consequently, multiplication activities started earlier. In correspondence with this the rate of multiplication of 7-day-old parent insects was 10 to 16-fold compared to the control, especially in the case of ABA^{-2} and ABA^{-3}. Marked differences were observed even in three-week-old V_1. These produced up to 15 young more than the control insects.

At present preparations are in practical use which, if applied at a specific time, initiate the process of ageing and in this way cause similar physiological reactions in the plant. In this regard we treated certain portions of the plant with retardants and observed not only a growth in the size of the animals but also a shortening of the multiplication period and at the same time an increase of the rate of multiplication. Similar effects may certainly be observed when defoliants are used (Tamaki and Weeks, 1968). Such results may indicate an explanation for certain autumnal outbreaks but at the same time they show that these processes may be accelerated by the use of certain products. However, the same preparation when applied at a much earlier date may delay the process of sprouting (predominantly retardants and other substances; ABA may be excluded for reasons of economy and poor penetrating capacity). In this manner we can modify the various developmental stages of a plant while the insect pest continues its normal development. Thus, by a chemically retarded plant development it is possible to deprive certain insect pests as aphids or tortricids of their natural food reserves.

Auxines and gibberellines acting as phytohormones increase the protein synthesis in the plant; they presumably control the production of amino acids as well as the synthesis of RNA and influence the activation and metabolism of enzymes. Although many informations have been collected on the direct effect of GA_3 on insects (Alonso, 1971; Carlisle et al., 1969; Nation and Robinson, 1966), not much is known yet about the indirect effects on these animals by way of the plant. Our experiments showed that on GA_3-incubated *Vicia* isolates the number of progeny was doubled in comparison to the control up to the 21st day if a concentration of 10^{-2} per cent was used. This "lead" was equalized by the control animals in the days which followed the third week. The lighter colour of the host-plants brought about by GA_3 may have increased the aphid infestation rate. Phenoxy fatty acids represented by 2,4-D and similar compounds (growth regulating herbicides) which are with respect to their action related to the phytohormone auxine (IAA), if applied in the lower concentration range (10^{-2} to 10^{-4}) initially double or even triple the rate of multiplication of *Aphis fabae*. The further course of the multiplication curve showed an adjustment to the control. These results are the more important as the use of such growth regulating herbicides accelerates the decomposition of protein and starch in the plant. Lower-molecular compounds are produced which may have a supporting effect on the multiplication especially of sucking insects.

Cytokinins are phytohormones which delay the decomposition of proteins and chlorophyll in isolated leaves. Amino acids, if any, are consumed in the metabolic cycle. The synthesis of protein is stimulated at the expense of the free amino compounds, sugars are consumed as sources of energy. Intact plants which have been treated with kinetin through their roots respond with stimulated protein synthesis, the free amino acids are diminished, inorganic nitrogen is bound and low-molecular sugars are used as energy sources. Scholze (1971) proved earlier that aphids fed on kinetin-treated plant parts of *Vicia faba* weighed less; in our experiments we observed also that females of *Aphis fabae* which were reared on *Vicia* plants (foliar or root application) produced few larvae or did not even multiply.

Our investigations covered also the reaction shown by aphids to the changes caused in the plant by CCC, MH, BOH and endothal. Although the literature cites many instances (e.g. van Emden, 1969) when the infestation of aphids declined, our trials with CCC^{-3} and CCC^{-4} showed an increase by 50 to 90 per cent in the larval number produced by 7 to 15 days old females. Following this period a considerable adjustment of the multiplication curve to the control could be observed. It should be checked whether plant portions which had received CCC treatment contain more amino acids than uninjured plants.

BOH and MH used in the practice differ widely in their plant mediated effect upon the insects. Whereas in case of BOH^{-2} we observed not only a very distinct ageing and yellowing of the plants but also a visible increase in the multiplication activity of *Aphis fabae* (most probably as a result of the aforementioned processes of physiological decomposition in the plant), the effect of MH was entirely different. Here, the number of descendants of 12-day-old *Aphis fabae* females was 90 per cent below the control figure at a concentration of 5×10^{-3} whereas higher concentrations (10^{-2}, 10^{-3} per cent) turned out to be insecticidal. This effect presumably is based upon the known ability of MH to fix sugar and amino acids. The insecticidal properties of the preparation and its ability to delay the sprouting of trees, thus acting on the insect by way of asynchronism, make this compound very interesting, though toxicologically questionable.

In general, the work carried out with phytohormones and synthetic growth regulating substances opens new ways in recognizing the relations existing between host-plants and insects as the autumnal fluctuations of carbohydrate and nitrogen metabolisms are at least influenced by these substances and can be shifted or even changed by the additional use of certain products. Moreover, the use of growth regulators permits us either to stimulate or to inhibit the development processes of the plant which will result in a disturbed relationship between the host-plant and the pest. It is in the interest of the ecosystem research and of the protection of our environment that whenever new products of this type are developed their effects on the insects by way of the plant should be taken into proper consideration.

REFERENCES

ALONSO, C. (1971): The effects of gibberellic acid upon developmental processes in *Drosophila hydei. Ent. exp. & appl.* **14**, 73–82.

CARLISLE, D. B., ELLIS, P. E. and OSBORNE, D. J. (1969): Effects of plant growth regulators on locusts and cotton stainer bugs. *J. Sci. Food Agric.* **20**, 391–393.

EIDT, D. C. and LITTLE, C. H. A. (1970): Insect control through induced host-insect asynchrony: a progress report. *J. econ. Ent.* **63**, 1966–1968.

EL-IBRASHY, M. T. (1972): Über die Sterilisationsaktivität biologisch aktiver Verbindungen gegen *Spodoptera littoralis* Bois. (Lepidopt., Noctuidae). *Z. ang. Ent.* **71**, 326–332.

EMDEN, H. F. VAN (1969): Plant resistance to *Myzus persicae* induced by a plant regulator and measured by aphid relative growth rate. *Ent. exp. & appl.* **12**, 125–131.

NATION, J. I. and ROBINSON, F. A. (1966): Gibberellic acid: effects of feeding in an artificial diet for honeybees. *Science* **152**, 1765–1766.

SCHOLZE, P. (1971): Untersuchungen zum Einfluß trophischer Faktoren auf die Entwicklung der Blattläuse, speziell der Schwarzen Bohnenlaus, *Aphis fabae* Scop. (Homoptera, Aphidina). *Zool. Jb. Syst.* **98**, 455–510.

TAMAKI, G., WEEKS, R. E. (1968): Use of chemical defoliants on peach trees in integrated program to suppress populations of green peach aphids. *J. econ. Ent.* **61**, 431–435.

Symp. Biol. Hung. 16, pp. 261–266 (1976)

ON THE VARIABILITY OF CHEMOSENSORY INFORMATION

by

L. M. Schoonhoven

DEPARTMENT OF ANIMAL PHYSIOLOGY, AGRICULTURAL UNIVERSITY,
WAGENINGEN, THE NETHERLANDS

When the nerve impulse patterns induced by a natural stimulus, e.g. a plant sap, in the taste sensilla of some caterpillars of the same species are compared a discouraging great variability is usually observed. To explain such variabilities of complex impulse codes, it is essential to know (a) the degree of the variability of individual receptor cells and its origin and (b) whether or not the individual cells react independent of each other.

Three questions in this respect and their answers may be mentioned. 1) Does age affect the sensitivity of a receptor cell? Two receptor cell types, the salicin-sensitive cell as well as the inositol-sensitive cell of *Manduca sexta* show during the fifth larval stage optimal responses between the second and third day. The sensitivity immediately after the fourth moult and before the pupal moult is about 25% lower. 2) Does the composition of the diet affect chemoreceptor sensitivity? The sensitivity to their specific stimuli of both receptor types mentioned above may be depressed by adding these chemicals to the diet. The depression may amount to 40% reduction in impulse frequency under standard conditions. 3) Are receptor sensitivities of different cells in a particular insect correlated with each other or do they behave independently? Since the sensitivity levels appear to vary independently in individual cells, the impulse pattern arising in a small population of different cells when stimulated by a complex stimulus, may vary considerably. Such variations in sensory input to the CNS are probably one of the causes underlying the variation in behavioural responses.

INTRODUCTION

The behavioural physiologist seeks to explain animal behaviour patterns in terms of physiological processes and activities. The nervous system attracts special attention in such analyses. The student of food selection behaviour in phytophagous insects may pose the question whether the different behavioural reactions towards certain foodplants by different insect species are due to the fact that they receive different sensory information from the same plant or whether in different species identical sensory messages are processed in a different way within their central nervous systems and therefore lead to different behavioural reactions. Our present knowledge indicates that such interspecific behavioural differences are based upon peripheral as well as centrally located variations (Dethier, 1972; Schoonhoven, 1972).

When the same question is put with regard to the variation in food acceptance behaviour between various individuals belonging to the same species, quantitative data on both behavioural reactions and chemosensory sensitivities are needed. Combining both sets of data will help to elucidate the relationship between both phenomena.

The present paper describes some quantitative aspects of some taste receptors in lepidopterous larvae. When a natural stimulus, such as a plant sap, is applied to one of the two maxillary sensilla styloconica of a caterpillar, the resulting impulse pattern in the four chemoreceptory cells present, is often very variable when different individuals are compared. To explain such variations of complex impulse codes one would like to know (a) the degree of variability of individual receptor cells and its origin and (b) whether or not the individual cells react independently of each other. These questions are difficult to solve when using complex stimuli, such as plant saps, which activate several cells. Therefore the following experiments make use of chemicals which are known to stimulate specifically one cell in the sensillum.

EXPERIMENTAL METHODS

Electrophysiological procedures were as described previously (Schoonhoven and Dethier, 1966), the only difference being the use of a Bioelectric NF-1 preamplifier and a SE-3006 UV-recorder as the recording system. All impulse frequencies refer to the total number of action potentials arising in a sensillum between 0.2 and 1.2 seconds after the onset of stimulation. All stimuli were applied in standard 0.1 M NaCl solutions. Pure salt solutions at this concentration induce only small spike numbers in the sensilla under investigation. All results are derived from the medial or lateral sensillum styloconicum, which occur on the galea of the maxilla of lepidopterous larvae. These sensilla each contain four different chemoreceptive cells (Schoonhoven and Dethier, 1966). All insects were raised at 25 °C and starved for one hour at this temperature prior to experimentation. Only one determination was done on each sensillum, except when otherwise stated.

RESULTS AND DISCUSSION

A. Effects of age

To test the possibility that age affects the sensitivity levels of chemoreceptors, two cells have been investigated, namely the inositol-sensitive cell in the medial sensillum styloconicum of *Manduca sexta* and the salicin-sensitive cell in its lateral sensillum. Both cells show clearly an increase in sensitivity during the middle of the fifth instar, with a maximal response which is about 25% above the reaction levels immediately after the last larval moult or shortly before the pupal moult (see Figs 1 and 2). Consequently, age should be taken into account when comparing impulse patterns or frequencies in different experimental insects.

B. Effects of feeding history

As has been shown previously (Schoonhoven, 1969) the type of food to which larvae of *M. sexta* have been exposed may drastically affect the sensitivity of various chemoreceptors, with important consequences for their food selection behaviour. Since these experiments were performed with only a general control for age, it seemed appropriate, in view of the results of the foregoing section, to check the effects of some dietary additives on

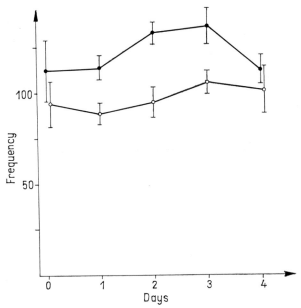

Fig. 1. Impulse frequencies elicited by 0.01 M salicin in the lateral maxillary sensillum styloconicum of *M. sexta* larvae at various time intervals after the last larval moult. Mean frequencies are shown ± 2 S.E. Each dot is based on values obtained from 12–30 different sensilla. Black dots represent larvae kept continuously on a standard diet. Open circles represent insects kept on a diet to which 0.01 M salicin was added

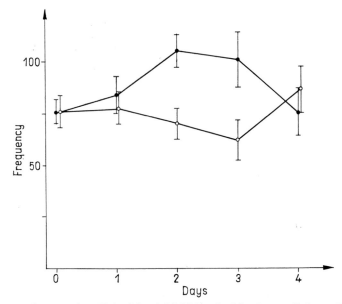

Fig. 2. Impulse frequencies elicited by 0.05 M inositol in the medial maxillary sensillum styloconicum of *M. sexta* larvae at various time intervals of the last larval moult. Mean frequencies are shown ± 2 S.E. Each dot is based on values obtained from 8–49 different sensilla. Black dots represent larvae kept on a standard diet, open circles represent insects which were kept on a diet containing 0.01 M inositol

the sensitivity levels of the chemoreceptors involved, at exact time intervals during the last larval stage. Results on inositol and salicin, which both depress receptor sensitivities considerably in the mid-instar period (but not shortly after the fourth moult or before the pupal moult) are represented in Figs 1 and 2. Thus age as well as the type of food may alter sensory responses. Dietary effects have not been found in some other caterpillar species. The sensitivity level of the "sinigrin-receptor" in *Pieris brassicae* is identical for larvae reared on cabbage and on an artificial diet either with or without $3 \cdot 10^{-3}$ M glucocapparin. The response intensity of the inositol receptors in *Adoxophyes orana* larvae, likewise, is not influenced by the presence or absence of this chemical in the diet. The marked receptor changes in *M. sexta*, of which the possible consequences to feeding behaviour have been discussed elsewhere (Schoonhoven, 1969), and which were also observed by Städler and Hanson (pp. 267–273 of the present volume), thus cannot be considered to represent a widespread phenomenon.

C. Relationship between two symmetrical sensilla

Since the sensilla discussed occur in symmetrical pairs one may wonder whether or not a specific cell in the left maxilla shows a sensitivity level which is related to the sensitivity of its contralateral counterpart. In Table 1 data obtained with different insect species are given, indicating that in some cases the sensitivity of the left hand and right hand side are correlated. No explanation is available at present for the fact that such correlation is only found with regard to inositol and some sugar cells.

TABLE 1

Correlations between receptor sensitivities of maxillary sensilla styloconica and their contralateral counterparts

Insect species	Stimulus	No. of obser- vations	Correlation coefficient	Significance	Avg. impulse frequency + stand. dev.
Manduca sexta	0.01 M salicin	40	0.04	n.s.	99 ± 22.1
M. sexta	0.03 M sucrose	14	0.80	<0.01	47 ± 17.1
M. sexta	0.1 M sucrose	19	0.92	<0.001	65 ± 19.9
M. sexta	0.05 M inositol	40	0.37	<0.05	121 ± 21.7
Philosamia cynthia	0.01 M inositol	19	0.65	<0.01	34 ± 8.1
Adoxophyes orana	0.01 M inositol	19	0.68	<0.01	44 ± 10.5
Pieris brassicae	0.01 M sucrose	15	0.09	n.s.	45 ± 10.4
P. brassicae	0.01 M glucocapparin	37	0.09	n.s.	50 ± 11.7

D. Relationship between cells within a sensillum

It is conceivable that the various chemosensory cells of a sensillum do not vary in their sensitivity levels independently of each other, but rather show concerted fluctuations of their response intensity. For some instances the degree of correlation between different cell types located in the same sensillum has been determined (Table 2). Since no correlation between sensitivity levels of different receptors appears, it is concluded that their response intensities vary randomly and independently of each other.

TABLE 2

Degree of correlation of sensitivity levels of different chemoreceptors located in the same sensillum

Insect species	Receptor types	No. of observations	Correlation coefficient	Significance
M. sexta	salicin and sucrose	24	0.07	n.s.
P. brassicae	sinigrin and proline	31	0.15	n.s.
P. brassicae	sinigrin and sucrose	31	0.11	n.s.

E. Correlation of receptor sensitivity and behaviour

Forty larvae of *M. sexta* were individually offered elder pith discs saturated with 0.005 M, 0.01 M and 0.05 M salicin solutions in a triple choice situation. When each larva had eaten 50% of the substrate with the lowest concentration of salicin, the amount eaten from the higher concentrations was determined (Jermy et al., 1968). The salicin receptor sensitivity was tested in eight larvae which had eaten only 3.5% of the 0.05 M salicin elder pith and in six larvae which on the average had eaten 49.8% of the 0.05 M salicin substrate. One day later (to neutralize possible receptor changes of the receptor sensitivity by the salicin) the insects were tested for their receptor sensitivity. The average action potential frequency, elicited by 0.01 M salicin between 1.0 and 2.0 seconds after the onset of the stimulus was 97 impulses/second (S.E. 16.7) in the former group of larvae, but only 67 impulses/second (S.E. 11.6) in the latter group. This result strongly indicates that the larvae tolerating a relatively high level of salicin in their diet did so because of a low receptor sensitivity to this compound.

CONCLUSIONS

In physiology, like in any other biological discipline, one is confronted with variations between different individuals. By a careful design of his experiments the physiologist tries to reduce these variations to the minimum and subsequently tends to ignore their existence. Since, however, such variations do occur, their negligence is condemnable. We rather should, after determining their extent, ascertain the consequences of the variability of the phenomenon studied to the processes related with it. In our case this means that the natural variability of the sensitivity of insect chemoreceptors is related to behavioural activities.

The two pairs of maxillary styloconica of lepidopterous larvae, which play a major role in foodplant recognition (Schoonhoven and Dethier, 1966) seem an ideal object for quantitative chemoreceptor studies, since in contrast to the situation with large populations of chemoreceptors in other insects, a particular chemoreceptive cell is easily located in different individuals.

Complex chemical stimuli, which plant leaf contents undoubtedly are, may elicit fairly different action potential patterns in different individuals or even in the two identical contralateral sensilla of the same individual. However, in view of the considerable variation of sensitivity of each receptor and the fact that, as described in the foregoing paragraphs, the various cells may fluctuate independently of each other, one may expect to obtain fairly variable spike patterns in different insects.

Striking differences exist between individuals of the same species with regard to food acceptance, as has been observed by various authors (e.g. Merz, 1959). Great individual variations in food uptake are also shown when insects are exposed to simple chemical stimuli. One example may be mentioned. When hungry larvae of *P. brassicae* grown on an artificial diet are exposed to a suboptimal feeding substrate, consisting of 4% agar, 4% cellulose and 0.006 M sucrose, the quantity of fecal pellets produced amounts to 7.8 mg (dry weight) per larva per day (mean of 18 larvae), with a standard deviation as high as 5.6.

In view of the considerable variability of individual receptors (see standard deviations of spike frequencies in Table 1) it seems likely that the above mentioned behavioural variations are due, at least partly, to variations in the sensory input. This idea is supported by the results of an experiment in which the behavioural aversion towards salicin in *M. sexta* larvae could be correlated with an above average sensitivity to this substance.

In conclusion it appears that the natural variability of chemoreceptor sensitivity present in insect populations and in some cases modified by experience or age, seems to be one of the causes of the variability in feeding behaviour.

REFERENCES

DETHIER, V. G. (1972): Electrophysiological studies of gustation in lepidopterous larvae. II. Taste spectra in relation to foodplant discrimination. *J. comp. Physiol.* **82**, 103–134.

JERMY, T., HANSON, F. E. and DETHIER, V. G. (1968): Induction of specific food preference in lepidopterous larvae. *Ent. exp. & appl.* **11**, 211–230.

MERZ, E. (1959): Pflanzen und Raupen, über einige Prinzipien der Futterwahl bei Großschmetterlingsraupen. *Biol. Zentr.* **78**, 152–188.

SCHOONHOVEN, L. M. (1969): Sensitivity changes in some insect chemoreceptors and their effect on food selection behaviour. *Proc. K. Ned. Akad. Wetensch. Amsterdam*, (C) **72**, 491–498.

SCHOONHOVEN, L. M. (1972): Plant recognition by lepidopterous larvae. In: *Insect/plant relationships* (Ed. H. F. van Emden). *Proc. Roy. Ent. Soc. Lond.* **6**, 87–99.

SCHOONHOVEN, L. M. and DETHIER, V. G. (1966): Sensory aspects of host-plant discrimination by lepidopterous larvae. *Arch. Néerl. Zool.* **16**, 497–530.

Symp. Biol. Hung. 16, pp. 267–273 (1976)

INFLUENCE OF INDUCTION OF HOST PREFERENCE ON CHEMORECEPTION OF *MANDUCA SEXTA:* BEHAVIORAL AND ELECTROPHYSIOLOGICAL STUDIES

by

E. STÄDLER* and F. E. HANSON

DEPT, OF BIOLOGICAL SCIENCES, UNIVERSITY OF MARYLAND,
CATONSVILLE, MD. 21228, USA

Glass fiber filter discs impregnated with leaf extracts of *Lycopersicon esculentum* (tomato) and *Solanum pseudocapsicum* were biossayed with tobacco hornworm (*Manduca sexta*) which had been reared on either of the two hostplants. The choice behavior was compared with that on leaves and it could be concluded that nonpolar compounds (soluble in hexane) of low volatility must be mainly responsible for host selection by induced larvae. Polar extracts were stimulating too but, like physical factors, seemed not to be important for discrimination between the two plants. It is likely that water soluble deterrents contribute also to this induced host preference.

Tip and sidewall recordings from the maxillary sensilla styloconica were performed and analysed by total spike counts of the first second of stimulation. Evidences were found for a change of the sensitivity of the lateral sensillum due to the prior rearing on either tomato or *S. pseudocapsicum*. In the medial sensillum no such effect was observed, but regardless of the induced food preference of the larvae, the sensory response to stimulation of tomato was different from that of *S. pseudocapsicum*.

INTRODUCTION

The induction of a host preference in the larvae of *Manduca sexta* as a result of prior feeding experience has been observed first by Yamamoto and Fraenkel (1960). Later Jermy et al. (1968) studied and verified this phenomenon extensively in *Heliothis zea* and *Manduca sexta*. Schoonhoven (1969) found changes of sensitivity in the sensilla styloconica on the maxillae of *Manduca sexta* due to feeding on artificial diets which could explain the change of feeding behavior observed earlier (Schoonhoven, 1967). Hanson and Dethier (1973) studied consecutively the role of the different chemoreceptors in the host selection of induced larvae. The authors concluded that the maxillary sensilla styloconica as well as the maxillary palpae and antennae must be involved in the induction of preference and host selection.

The investigation of the involvement of plant chemicals in the induction of preference can be attempted by studying either their influence during the feeding and induction period or the subsequent host selection behavior. Only the second approach was successful, probably because artificial diets, in which plant compounds could be incorporated, have been shown to be not "neutral" for the larvae (Städler and Hanson, in preparation). The electrophysiological study was aimed at finding the sensory correlation to the induction of preference shown in the host selection behavior.

* Present address: Swiss Federal Research Station, CH–8820 Wädenswil/Switzerland.

MATERIALS AND METHODS

Insects: Tobacco hornworms (*Manduca sexta* Johanssen) were reared from eggs supplied by Dr. J. N. Kaplanis (USDA Beltsville) on cut leavet of either tomato (*Lycopersicon esculentum*) or *Solanum pseudocapsicum*.*

Extraction: Fresh leaves were extracted with the solvents in a blender. For freeze drying the leaves were ground at 4 °C, frozen and then transferred to a commercial drier. The obtained leaf powder was extracted in a soxhlet apparatus. For fractionation the MeOH extract of tomato was concentrated by distillation under vacuum. The residue was extracted first by hexane and the residue dissolved completely in diethyl-ether and distilled H_2O. The ether was separated from the H_2O and dried over anhydrous Na_2SO_4. The residue of the hexane extraction of freeze dried tomato and *Solanum* leaves was again extracted in the soxhlet apparatus with diethyl-ether and later with distilled H_2O.

Behavioral test: Basically the same method as described by Jermy et al. (1968) was used. 0.1 ml of the extracts or of the solvent (control) were applied on each glass fiber filter paper (Whatman GFA) disc (diam. = 17 mm) and the solvent (except water) allowed to evaporate. The use of glass filter paper has the following advantages over other feeding substrates: it is chemically pure, a good absorbent for all solvents, readily eaten by the larvae when wet (0.1 ml H_2O per disc) and not toxic. The test was terminated at the time when the animal has eaten approximately 50% of the discs containing one extract (t_{50}). The proportion eaten of each of the three kinds of discs was then estimated (for further description see Jermy et al., 1968). The obtained percent values (sum of 4 discs) were transformed (arc sin x) and analysed by an analysis of variance or t-test.

Electrophysiology: Isolated heads with a neck ligature were mounted on the indifferent electrode. Tip recordings were performed as described by Schoonhoven and Dethier (1966); the sidewall recording technique using tungsten electrodes was developed by Hanson (1970) and eliminates most of the loss of data due to contact artifacts. Plant saps for stimulation were prepared with a kitchen juicer at 4 °C, centrifuged in the cold and stored in the refrigerator. For the sidewall-recordings freshly cut small leaf pieces were mounted on the tip of a glass electrode filled with 0.05 M NaCl.

RESULTS AND DISCUSSION

Behavioral study

Table 1 shows the results of a first attempt to extract and characterize the feeding stimulants from tomato leaves. The whole extract as well as its fractions contain feeding stimulants which can be tested successfully with the bioassay using glass filter paper as a neutral support. Obviously no stimulants were lost during the concentration of the methanol extract because no activity was found in the collected distilled MeOH. It seems therefore that very volatile compounds of the leaves are not stimulating. Further fractionation of the methanol extract also proved interesting. The

* In this paper referred as "*Solanum*".

TABLE 1

Feeding stimulation by a methanol-extract
and its fractions of fresh tomato leaves
(Larvae raised [induced] on tomato leaves)

Choice	N	% eaten at $t_{5\bullet}$
MeOH-extract Control	21	48.0 *** 13.4
MeOH distilled Control	16	47.3 0 47.4
Hexane-fraction Control	22	49.9 *** 25.2
Et$_2$O-fraction Control	9	49.5 *** 23.2
H$_2$O-fraction Control	22	48.1 * 41.6

*** $P < 0.001$.
* $P < 0.05$.
0 $P > 0.05$.

major fraction of the chemicals responsible for feeding stimulation are either non-polar (Table 1: hexane fraction) or weakly polar (Table 1: ether fraction), since the animals clearly selected these fractions in preference to the control ($P < 0.01$). Polar compounds, which include sugars and amino acids, yielded surprisingly little stimulation ($P < 0.05$), indicating these are either not as important stimulants as the nonpolar, and slightly polar compounds soluble in hexane and ether, or else their effect is masked by deterrents.

These results are interesting for their contribution to the knowledge of the chemical basis of host selection. They are partially similar to those reported about other Lepidoptera larvae (reviewed by Ma, 1972; Schoonhoven, 1972). We subsequently investigated whether these plant extracts could be responsible for the host preference of induced larvae. For this study we raised (induced) larvae on one of two host plants, either tomato or *Solanum*. The response of hornworms induced on tomato or *Solanum* to leaves and extracts of tomato and *Solanum* and control is summarized in Fig. 1. The test of leaves (with a control to make it compatible with the extract tests) is a repetition of the experiments reported by Jermy et al. (1968) and Hanson and Dethier (1973), and confirmed the induction of preference in *Manduca sexta;* namely that larvae raised on a specific plant prefer it later in a choice test over other plants within its host range.

Feeding choice tests using methanol and hexane extracts were similar (not significantly different) to the feeding choice tests where leaves themselves were used. The discrimination between tomato and *Solanum* leaves seems to be caused only by chemical factors. Even if not all the significant compounds responsible for the specific food choice could be extracted, methanol and hexane extracts must contain the important substances which cause the selective behavior of induced larvae. (It is noted that the control is always eaten to a greater proportion in the tests using extracts than those in tests using leaves. Perhaps this could be explained by the preference of the larvae for the physical properties of leaves.)

Diethyl-ether and water extracts resulted in food choices completely different from the above. Ether extracts of both tomato and *Solanum* contained stimulants but the larvae did not discriminate between them. The

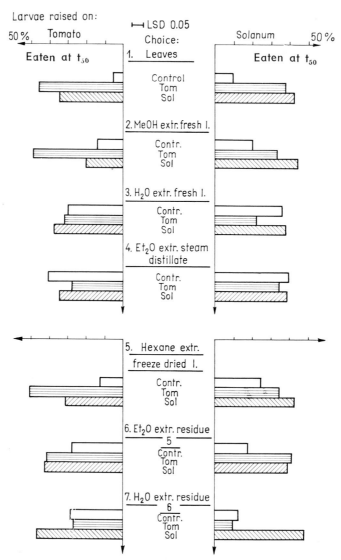

Fig. 1. Leaves compared with different leaf extracts as factors in food choices of larvae induced on two host plants. Tom = tomato (*Lycopersicon esculentum*); Sol = *Solanum pseudocapsicum*

two water extracts stimulated feeding, but only with *Solanum*. Feeding on H_2O extracts of tomato were similar to or *less* than the controls (deterrent effect). This is surprising but in accordance with the low activity in the earlier extraction procedure (Table 1). It seems likely that the exclusive inactivity of the tomato water extracts is mainly the result of deterrents which might be more potent than in *Solanum*. This is further suggested by the fact that this deterrent effect of water extracts of tomato is significantly stronger in *Solanum* reared animals. The proposed water soluble deterrents could contribute also (in the negative sense) to the preference behavior of induced larvae by adaptation of the larvae to deterrents occurring in their host plants. Such an effect of deterrents in host selection has already been proposed by Jermy (1966) and Schoonhoven (1969).

Further information is provided by the negative results of the diethylether extracts of steam distillates. The chemicals responsible for the host selection of induced larvae are not steam distillable and therefore not very volatile. On the basis of these data (Fig. 1) and from the results in Table 1, we conclude that the main chemicals involved in inducing preferences are non-volatile, non-polar compounds, possibly lipids. These chemicals also appear to be involved in host selection *per se*, but slightly polar and, to some extent, polar compounds contribute as well.

Electrophysiology

The total counts of action potentials from chemoreceptors in the maxillary sensilla styloconica in the first second of stimulation (Fig. 2) suggest a possible sensory basis of induction of preference. In the lateral sensilla we found a significant difference in the overall response between receptors of tomato and *Solanum* raised animals. More spikes were elicited during stimulation with the plant with which the larvae had no contact during rearing. It is likely that the "deterrent receptor" described by Schoonhoven

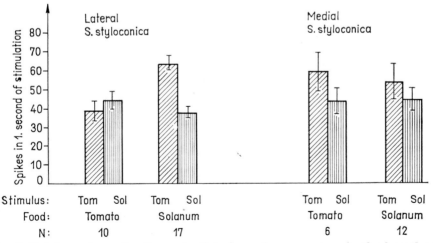

Fig. 2. Total numbers of action potentials from chemoreceptors in the lateral and medial sensilla styloconica after stimulation with leaf extracts or leaves

and Dethier (1966) is involved in this response. Basically the same effect was observed by Schoonhoven (1969) comparing the response of the sensilla styloconica of larvae which were raised either on tomato leaves or on an artificial medium.

In contrast to Schoonhoven's results we found no such difference in the medial sensilla. The relationship of the responses to tomato and *Solanum* stimulation from these sensilla were the same regardless of the prior feeding experience. Thus tomato elicited significantly more spikes than did *Solanum*. This means first that the sensory information of the medial sensilla would be sufficient to allow the animals to discriminate between the two plants, and second that the induction of preference had no obvious effect (across-fiber patterning [Dethier, 1973] not considered) on the response of the chemoreceptors in these sensilla. Our preliminary investigation supports therefore the assumption that the induction of preference involves probably both central and peripheral modifications of the nervous system.

Future electrophysiological studies should be concentrated on the side-wall recording technique. This method not only improves the possibility to discriminate between different spikes of the sensory cells but also allows the analysis of the response prior to and after stimulus contact. This has been proven to be important for the understanding of some aspects of chemoreception in these sensilla styloconica (Städler and Hanson, in preparation). Further studies of host selection, induction of preference, and chemoreception using pure compounds and meridic artificial diets are in progress. These investigations, along with a detailed chemical analysis of the natural feeding stimuli of the plants, may further our understanding of host selection in lepidopterous larvae.

SUMMARY

The physiological basis for host discrimination by the tobacco hornworm (*Manduca sexta*) was studied using techniques adapted from Jermy et al. (1968). Larvae reared on two different host plants (*Lycopersicon esculentum* [tomato] and *Solanum pseudocapsicum*) scored differently in preference tests with leaf discs of the two host-plants as well as discs of an artificial substrate (glass fiber filter paper) impregnated with extracts of these plants. The food choice behavior elicited by the various extracts was compared with that on leaves and it could be concluded that nonpolar compounds (soluble in hexane) of low volatility must be mainly responsible for host selection by induced larvae. Polar extracts were stimulating too but, like physical factors, seemed not to be important for discrimination between the two plants. Water soluble deterrents may also contribute to this induced host preference.

Electrophysiological recordings from the maxillary sensilla styloconica were performed and analysed by total spike counts of the first second of stimulation. Evidences were found for a change of the sensitivity of the lateral sensillum due to the prior rearing on either tomato or *Solanum*. In the medial sensillum no such effect was observed, but regardless of the prior feeding experience of the larvae, the sensory response to stimulation of tomato was different from that of *Solanum*.

REFERENCES

DETHIER, V. G. (1973): Electrophysiological studies of gustation in lepidopterous larvae. II. Taste spectra in relation to food-plant discrimination. *J. comp. Physiol.* **82**, 103–134.

HANSON, F. E. (1970): Sensory responses of phytophagous Lepidoptera to chemical and tactile stimuli, 81–91. In: Wood, D. L., Silverstein, R. M. and Nakajima, M. (eds.): *Control of insect behavior by natural products.* Academic Press, New York.

HANSON, F. E. and DETHIER, V. G. (1973): Role of gustation and olfaction in food plant discrimination in the tobacco hornworm, *Manduca sexta. J. Insect Physiol.* **19**, 1019–1034.

JERMY, T. (1966): Feeding inhibitors and food preference in chewing phytophagous insects. *Ent. exp. & appl.* **9**, 1–12.

JERMY, T., HANSON, F. E. and DETHIER, V. G. (1968): Induction of specific food preference in Lepidopterous larvae. *Ent. exp. & appl.* **11**, 211–230.

MA, WEI-CHUN (1972): Dynamics of feeding responses in *Pieris brassicae* Linn. as a function of chemosensory input: A behavioural, ultra-structural and electro-physiological study. *Meded. Landbouw. Wageningen,* **72–11**, 1–162.

SCHOONHOVEN, L. M. (1967): Loss of host plant specificity by *Manduca sexta* after rearing on an artificial diet. *Ent. exp. & appl.* **10**, 270–272.

SCHOONHOVEN, L. M. (1969): Sensitivity changes in some insect chemoreceptors and their effect on food selection behaviour. *Proc. K. ned. Akad. Wet.* (C) **72**, 491–498.

SCHOONHOVEN, L. M. (1972): Plant recognition by lepidopterous larvae. In: van Emden, H. F. (ed.): *Insect/plant relationships.* Blackwell Scientific Publications, Oxford, 87–99.

SCHOONHOVEN, L. M. and DETHIER, V. G. (1966): Sensory aspects of host-plant discrimination by lepidopterous larvae. *Arch. Néerl. Zool.* **16**, 497–530.

YAMAMOTO, R. T. (1974): Induction of host plant specificity in the tobacco horn-worm, *Manduca sexta. J. Insect Physiol.* **20**, 641–650.

YAMAMOTO, R. T. and FRAENKEL, G. (1960): The physiological basis for the selection of plants for egg-laying in the tobacco hornworm, *Protoparce sexta* (Johan.). *Proc. 11th Int. Cong. Ent.* **3**, 127–133.

NOTE ADDED IN PROOF

Yamamoto (1974) confirmed the induction of preference in *Manduca sexta* due to prior feeding on Solanacea (tomato) and to some extent on plants belonging to other taxa. The author found further that more tomato-reared larvae chose a diet with a purified extract of tomato than did diet-reared larvae.

18

Symp. Biol. Hung. 16, pp. 275–281 (1976)

THE EFFECT OF THE AMPUTATION OF HEAD APPENDAGES ON THE OVIPOSITION OF THE BEAN WEEVIL, *ACANTHOSCELIDES OBTECTUS* SAY (COLEOPTERA: BRUCHIDAE)

by

Á. Szentesi

RESEARCH INSTITUTE FOR PLANT PROTECTION, H–1525 BUDAPEST, PF. 102, HUNGARY

The oviposition behaviour of amputated bean weevil females were investigated, providing different oviposition stimuli (chemical and tactile). All the head appendages have chemosensory functions, however, the maxillary palpi seem to be most significant. The ovipositor is also able to perceive chemical stimuli. Its preliminary morphological examination showed different types of hairs on it, some of them with possible chemosensory functions.

INTRODUCTION

The selection and recognition of the oviposition site and the induction of oviposition in most insects is the result of different reflex mechanisms in which chemical and tactile stimuli play an important role. Different oviposition-behavioural patterns can be observed among insect species regarding the importance of head appendages and/or the ovipositor, in selecting the oviposition site.

In some insects the choosing of the oviposition site is a function of head appendages only, and the misfunction or removal of the antennae will result inadequate responses in oviposition as it has been shown in *Manduca* by Yamamoto and Fraenkel (1960). It is probable that if there are any chemoreceptors on the tip of the abdomen they do not play any role in the oviposition behaviour in *Manduca*.

However, many other insects need the simultaneous performance of the antennae, palpi and the ovipositor during this process. Stimuli, received by the ovipositor, will modify the information entering the central nervous system throughout the sensors of head appendages. There are some good examples in parasitic Hymenoptera, where the role of the ovipositor is equally important with that of the antennae. The localization of the host is the function of the antennae, but egg laying is evoked by stimuli perceived by the ovipositor on the tip of which different types of chemoreceptors were found (van Lenteren, 1972; Hawke et al., 1973). Scanning electron microscope examination showed the presence of chemoreceptors on the ovipositor of *Musca autumnalis*, too (Hooper et al., 1972).

Salama and Ata (1972) refer to the oviposition site selection of *Culex pipiens*, where neither the head appendages, nor the ovipositor play any significant role, because their removal did not cause any loss in sensitivity. It is probable that receptors on the legs guide the selection of oviposition site.

There are several stimuli originating from the host-plant and influencing oviposition in the bean weevil. It has been shown by Labeyrie (1961)

18*

that there are two strains in respect to oviposition response. One of them lays eggs only in the presence of bean. Pouzat (1970) examining the role of head appendages in the oviposition of the bean weevil found the maxillary palpi as the most important ones in affecting egg laying. When the antennae were amputated, there was a high oviposition response. This was explained by the author as the release of oviposition from an inhibitory state. However, Nakamura (1971) did not observe any significant effect on egg laying after cutting off the antennae of the females in two *Callosobruchus* species.

Not only contact, but air-borne stimuli originating from the host-plant elicit searching, locomotion, attraction toward the direction of the odour source in inseminated bean weevil females (Halstead, 1973; Pouzat, 1974), while in a choice experiment using dry bean odour as stimulus and glass beads as substrate for oviposition egg laying occurred anywhere randomly (Jermy, unpublished).

Besides chemical stimuli from the bean (pod, seed) tactile stimuli are supposed to be also important. Inseminated females of *Ceutorrhynchus maculaalba* will not lay eggs on the flattened strips of poppy capsule, but there is an immediate ovipositional response if the strips are bent (Sáringer, 1975, in this volume). Similar behaviour was shown in pea weevil (*Bruchus pisorum*) where females preferred laying eggs on normal pea pods to flattened ones (Jermy, 1972). Shape sensation is supposed to be an important element in the complex oviposition behaviour in the bean weevil, too. The perception of shape stimuli is probably a function of antennae or palpi, or some proprioreceptors possibly localized on the legs.

In order to analyze the importance of different stimuli and to find out what sort of receptors are involved and where they are localized in the bean weevil in relation to oviposition behaviour, amputation and oviposition inhibition experiments as well as SEM examinations have been carried out.

MATERIAL AND METHODS

Amputation experiments

One to three days old bean weevil females originating from the laboratory mass rearing (Szentesi, 1972) and immobilized at low temperature ($+0.5$ to $+4\,°C$) were amputated to different degrees. After cutting off the antennae or the palpi or all of them, the females were kept individually in 6 ml glass vials with one male in each until their deaths at $23\,°C$. In the vials: (1) white dry beans (complete oviposition stimuli), (2) glass beads, 5 mm in diameter (tactile stimuli only), and (3) ground dry bean pods (chemical stimuli only), respectively, were provided as oviposition sites. The replicates of the series of amputations were 50. After the death of all adults the number of eggs was counted.

Oviposition inhibition in normal and amputated females

White dry beans were dipped into 0.05 M $CuSO_4$ solution for a few minutes then taken out and dried immediately in warm airstream. Dry beans treated with water in the same way served as control. The bottoms of

10 cm petri dishes were divided into four sections by paraffin wax walls: two sections for the untreated and two for treated beans. Equal number of bean seeds was placed into each petri dish. There were 5 replicates in each of the following treatments: (1) females without amputation (control), (2) females having only maxillary palpi, (3) totally amputated females. After 20 days the number of eggs laid into the treated and untreated sections was counted. The experiment was conducted at 23 °C, and at about 50% rel. hum.

Morphological investigations on the ovipositor of female
bean weevil using scanning electron microscope

The ovipositor of freshly emerged virgin bean weevil females was used. By slightly pressing the abdomen of a female, the ovipositor protruded and was cut off, dried in desiccator above conc. sulphuric acid and prepared for SEM examination. Ovipositors were stuck to a specimen holder with conductive silver paint and coated with gold in high vacuum evaporator. The examinations were carried out by a JEOL JSM 50A SE microscope at 20 kV.

RESULTS AND DISCUSSION

Amputation experiments

Sandner and Pankanin (1973) observed an increasing rate of oviposition in normal, inseminated bean weevil females when introducing them to gradually increasing amounts of dry beans. On the contrary, there was a high difference in the number of eggs laid if only indifferent stimuli (empty box, glass beads) were present.

In our experiments the oviposition behaviour of normal and amputated females showed unambiguous preference to bean seeds (Fig. 1). The number of eggs laid was the highest, because the stimuli provided were complete and complex. The ovipositional response was weaker, though the difference was not significant, if only one of the stimuli was provided. In case of normal, inseminated females the importance of shape and chemical stimuli, respectively, seems to be equally necessary and important. There was a considerable number of eggs laid even without any oviposition stimuli present. This finding may refer to the possible occurrence of individuals ovipositing in the absence of bean (Labeyrie, 1961).

Any other combination of the amputation gave results similar to one another, and the most preferred oviposition site was the dry bean. However, the tendency of oviposition showed that the maxillary palpi had the most important role in chemical sensation, while the antennae may function both in chemical and shape recognition, though to a smaller extent. These results are in good accordance with those of Pouzat (1970).

There were some ovipositional responses in the case of total amputation. We concluded that both spontaneity in oviposition and the role of ovipositor were involved in this reaction.

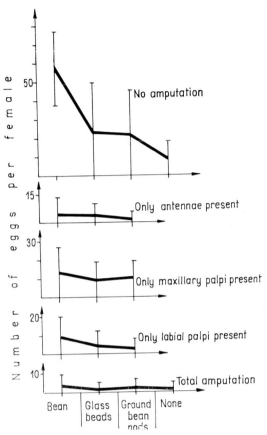

Fig. 1. Ovipositional responses given by normal and amputated bean weevil females
in the presence or absence of different egg-laying stimuli

Oviposition inhibition experiment

In order to clear the importance of the ovipositor in the egg-laying behaviour, attempts were made to examine its role alone. Head appendages were cut off in different degrees while the ovipositor always remained intact. $CuSO_4$ has a definite and strong inhibitory effect on oviposition, disturbing the perception of the natural stimulating effect of dry bean. It was supposed that after a total amputation there would be no perception of stimuli either positive or inhibitory unless the ovipositor had chemoreceptors. According to the results (Table 1) amputated females could distinguish treated and untreated surfaces, and this fact indicated the presence of chemoreceptors on the tip of the ovipositor.

Morphological examinations on the ovipositor

By morphology the bean weevil ovipositor is a dorsoventrally flat organ. Its tip bears a great number of hairs of different lengths (Fig. 2). Pre-

TABLE 1

Oviposition preference of amputated bean weevil females

Degree of amputation	Percentage of eggs laid on		S.D.
	untreated	0.05 M CuSO$_4$ treated	
	beans		
No amputation	90.48	9.51	1.5[a]
Antennae + labial palpi cut off (only maxillary palpi present)	70.96	29.05	12.0[b]
Total amputation	70.42	29.57	11.5[c]

a−b: P < 0.1%.
a−c: P < 0.1%.
b−c: N.S.

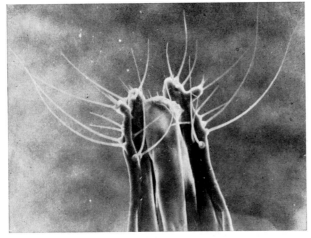

Fig. 2. General view of the tip of the ovipositor of *Acanthoscelides obtectus* Say showing different types of hairs. SEM photo (×300)

liminary investigations showed that most of them are long mechanoreceptor hairs, greatly varying in length and with probable tactile functions. They may play an important role in sensing distance between adjacent dry beans or the position and size of small holes gnawed on dry bean pods. Their functions may be responsible for the spatial distribution of eggs laid.

Besides the mechanoreceptors, scattered among them, there are shorter hairs with the shape of sensilla chaetica (Flower and Helson, 1971; Schafer, 1971; Steinbrecht and Müller, 1971).

There are two small warts on the lateral-distal parts of the ovipositor (Fig. 3). There are three or four (?) hairs raising from each. Two of them, though varying in size, must be a pair of long tactile hairs, the others rooting side by side seem to be a pair of sensilla chaetica type chemoreceptors.

Hairs distributed around the warts and in general on the surface of the ovipositor having different length are supposed to have similar functions.

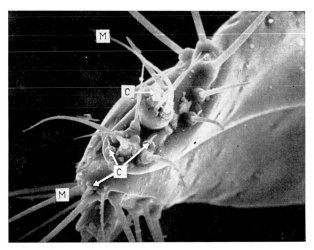

Fig. 3. The lateral-distal part of the bean weevil's ovipositor. One of the warts can be seen holding hairs thought to be mechanoreceptors (M) and chemoreceptors (C). SEM photo (×1000)

CONCLUSIONS

Studying the oviposition behaviour of the bean weevil it was proved:

1. Chemical and tactile (shape and mechanical) stimuli play a decisive role in choosing the oviposition site. Untreated females show the strongest oviposition response.

2. Dry beans are the most preferred oviposition site. There were fewer eggs laid if only chemical or only shape (tactile) stimuli were supplied.

3. In spite of the importance of the antennae and palpi in the recognition of the oviposition site, the role of the maxillary palpi seems to be the most significant.

4. In the oviposition inhibition experiment combined with the amputation of head appendages, it was proved that there must be chemoreceptor(s) on the tip of the ovipositor. These receptors take part in selecting the oviposition site. The oviposition behaviour is the result of a physiological state influenced, among others, by information given by the receptors on the head appendages and the ovipositor in the bean weevil.

5. Preliminary morphological investigations showed the presence of chemosensory sensilla on the ovipositor, however, further histological and physiological experiments are needed to clarify their exact functions.

ACKNOWLEDGEMENTS

I am indebted to Dr. T. Jermy (Director of the Research Institute for Plant Protection, Budapest) for his helpful criticism of the manuscript and to Dr. M. Sass (Department of General Zoology, Eötvös Loránd University, Budapest) for performing the SEM examinations.

280

REFERENCES

FLOWER, N. E. and HELSON, G. A. H. (1971): The structure of sensors on the antennae and proboscis of *Heliothis armigera conferta* Hubn. *New Zeal. J. Sci.* **14** (4), 810–815.

HALSTEAD, D. G. H. (1973): Preliminary biological studies on the pheromone produced by male *Acanthoscelides obtectus* (Say) (Coleoptera, Bruchidae). *J. stored Prod. Res.* **9**, 109–117.

HAWKE, S. D., FARLEY, R. D. and GREANY, P. D. (1973): The fine structure of sense organs in the ovipositor of the parasitic wasp, *Orgilus lepidus* Muesebeck. *Tissue and Cell*, **5** (1), 171–184

HOOPER, R. L., PITTS, C. W. and WESTFALL, J. A. (1972): Sense organs on the ovipositor of the face fly, *Musca autumnalis. Ann. ent. Soc. Am.* **65** (3), 577–586.

JERMY, T. (1972): A növényevő rovarok táplálékspecializációjának etiológiája (Behavioural aspects of host selection in phytophagous insects) (MS).

LABEYRIE, V. (1961): Longevité et capacité reproductrice de lignées d'*Acanthoscelides obtectus* selectionnées en fonction de la réponse aux stimuli de ponte. *C. R. Soc. Biol.* **155** (6), 1366–1369.

VAN LENTEREN, J. C. (1972): Contact-chemoreceptors on the ovipositor of *Pseudeucoila bochei* Weld (Cynipidae). *Neth. J. Zool.* **22** (3), 347–350.

NAKAMURA, H. (1971): Effect of the amputation of female's antennae on the oviposition in *Callosobruchus chinensis* and *C. sp. (C. rhodesianus?). Jap. J. Ecol.* **21** (3–4), 167–169.

POUZAT, J. (1970): Rôle des organes sensoriels céphaliques dans l'ovogenèse et l'émission chez la Bruche du Haricot *Acanthoscelides obtectus* Say. In: *L'influence des stimuli externes sur la gamétogenèse des insectes.* Colloque International du C.N.R.S. **189**, 381–400.

POUZAT, J. (1974): Comportement de la Bruche du Haricot femelle, *Acanthoscelides obtectus* Say (Col., Bruch.) soumise à différents stimulus olfactifs: male, plantehôte. *C. R. Acad. Sci. Paris*, **278**/D, 2173–2176.

SALAMA, H. S. and ATA, M. A. (1972): Reactions of mosquitoes to chemicals in their oviposition sites. *Z. ang. Ent.* **71** (1), 53–57.

SANDNER, H. and PANKANIN, M. (1973): Effect of the presence of food on egg-laying by *Acanthoscelides obtectus* (Say) (Coleoptera, Bruchidae). *Polsk. Pismo Ent.* **43**, 811–817.

SÁRINGER, G. (1975): Oviposition behaviour of *Ceutorrynchus maculaalba* Herbst. (Col.: Curculionidae) *Symp. Biol. Hung.* **16**, 241–245.

SCHAFER, R. (1971): Antennal sense organs of the cockroach, *Leucophaea maderae. J. Morphol.* **134** (1), 91–104.

STEINBRECHT, R. A. and MÜLLER, B. (1971): On the stimulus conducting structures in insect olfactory receptors. *Z. Zellforsch.* **117**, 570–575.

SZENTESI, Á. (1972): Studies on the mass rearing of *Acanthoscelides obtectus* Say (Coleoptera: Bruchidae). *Acta Phytopath. Acad. Sci. Hung.* **7**, 453–463.

YAMAMOTO, R. T. and FRAENKEL, G. S. (1960): The specificity of the tobacco hornworm, *Protoparce sexta*, to solanaceous plants. *Ann. ent. Soc. Am.* **53**, 503–507. Cited by Engelmann, F. (1970): *The Physiology of Insect Reproduction*, Pergamon Press, Oxford.

Symp. Biol. Hung. 16, pp. 283–285 (1976)

A PRELIMINARY STUDY OF HOST SELECTION AND ACCEPTANCE BEHAVIOUR IN THE CABBAGE APHID, *BREVICORYNE BRASSICAE* L.

by

W. F. TJALLINGII

DEPARTMENT OF ANIMAL PHYSIOLOGY, AGRICULTURAL UNIVERSITY, WAGENINGEN, THE NETHERLANDS

The present study investigates the role of the chemical senses in host selection and acceptance behaviour of the cabbage aphid *Brevicoryne brassicae* L. In order to determine the composition of this behaviour we observed aphids on hosts, non hosts and intermediate plants on which the aphid incidentally occurs (Markkula, 1953): non-cruciferous plants which contain mustard oil glucosides. About fifteen different behavioural components were distinguished and scored during a period of 30 minutes after an aphid had benn put on a plant. On formal grounds we used the term "proboscis contact" (p.c.) instead of "probe". Some authors (cf. Klingauf, 1971) use the duration of the first walk or the first probe as parameters in host selection behaviour. In our experiments the frequency of p.c., its mean duration and the percentage of time spent in p.c. were used. These parameters appeared to be less sensitive to pretreatments and reflected greater differences between the various plant species than the duration of the first walk or first probe. Apterous adults showed (Fig. 1) on the host *Brassica oleracea* a relatively low frequency of p.c. compared to aphids on the non host *Vicia faba*. The mean duration of p.c. and the percentage of time spent in p.c. were relatively high on host-plants. Compared with *Brassica*, *Tropaeolum majus* was, under our experimental conditions, an equally attractive host for aphids. Similar results were found with alate insects (Fig. 2). Aphid behaviour on *Reseda odorata* may be described as intermediate as compared to *Brassica* and *Vicia*.

Aphids migrated voluntarily from densely populated *Brassica* to *Reseda*, *Tropaeolum* or *Vicia*, but they died within three days after these plants had been placed in separate cages. Although further data are needed we found no significant differences in mortality rates during the first two days between aphids on these latter three plant species and aphids placed on filter paper.

Preliminary observations on locomotion of the aphids on these plants showed that aphids on *Vicia* soon become restless and leave the plant. On *Tropaeolum* and *Brassica* they stayed more often at the same place. Fig. 3 represents the number of aphids found on the same place of the leaf at the previous observation. In a similar experiment *Reseda* elicited an intermediate response between *Tropaeolum* and *Vicia*. Within three days, the number of aphids becoming restless increased and they either died or left the plant. Increasing restlessness was also observed in aphids on cut *Vicia* leaves kept in solutions of sinigrin (Wensler, 1962) or glucotropaeolin (both 25 mMol) or distilled water as a control. After four hours most aphids

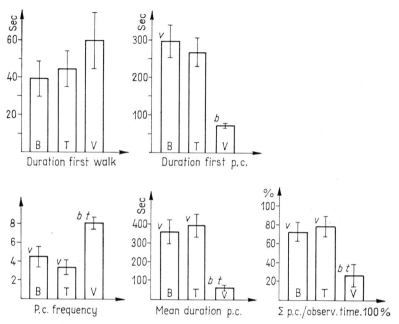

Fig. 1. First 30 min of acceptance behaviour. Means of 10 adult apterous aphids on different plants and standard errors. B = *Brassica oleracea;* T = *Tropaeolum majus;* V = *Vicia faba;* b, t or v = significantly different from B, T or V; P ≤ 0.05; Wilcoxon two sample test

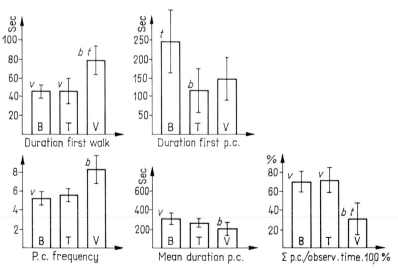

Fig. 2. First 30 min of acceptance behaviour on different plants. Means of 10 adult alate aphids and standard errors

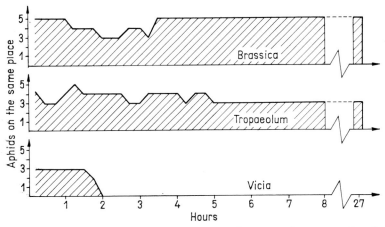

Fig. 3. Locomotion on different plants after transplantation from *Brassica*. 5 apterous
adults per plant; observations every 15 min

had left the leaves except the sinigrin treated, on which they stayed for
about one day.

Although the information available is still incomplete, some conclusions
may be drawn. The occurrence of p.c. was not affected by plant species.
Possibly chemicals present in the first sap sample(s) determine the *duration*
of the p.c. The experiments presented, however, give no information of
what occurs during p.c. and which differences between *Brassica* and
Tropaeolum influence p.c. characteristics. At the moment, therefore, it is
difficult to differentiate between selection and acceptance, two essential
behavioural phases of host finding in aphid/plant relationships.

REFERENCES

KLINGAUF, F. (1971): Die Wirkung des Glucosids Phlorizin auf das Wirtswahlsver-
    halten von *Rhopalosiphum insertum* (Walk.) und *Aphis pomi* DeGeer (Homoptera,
    Aphididae). *Z. ang. Ent.* **68**, 41–55.
MARKKULA, M. (1953): Biologisch-ökologische Untersuchungen über die Kohlblatt-
    laus, *Brevicoryne brassicae* L. (Hem., Aphididae). *Ann. Zool. Soc. "Vanamo"*, **15**,
    1–113.
WENSLER, R. J. D. (1962): Mode of host selection by an aphid. *Nature* **195**, 830–831.

Symp. Biol. Hung. 16, pp. 287–290 (1976)

HOST SELECTION AND COLONIZATION
BY SOME SPRUCE BARK BEETLES

by

G. I. Vasetschko

UKRAINSKIY INSTITUT ZASHTSHITY RASTENIY, KIEV-22, USSR

Oleoresin and turpentine, obtained from resistant or susceptible spruce trees, excite equal responses of *Ips typographus* L., *Dendroctonus micans* Kug. and *Hylurgops glabratus* Zett. depending on insect species only. There is no significant difference between resistant and susceptible trees in turpentine composition. There in no correlation between the intensity of oleoresin odour and tree attractiveness. The explanation of primary attractiveness of odour of oleoresin or other substances is unconvincing. *I. typographus* beetles select weakened trees responding to a complex of stimuli including visual, some olfactory and, possibly to biophysical ones. There is no "contact" pheromone in this species. Mass aggregation of *I. typographus* males is caused by host stimuli. After suppressing oleoresin exudation by the males they begin to produce a frass pheromone that attracts mainly the females. *D. micans* has no aggregative pheromone. Perhaps, the odour of oleoresin is important in host selection by the fertilized females of this species. The factor of spruce resistance to bark beetle attack is solely oleoresin flow.

MATERIALS AND METHODS

The comparison of resistant and susceptible trees was the aim of this work. The survival of introduced bark beetles or the success of their natural attack served as a criterion of susceptibility, and destruction of the insects was a criterion of resistance. The responses of the beetles to many objects including different substances, alive and crushed beetles, hindguts were also tested. In laboratory experiments responses were studied by using a multiple-choice olfactometer of an original construction. The objects were tested *per se* in quantities between traces (below 0.5 mg) and up to 50 mg and as solutions. The olfactometer tests were conducted in the dark. In field experiments responses of bark beetles to the same objects and trees which had been treated by different ways were studied. Response of flying beetles to an iron screen cylinder (3 m long and 0.3 m in diameter) imitating tree trunk, was observed. The trees were characterized by the following data: oleoresin exudation pressure (o.e.p.) determined by capillary glass tubes with one end sealed, rate of oleoresin flow (r.o.f.; Polozhentsev's method), turpentine content (steam distillation), turpentine composition (gas-liquid chromatography), toxicity of oleoresin and turpentine (biotests in petri dishes and probit-analysis), osmotic pressure (plasmolysis), relative turgidity (Weatherley's method), water content (drying to constant weight).

Attractiveness; laboratory experiments

For *I. typographus* α-pinene and β-pinene were repellent in all quantities; borneol and α-terpineol always attracted the beetles; acetone, oleoresin, turpentine and the majority of other terpenoids were attractant only at low quantities and repellent at large quantities, but attractiveness of oleoresin was insignificant; ethanol and vinegar were strong attractants but methyl ethers of fatty acids were repellent in all cases. Oleoresin, turpentine and terpenoids were strong attractants for *D. micans* and repellents for *H. glabratus* in all quantities. It seems to be important that oleoresin and turpentine, obtained from resistant or susceptible trees, elicited equal responses depending on insect species only. When responding to α-terpineol and turpentine the sex ratio (♀♀ : ♂♂) of *I. typographus* was 0.58 : 0.42 and 0.61 : 0.39, respectively. Male frass of *I. typographus* and ethanol extracts of this product were attractive for this species. The sex ratio was 0.72 : 0.28. Female frass was unattractive. Mixture of frass and oleoresin collected near initial attacks of *I. typographus* males on spruce trunk, that had weak oleoresin exudation, was unattractive. Alive and crushed *I. typographus* beetles, which had been in contact with spruce oleoresin, and hindguts of such beetles had the same attractiveness as the check objects. The frass of *D. micans* did not elicit positive reaction of these beetles. *H. glabratus* frass served as an attractant in spring, as a repellent during summer and in autumn it was not active.

Attractiveness; field experiments

Oleoresin, turpentine, terpenoids, other substances, frass and ethanol extract of frass *per se* were unattractive for any insect. Exposition of these objects on resistant trees did not elicit attacks. In the tests bark beetles attacked cut trees only. However, different treatments clearly influenced attractiveness. E.g. spraying the whole surface or cut trees with insecticide (solution of BHC and DDT in solar oil) attracted many bark beetles. The sex ratio was determined as follows: *I. typographus* 0.39 : 0.61; *Pityogenes chalcographus* L. 0.05 : 0.95; *Trypodendron lineatum* Ol. 0.64 : 0.36. In the check (trap tree) the sex ratio of these species was as follows: 0.70 : 0.30; 0.82 : 0.18; 0.50 : 0.50, respectively. Exposition on sprayed trees of ethanol extract of *I. typographus* frass to be released from small-bore tubings did not increase the trees' attractiveness. The normal sex ratio and a huge number of attracted bark beetles were observed when 10–20% of the surface of cut trees was left unsprayed in different parts of the trunk. In such cases the beetles were attracted by host stimuli and pheromone over the flight period. The adults of *I. typographus* were attracted by the iron-screen cylinder. But their number per time-unit was 3 times less than on a completely sprayed cut tree. The sex ratio was 0.40 : 0.60. The attractiveness was increased as soon as a spruce log was inserted into the cylinder.

The characteristics of resistant trees were the following: o.e.p. = more than 0 atm and up to 9 atm; indicator of r.o.f. = 4 or 3; turpentine content = 0.42% to dry weight; lethal concentration–50 (LC-50) of turpentine for *I. typographus* beetles = 0.2714 mg/cm³; LC-50 of oleoresin for these beetles = 0.5124 mg/cm³; water content ± 57–58%; osmotic pressure = = 13–14 atm; relative turgidity = 71–72%. The characteristics of the susceptible trees were: o.e.p. = 0 atm; indicator of r.o.f. = 0–2; turpentine content = 0.38–0.57%; LC-50 of turpentine for *I. typographus* beetles = = 0.2538 mg/cm³; LC-50 of oleoresin for these beetles = 0.4898 mg/cm³; water content = 51–59%; osmotic pressure = 10 – 20 atm; relative turgidity = 58–72%. There was no significant difference between resistant and susceptible trees in turpentine composition. Turpentine did not show toxicity for beetles and the half-grown larvae of *D. micans*. When insects were introduced into the phloem, the ranges of o.e.p. and r.o.f., respectively, at which they could survive, where as follows: *D. micans* beetles = 0–9 atm, 0–4; *D. micans* larvae = 0 atm, 0–2; *I. typographus* beetles = 0 atm, 0–2; *I. typographus* larvae = 0 atm, 0–1; *H. glabratus* beetles = 0 atm, 0–1; *H. glabratus* larvae = 0 atm, 0. At high water content, in the phloem of a cut tree, *I. typographus* larvae survived poorly, but progenies of *D. micans* and *H. glabratus* developed under such conditions normally. The progeny of *I. typographus* survived better, *D. micans* and *H. glabratus* larvae dried out when the water content fell below the level characteristic of resistant trees.

DISCUSSION

In course of the above investigations no "contact" pheromone, but only a "food" pheromone has been found in *I. typographus*. Hence, host selection and colonization by the beetles is not identical with that of *Dendroctonus* species as Vité et al. (1972) claimed in respect of *Ips calligraphus* Germ. and other species of this genus. In my opinion a right concept was proposed by Vité and Pitman (1968) about different evolutionary ways in the two genera of bark beetles: *Dendroctonus* species release pheromones into the air at the contact with oleoresin but pheromones of *Ips* species are released from frass. This concept has an important corollary, namely, mass attack of *Ips* species is affected by host-plant stimuli and the pheromones in this genus have the properties of sex attractants. Indeed, *I. typographus* is an aggressive pest attacking trees with significant oleoresin exudation. When attacking such trees pheromones cannot be released. Therefore, the beetles respond to host stimuli, and host (primary) attractiveness is strong. Oleoresin, obtained from resistant or susceptible trees, is a repellent or a very weak attractant. A healthy tree is unattractive even if its trunk is overlaid with oleoresin that exuded from wounds. Hence, host attractiveness cannot be explained simply by the alteration of turpentine composition or by an increasing intensity of oleoresin odour. There is no possibility to connect host attractiveness with other substances. I do not agree with Chararas (1958), Adlung (1958), Rudinsky et al. (1971) who have supposed that host stimuli for *I. typographus* are represented by methyl ethers or oleoresin odours. Perhaps, in primary attractiveness biophysical, visual and olfactory stimuli play a role. In *I. typographus* a pheromone is released only from such trees whose defence was overcome by males. Many of them are killed by oleoresin flow. But it is unessential for a population, because males develop in redundancy. The surviving males are quite enough for the survival of this polygamous species. Normal sex ratio and strong response of flying beetles appear when host stimuli and the pheromone act simultaneously. The spraying of trap trees with insecticides before the flight period, proposed by Rudnev (1965), is a very effective means for bark beetle control including such genera as *Ips*, *Pityogenes*, *Trypodendron*, and also some others. *I. typographus* beetles especially females respond positively to low quantities of turpentine and all quantities of α-terpineol in laboratory tests. It is out of question that terpene alcohols appear in turpentine during steam distillation. The identified pheromones of bark beetles are just terpene alcohols. The testing of substances in the form of solutions, creating artefacts, is not permissible. The pheromone of *Ips para-confusus* Lanier was identified as early as 1966. But a report about a successful application of these substances for the suppression of the beetles' population has not been published yet, which is probably due to the inadequate knowledge of host stimuli. When studying primary attractiveness most scientists pay attention almost exclusively to chemical stimuli. Thus, these works are unable to progress. It is important to study hypothetical biophysical stimuli, e.g. infra-red radiation and Kirlian effect.

D. micans has no aggregative pheromone. Inseminated females of this species when selecting weakened trees are guided by host stimuli including the odour of the oleoresin.

The factor of spruce resistance to bark beetle attacks is oleoresin flow which manifests itself as o.e.p. or r.o.f. Water conditions in the phloem do not play a role in resistance but are important factors in influencing the fate of the progeny.

ACKNOWLEDGEMENTS

My grateful acknowledgements are due to Prof. Dr. D. F. Rudnev for his guidance while preparing my work. I wish to thank Dr. V. S. Karasiov for making the gas-liquid chromatography analysis of the turpentine.

REFERENCES

ADLUNG, K. G. (1958): Die Lockwirkung von Methylestern der Leinölfettsäuren aus Borkenkäfer. *Naturwiss.* **45** (24), 624–627.

CHARARAS, C. (1958): Rôle attractive de certains composants des oléorésines à l'égard des Scolytidae des résineaux. *C. R. Acad. Sci.* **247** (18), 1653–1654.

RUDNEV, D. F. (1965): *Guide for bark beetle control in spruce stands of the Carpathians.* Harvest Publ., Kiev (in Ukrainian).

RUDINSKY, J. A., NOVÁK, V. and SVICHRA, P. (1971): Attraction of the bark beetle *Ips typographus* L. to terpenes and a male-produced pheromone. *Z. ang. Ent.* **67** (2), 179–188.

VITÉ, J. P. and PITMAN, G. B. (1968): Bark beetle aggregation: effects of feeding on the release of pheromones in *Dendroctonus* and *Ips. Nature, Lond.* **218** (5137), 169–170.

VITÉ, J. P., BAKKE, A. and RENWICK, J. A. (1972): Pheromones in *Ips* (Coleoptera: Scolytidae), occurrence and production. *Can. Ent.* **102** (12), 1967–1975.

Symp. Biol. Hung.   16_2,pp91–300 (1976)

THE OLFACTORY COMPONENT IN HOST-PLANT SELECTION IN THE ADULT COLORADO BEETLE (*LEPTINOTARSA DECEMLINEATA* SAY)

by

J. DE WILDE*

DEPT. OF ENTOMOLOGY, AGRICULTURAL UNIVERSITY, WAGENINGEN, THE NETHERLANDS

In many oligophagous insects it is known that specific olfactory stimuli emerging from the host-plant orientate locomotion, and specific contact chemical stimuli release feeding responses. Many of the substances concerned have now been defined. The Colorado beetle is an exception, as no specific chemicals in solanaceous plants are known which orientate its locomotion, and neither have specific phagostimulants been described. Experiments to isolate specific gustatory phagostimulants have failed.

In this paper arguments will be given in favour of the idea that in *Leptinotarsa*, long and short-distance orientation as well as feeding responses are enhanced by a specific volatile factor emerging from potato foliage. This factor while carried in a stream of air, most probably is directing flight behaviour, but clearly directs walking activity. It will be discussed how far the orientation of flight and of walking activity may involve the same chemo-anemotactic complex.

INTRODUCTION

It is generally presumed that phytophagous insects are orientated towards their host-plants by optical and/or olfactory stimuli, while host-plant acceptance or rejection is the result of the presence or absence of contact feeding stimulants and feeding deterrents, the specificity of which may vary greatly.

We now know that the information which an insect obtains from the chemical composition of its host-plant may be extremely complex, and may lead to an even more complicated set of responses (Schoonhoven, 1968).

In insects reaching their host-plant during flight, the study of these relations has often been hampered by the difficulty that flight may only take place under very specified circumstances. The Colorado beetle belongs to this type, its pre-reproductive dispersal flight taking place under very special physiological and environmental conditions (Johnson, 1969; LeBerre, 1952). During these typical dispersal flights, the beetles are passively transported by prevailing winds, and may even drift above the sea, being washed ashore later in massive numbers.

On June 8, 1946, while walking in the Belgian greenhouse district of Hoeylaert, it happened that the author was passing alongside a large potato field bordered by greenhouses, a roadside and a railroad track grown with grasses and shrubs. The weather was warm and a thunderstorm was approaching. Extensive flights of Colorado beetles occurred and very soon one

---

* The experiments were made by Mrs. K. Hille Ris Lambers-Suverkropp and Miss A. van Tol during their graduate study.

could observe many growers and their families actively collecting hundreds of beetles which had apparently just landed. The author repeatedly observed a beetle falling into the potato fields or in the surrounding vegetation, often from a considerable height of several metres. These beetles apparently arrested their flight when above or near potato plants.

Dispersal flights merely occurring during limited periods, the beetle in most cases will have to rely on walking to reach its host plants. This occurs during two phases of its life cycle: in the newly-emerged beetle shortly after adult ecdysis and in the post-diapause beetle which has left the soil after hibernation. In the first instance, the beetle will usually emerge in a potato field, in the second case, crop rotation prevailing, the beetle will emerge in a non-host-plant crop, mostly strewn with some subsisting potato-plants sprouted from last year's crop. Experience learns that these plants are mostly the first to be attacked.

In the experiments to be described here we have tried to answer the following questions:

1. Does olfaction play a role in host-plant finding by the adult beetle?
2. Does olfaction play a role in host-plant discrimination?

PREVIOUS OBSERVATIONS

McIndoo (1926) introduced the well-known Y-tube insect olfactometer and established the fact that the adult Colorado beetle is attracted by the odour of the potato plant. In this apparatus, the beetles are faced with the rather abnormal situation of having to choose between two airflows, only one of which carries the odour of the plant.

Schanz (1953) using a similar olfactometer, has made some significant observations which are still of basic value. Although she describes that the adult shows positive anemotactic responses, she does not incorporate these responses in the evaluation of the tests, but applies the positive phototactic response of the beetles to lead them to the point of choice. By adequate experimentation air humidity is excluded as a causal factor in the positive response the beetle shows towards air having passed along potato leaves. Amputation of the four terminal antennal segments eliminates the olfactory response which is still intact when one, two or three segments are removed. Morphological studies show that on the five terminal antennal segments, thin-walled trichoid chemoreceptors are located. Only 1% of these receptors is present on the fifth segment, which explains why this segment alone does not suffice to elicit the response.

Bongers (1970) has demonstrated that adult beetles respond to potato leaf odour in still air in a "screen test" which, however, also does not provide a situation comparable with natural choice conditions. By combining the screen test with the optical image of potato leaves, he could show that the movements of the beetle may also be directed by optical plant stimuli.

From both field and laboratory experiments, Jermy (1958) concludes that neither post-diapause nor newly emerged young beetles make use of olfactory stimuli in finding the potato plant under field conditions. Post-diapause beetles react positively to potato leaf odour in a modified McIndoo olfactometer, but newly emerged young beetles only react after having been fed for some time on potato foliage. It is concluded that contact chemoreception is the principal factor in host-plant recognition. De Wilde et al. (1969) making use of a wind-tunnel described by Haskell et al. (1962) have shown that potato leaf odour enhances the positive anemotactic response which is proper to adult beetles but absent in larvae. Some of these experiments, as well as an extension, are discussed in the next chapter.

WIND-TUNNEL EXPERIMENTS

The structure and dimensions of the wind-tunnel are shown in Fig. 1. The tunnel was lined with glass plates which were periodically removed and washed with alcohol. A silicone wax barrier prevented the beetles from climbing the vertical plates. Air speed was roughly 2 m/sec. The tunnel was placed in a climatized room at 26 °C and R. H. 65%. Beetles were released in groups of 20–40 at the centre of the tunnel. Choices were repeated 10 times. They were considered positive or negative when after a delay of 5 min the beetles were within a limit of 5 cm of the upwind, resp. downwind screen. Differences were tested for significance by means of the Wilcoxon test. The difference was considered significant if $P \leq 0.01$.

Tests were made with both males and females 2–3 weeks after adult emergence, reared at an 18-hr photoperiod. The responses were essentially the same, but the variability was larger in the male due to sexual behaviour.

In the experiment shown in Fig. 2, the effect of potato leaves is compared with that of three non-solanaceous plants. Alder (*Alnus incana*), dandelion (*Taraxacum officinalis*) and grass (*Poa annua*) significantly reduced, while potato leaves enhanced the positive anemotactic response. In the experiment of Fig. 3, the effect of removal of terminal antennal segments on the anemotactic response was studied in an experiment in which the plant reservoir of the tunnel was filled with glass wool. Removal of the segments 1–3 still allows a normal response, but removal of the fourth segment eliminates the response. The same is true for the experiment of Fig. 4, in which the stimulative effect of potato odour with respect to the anemotactic response is shown.

The above experiments were later extended to include newly emerged adults. 40 females, which had not received any food, were tested on the day of emergence.

The results are collected in Table 1.

It is obvious that both anemotactic and chemo-anemotactic responses occur in adults without any previous experience.

TABLE 1

Result of an experiment with 40 newly emerged, non-fed female beetles

	Mean choice			Mean choice difference	Effect of odour	Significance
	neg.	indiff.	pos.			
Control	0.8	23.5	15.7	14.9		
Potato	1.0	15.5	23.5	22.5	7.6	P<0.01

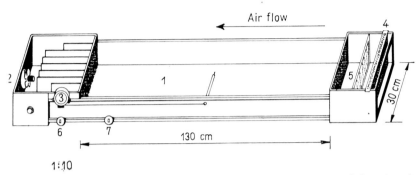

Fluorescent tubes

Air flow

1:10

Fig. 1. Wind-tunnel according to Haskell et al. (1962). 1. Glass inner lining; 2. exhaust ventilator; 3. thermostat connected with thermocouple; 4. reservoir containing plant material; 5. heating device for regulating temperature; 6 and 7. switches for ventilator and thermoregulation

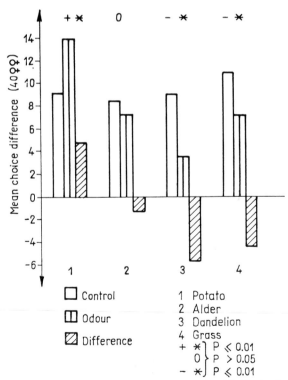

Fig. 2. Effect of non-host-plant odour compared with that of potato foliage

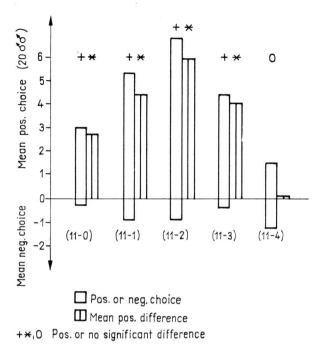

Fig. 3. Role of antennal segments in anemotactic response. (11–0), (11–1), (11–2) =
= 0, 1, 2 etc. terminal antennal segments removed

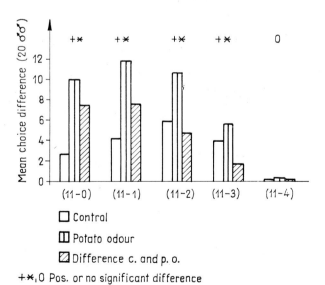

Fig. 4. Role of antennal segments in chemo-anemotactic response. Explanation as in
Fig. 3

A series of beetles were placed in succession in the centre of a 3 m broad asphalt road, alongside an experimental field partly planted with potato rows. The observations were made in July, when the plants were full-grown. Male and female beetles were studied in separate series. They were placed at different distances from the potato plants, with different wind directions and wind speed and at different times of the day. Wind speed and direction were measured with a Füss rotating anemometer and wind vane. The track followed by the beetles was drawn on the road by means of chalk, and marks were set at 1 minute intervals to obtain an impression of walking speed.

In the behaviour of the beetles, two points were remarkable.

1. While walking, the beetles often stopped, showing the clinotactic behaviour described by Schanz (1953). Head and thorax were elevated, the head was turned to and fro, the antennae waving alternately.

2. With sufficient wind speed (1–2 m/sec at a minimum) and when the distance of the potato rows did not exceed ± 6 m, the beetles mostly maintained their walking direction (against wind or at a limited angle with the wind direction) for several minutes, a behaviour also described by Jermy (1958). One could not escape the impression that the angle of locomotion was determined by both the wind direction and the direction of the nearest potato rows. Examples of this behaviour are shown in Figs 5–8. The direction of the sun apparently had no influence. This behaviour persists when one

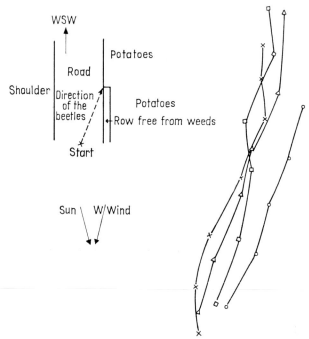

Fig. 5. Walking tracks of four beetles in relation to sun and wind direction. Distance from potato rows 4 m

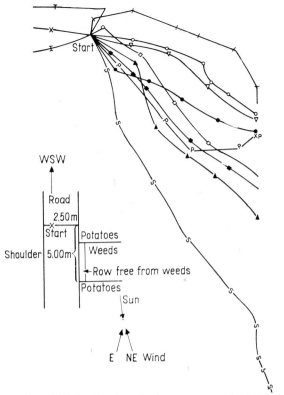

Fig. 6. Walking tracks of 12 beetles in relation to sun and wind direction. Temperature 26.0–26.4 °C; R.H. 56–60%. Wind speed 1.5–2.3 m/sec

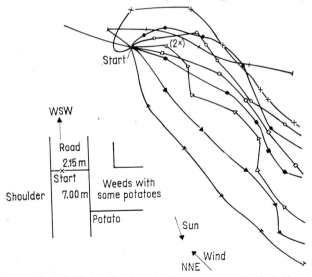

Fig. 7. Walking tracks of 10 beetles in relation to sun and wind direction. Temperature 25.4–25.6 °C; R.H. 51–60%. Wind speed 1.4–1.6 m/sec

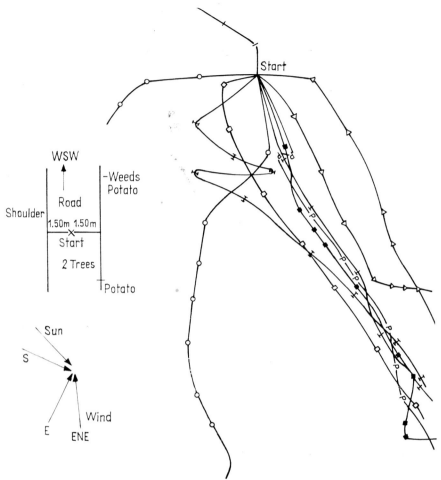

Fig. 8. Walking tracks of 10 beetles in relation to sun and wind direction. Right antennal flagellum amputated. Temperature 24.8–25.2 °C; R.H. 52–59%. Wind speed
0.9–1 m/sec

antenna is amputated, only the antenna at the most stimulated side remaining. When the "leeward" antenna was removed, the track became more
irregular.

If potato rows were absent or at a long distance, the tracks often bore no
relation to the wind direction, but were directed towards the nearest green
vegetation.

DISCUSSION

The results of Y-tube olfactometer tests (McIndoo, 1926; Schanz, 1953;
Jermy, 1958) and of screen tests (Bongers, 1970) as well as wind-tunnel
experiments show that the adult Colorado beetle reacts positively to the

298

odour of potato and some other solanaceous plants, and can discriminate between this odour and the scents emanated by several non-solanaceous species.

As the walking adult Colorado beetle has a clear anemotactic response, the wind-tunnel method as deviced by Haskell et al. (1962) provides by far the most natural conditions to test its behaviour with respect to odours in an air flow. We would like to use the term chemo-anemotaxis to indicate this type of orientative behaviour. As the orientation does not require a symmetric distribution of the stimuli over both sides of the body, but is retained after one antenna is amputated and disappears when both antennae are removed, we consider this response to be menotactic according to the definition given by Kühn (1919). This menotactic orientation is periodically corrected by clinotactic re-orientation. The role of optical stimuli in host-plant detection is certainly not negligible and becomes more predominant when air flow is at a minimum or potato scent rare.

For both anemotactic and chemo-anemotactic responses the presence of the four terminal antennal segments is essential. As 99% of the trichoid chemoreceptors are located on these four segments (Schanz, 1953) it is intriguing to think that these sensillae have at the same time an anemo-receptor function.

Many authors (Raucourt and Trouvelot, 1933; Ritter, 1967; Hesse and Meier, 1950; Hsiao and Fraenkel, 1968) have tried in vain to isolate from potato leaves a specific taste substance, which by its effect as a selective feeding stimulant could determine the oligophagous feeding behaviour of the Colorado beetle and its specific choice of solanaceous foodplants. Jermy (1958) even went so far as to suppose that this food choice can only be determined by the absence of repellent or deterrent substances, while Ritter (1967) denies the presence of a specific token feeding stimulant for the Colorado beetle.

Many of the above-mentioned experiments were only performed with larvae, but the conclusion remains that also for the adult beetle, no gusta-tory token feeding stimulants specific for solanaceous plants have been found.

As all efforts have up till now been directed towards contact chemorecep-tion (as detected e. g. by the elder pith test), and no experiments have been made to extract volatile token factors, it now seems an attractive hypothesis that in solanaceous plants, a volatile factor acts both as an attractant and a feeding stimulant towards the Colorado beetle. Preliminary experiments with low-temperature distillation give us reason to optimism in this respect.

SUMMARY

1. Walking Colorado beetles in a wind-tunnel show positive anemotactic responses.

2. These responses are enhanced by potato leaf odour, but reduced by the odour of several non-host-plants.

3. Removal of the four terminal antennal segments eliminates both anemotactic and chemo-anemotactic responses.

4. Under field conditions walking Colorado beetles find potato plant from a distance of 6 m when these plants are located above wind.

5. This also occurs when the "leeward" antenna is amputated. The orientation is menotactic with intermittent clinotactic re-orientation.

6. From some field observations on flying Colorado beetles it would seem that the odour of potato plants may have an arrestant effect.

7. The hypothesis is brought forward that both attractant and phagostimulant effect of solanaceous plants may be due to volatile factors.

REFERENCES

BONGERS, W. (1970): Aspects of host-plant relationship of the Colorado beetle. *Meded. Landbouwhogeschool, Wageningen* **70-10**, 77.

HASKELL, P. T., PASKIN, M. W. J. and MOORHOUSE, J. E. (1962): Laboratory observations on factors affecting the movements of hoppers of the desert locust. *J. Insect Physiol.* **8**, 53–78.

HESSE, G. and MEIER, R. (1950): Über einen Stoff, der bei der Futterwahl des Kartoffelkäfers eine Rolle spielt. *Angew. Chem.* **62**, 502–506.

HSIAO, T. H. and FRAENKEL, G. S. (1968): Selection and specificity of the Colorado potato beetle for solanaceous and non-solanaceous plants. *Ann. ent. Soc. Am.* **61**, 493–503.

JERMY, T. (1958): Untersuchungen über Auffinden und Wahl der Nahrung beim Kartoffelkäfer. *Ent. exp. & appl.* **1**, 197–208.

JOHNSON, C. G. (1969): *Migration and dispersal of insects by flight.* Methuen, London.

KÜHN, A. (1919): *Die Orientierung der Tiere im Raum.* Gustav Fischer, Jena.

LeBERRE, J. R. (1952): Contribution à l'étude du déterminisme de l'envol du Doryphore. *C. R. Acad. Sci. Paris* **234**, 1092–1094.

McINDOO, N. E. (1926): An insect olfactometer. *J. econ. Ent.* **19**, 545–571.

RAUCOURT, M. and TROUVELOT, B. (1933): Recherches sur les constituents des feuilles de pomme de terre determinant la choix de nourriture des larves de *Leptinotarsa decemlineata*. *C. R. Acad. Sci. Paris* **197**, 1153–1154.

RITTER, F. (1967): Feeding stimulants for the Colorado beetle. *Meded. Rijksfac. Landb. wetensch. Gent* **32**, 291–305.

SCHANZ, M. (1953): Der Geruchssinn des Kartoffelkäfers. *Z. vergl. Physiol.* **35**, 353–379.

SCHOONHOVEN, L. M. (1968): Chemosensory basis of host plant selection. *Ann. Rev. Ent.* **13**, 115–136.

DE WILDE, J., HILLE RIS LAMBERS-SUVERKROPP, K. and VAN TOL, A. (1969): Responses to air flow and air borne plant odour in the Colorado beetle. *Neth. J. Pl. Path.* **75**, 53–57.

Symp. Biol. Hung. 16, p. 301 (1976)

HOST SELECTION BY BARK BEETLES
IN THE MIXED-CONIFER FORESTS OF CALIFORNIA

by

D. L. Wood

DEPT. OF ENTOMOLOGICAL SCIENCES, UNIVERSITY OF CALIFORNIA,
BERKELEY, CALIFORNIA 94720, USA

(Summary only)

The capacity for bark beetles to discriminate between living host (*Pinus ponderosa*), dead host, and non-host trees (*Abies concolor, Libocedrus decurrens*), and between resistant and susceptible host trees prior to landing on the bole, was examined by measuring the landing rates of *Dendroctonus brevicomis* and *D. ponderosae* over month-long periods. Sticky traps were placed on the bole near the first living branches. Ponderosa pine exhibiting advanced symptoms of photochemical oxidant injury and root infection by *Verticicladiella wagnerii* were selected as the susceptible host trees. Some of these were screened over most of the bole to prevent initiation of boring activity and subsequent pheromone production and mass attack. Severe water stress was induced in another series of ponderosa pines by freezing, and injection of the herbicide, cacodylic acid. No significant differences in the landing rates of these bark beetle species were observed among the six classes of trees studied.

Thus, we find no support for the long-standing theory proposed by Person. Stimuli which elicit an arrestment, biting, and sustained feeding response must therefore explain the patterns of host tree discrimination observed for these species.

Symp. Biol. Hung. 16, pp. 303–306 (1976)

RESEARCH ON THE EFFECT OF ALPHA-TOCOPHEROL (VIT. E) ON THE GROWTH AND REPRODUCTION OF THE COLORADO POTATO BEETLE (*LEPTINOTARSA DECEMLINEATA* SAY)

by

Z. Zwolińska-Śniatałowa

INSTITUTE FOR PLANT PROTECTION, 60318 POZNAŃ, POLAND

The influence of alpha-tocopherol on the Colorado potato beetle has been investigated. Integrated biochemical, biological and histological studies were made. The biochemical studies embraced the gas-chromatographic analyses concerning the qualitative and quantitative composition of tocopherols in the leaves of *Solanum rostratum* and *Solanum tuberosum*, as well as in imaginal tissues, in the larvae and in the eggs. In the biological experiments Colorado potato beetles were reared on normal potato leaves and also on leaves to which the increased doses of alpha-tocopherol were applied. As criteria, fertility, mortality, the duration of beetles' activity and the consumption were considered. In the histological investigation the gonads of the larvae and adults were examined. The experimental results have shown that the most important differences occur in the reproduction of the insect. When increased doses of alpha-tocopherol were applied an increase in the quantity of eggs laid was recorded. Histological studies revealed that a higher quantity of sperm was present in the gonads of the males fed on increased doses of alpha-tocopherol and the gonads of the females were larger and the oocytes had thick follicular epithelium comprising large cells than in untreated adults. Generally it was established that alpha-tocopherol acted as a stimulant of fecundity in this particular insect.

INTRODUCTION

The problem concerning the influence of vitamin E on the physiology of the Colorado potato beetle arose as a specific question connected with the role of food in the development of this insect. The changing of food plant from *Solanum rostratum* to *S. tuberosum* as well as the changing of ecological and geographical conditions were the critical points in the life of this species. Since that moment the problems of unusual fecundity (from the physiological point of view) and catastrophical damages have arisen. Among many other factors which can influence the fecundity of this insect, undoubtedly one of the most important is vitamin E which acts directly on the reproductive system.

Recently many scientific centres in the world investigate vitamin E from different points of view because it appeared that this extremely interesting compound occurs in plant, human and animal tissues much more commonly than it was thought earlier.

It was also stated that it was responsible for a variety of biological functions and it participated in many metabolic processes.

The basic reaction of tocopherols responsible for their biological activity is oxidation, which occurs under the influence of different agents (Skinner and Parkhurst, 1971). Interests in studying the oxidation-reactions of the tocopherols are related to their use as antioxidants. Alpha-tocopherol

(5,7,8-trimethyltocol) is biologically the most active member of naturally occurring tocopherols and tocotrienols.

In entomology over a long period of time the role of fat soluble vitamins and their importance for insects were not really appreciated.

Recent investigations, particulary, House's works (1955, 1966), proved that fat soluble vitamins play a basic role in insect physiology, especially in those insects which show specific nutritional requirements (Vanderzant et al., 1957; Fraenkel, 1959; Dadd, 1963; Lipke and Fraenkel, 1956). House suggests that the main reason which makes the questions difficult is that there are different modes of accumulation of fat soluble vitamins in insect tissues and different manners of their transmission to the offspring.

Owing to the possibility of accumulation of these vitamins by insects the symptoms of deficiency may not appear even if the improper diet is applied for a long time.

Our investigation refers to the influence of alpha-tocopherol on the organism of the Colorado potato beetle. For this purpose integrated biochemical, biological and histological studies were carried out.

BIOLOGICAL EXPERIMENTS

Over many years experiments had been carried out in our laboratory for studying the formation of some compounds contained in leaves of several potato varieties during the vegetation period (Wojciechowski et al., 1957). The content of nitrogen, proteins, fats, carotenes, carbohydrates and some vitamins was determined. It allowed us to estimate the nutritional value of potato leaves from a biochemical point of view.

We assumed while studying the influence of only one component (vitamin E) added to the natural food of the Colorado potato beetle (potato leaves of known varieties) that the content of other substances remains approximately constant.

In our biological experiments the Colorado potato beetles were reared on potato leaves and on leaves supplemented with a known dose of alpha-tocopherol. The investigation was carried out in all stages of the insect.

The following observations have been made during the breeding treatments: fertility, mortality, the duration of beetles' activity and quantity of consumed food.

As the main criterion concerning the effect of vitamin E on the insects was the number of eggs laid.

In those experiments where the food with a higher amount of vitamin E was given, the females laid more eggs than the control ones (from 30% to 90%). This fact demonstrated the stimulating property of vitamin E in insect fertility. Considering the amount of consumed food we did not find any difference among the treatments. This means that the addition of alpha-tocopherol did not cause any stimulation in nutrition.

BIOCHEMICAL RESULTS

A large quantity of information has accumulated concerning the effect of tocopherols in plant and animal organisms (Herting, 1971). Their metabolic functions are so important that we may assume that they are present in almost all living organism. Recently we have been interested in the question which isomers are present and why the quantitative composition is so different in many cases (Booth, 1963). It is a well-known fact that alpha-tocopherol is a dominant form of the group of naturally occurring compounds possessing vitamin E activity and this is the most active biological form. The determination of the quantitative and qualitative composition of tocopherols in tissues is now a very important question because it could solve the problem of physiological activity of all compounds belonging to the tocopherol group.

By gas-chromatographic analyses the qualitative and quantitative composition of tocopherols in leaves of S. rostratum and S. tuberosum as well as in adult potato beetles, larval tissues and in eggs was determined.

We applied the gas liquid chromatograph Pye 104 with flame ionization detector, and glass column packed with 3% SE 30. The temperature of the column and the detector was 260 °C and 280 °C, respectively. Table 1 presents the content of tocopherols in the leaves of S. rostratum and S. tuberosum. Table 2 summarizes the content of tocopherols in eggs, larvae and adults bred on leaves enriched with vitamin E.

The quantitative analyses of plants indicated that in the leaves of S. tuberosum the content of tocopherols was higher than in S. rostratum. The quantitative analyses of eggs, larvae and adults showed that the content of the tocopherols was higher in those treatments in which the food was enriched with vitamin E. The qualitative analyses of plant extracts of S. rostra-

TABLE

The content of tocopherols in the leaves
of S. rostratum *and* S. tuberosum

Plant species	The content of tocopherols in μg/g of fresh weight	
	alpha	delta
Solanum rostratum	56	16
Solanum tuberosum	131	22

TABLE 2

The content of tocopherols in eggs, larvae
and adults enriched in vitamin E

Object	The content of tocopherols in μg/g of fresh weight		Percentage of tocopherols in breeding supplemented with vitamin E	
	alpha	delta	alpha	delta
Eggs	233	194	329	230
Larvae (L$_4$)	121	26	178	230
Adults	331	38	209	227

tum and *S. tuberosum* as well as extracts from eggs, larvae and adults showed similar results. By the chromatograms trimethyl derivatives were dominant, monomethyl derivatives occurred in lesser quantity, and no dimethyl derivatives appeared at all. These data indicate that these compounds may be stored in the insects' body.

HISTOLOGICAL INVESTIGATION

Simultaneously histological studies were made in order to elucidate the influence of vitamin E on the reproductive system.

It is necessary to emphasize that this investigation was only a contribution to the histological problem (Lipa et al., 1971). Histological studies showed that in the gonads of males fed with increased doses of alpha-tocopherol a higher quantity of sperm was present and sperm maturation occurred sooner than in the gonads of the control group. No differences in the size of male gonads were observed.

The gonads of females fed with increased doses of alpha-tocopherol were larger than those of the control. The oocytes of females fed with increased doses of alpha-tocopherol had a thick follicular epithelium formed by large cells.

CONCLUSION

The results prove that alpha-tocopherol has a positive effect in general on the organism of the Colorado potato beetle affecting it especially as a fecundity stimulant.

REFERENCES

BOOTH, V. H. (1963): Determination of tocopherols in plant tissues. *Analyst* **88,** 627–632.

DADD, R. H. (1963): Feeding behaviour and nutrition in grasshoppers and locusts. *Adv. Insect Physiol.* **1,** 47–109.

FRAENKEL, G. (1959): The chemistry of host specificity of phytophagous insects. Fourth International Congress of Biochemistry. *Biochemistry of Insects*. Pergamon Press Ltd. **12.**

HERTING, D. C. (1971): Introduction to the symposium on chemistry and biochemistry of tocopherols. *Lipids* **6,** 238–239.

HOUSE, H. L. (1955): Nutritional requirements and artificial diets for insects. *Ann. Rept. But. Soc. Ontario* **86,** 5–10.

HOUSE, H. L. (1966): Effects of vitamin E and A on growth and development, and the necessity of vitamin E for reproduction in the parasitoid *Agria affinis* (Fallen) (Diptera: Sarcophagidae). *J. Insect Physiol.* **12,** 409–417.

LIPA, J. J., ZWOLIŃSKA-ŚNIATAŁOWA, Z. and BARTKOWSKI, J. (1971): Wpływ witaminy E na gonady stonki ziemniaczanej (*Leptinotarsa decemlineata* Say). *Prace Naukowe IOR* **13** (2), 218–240.

LIPKE, H. and FRAENKEL, G. (1956): Insect nutrition. *Ann. Rev. Ent.* **1,** 17–44.

SKINNER, W. A. and PARKHURST, R. M. (1971): Reaction products of tocopherols. *Lipids* **6** (4), 240–244.

VANDERZANT, E. S., KERUR, D. and REISER, R. (1957): The role of dietary fatty acids in the development of the pink bollworm. *J. econ. Ent.* **50,** 606–608.

WOJCIECHOWSKI, J., GIEBEL, J., GŁOGOWSKI, K., SZYMAŃSKI, S. and ZWOLIŃSKA-ŚNIATAŁOWA, Z. (1957): Ksztaltowanie sie zawartosci niektórych skladników w lisciach odmian ziemniaków w czasie wegetacji. *Rocz. Nauk. Rol.* **74**-A, 2.

Symp. Biol. Hung. 16, pp. 307–309 (1976)

GENERAL CONCLUSIONS

(Symposium on "The Host-plant in Relation to Insect Behaviour
and Reproduction", Tihany 1974)

by

R. F. CHAPMAN[1] and J. S. KENNEDY[2]

[1]CENTRE FOR OVERSEAS PEST RESEARCH, COLLEGE HOUSE,
WRIGHTS LANE, LONDON; [2]IMPERIAL COLLEGE FIELD STATION,
SILWOOD PARK, ASCOT, UK

The papers given during the symposium have covered many aspects of plant/insect interactions. They may be grouped into broad categories in order to generalize on the changes in knowledge and thought which have occurred since the last symposium in Wageningen and also to indicate the obvious gaps in our present understanding of the insect/plant interaction. These conclusions are based mainly on points made by Drs Beck, Dethier, Schoonhoven, Städler, de Wilde and Wood during the final discussion and are an attempt at a consensus of opinion of the meeting as a whole.

Most of the papers were essentially behavioural in content, with a general lack of physiological background. Despite this, a significant change in thinking since the last symposium is in the field of sensory physiology with the realization that food selection is not always based on the recognition of a few key chemicals which stimulate specific receptor cells, but rather that many compounds may act together to elicit complex activity patterns in the receptors. It is necessary to consider quantitatively the pattern of changes occurring in the total input from all the receptors and not to restrict ourselves to a consideration of qualitative changes. This has important behavioural implications in questioning the concept of the token stimulus in plant recognition. At present, however, our knowledge is such that we must retain an open mind: while the concept of changes in pattern may be important in some situations it is not necessarily entirely at variance with the importance of specific qualitative changes in others.

No consensus was reached on the importance of chemical stimuli in attraction to host-plants from a distance. In an increasing number of examples it appears that selection does not occur until the insect reaches the host indicating that specific olfactory attraction is not important. There are, however, other instances where specific olfactory cues are known to guide the insects to their host-plants. The number of instances in which details of attraction are known to us is small, but it seems likely that there is truly considerable variation in the role of olfaction in distant recognition of the host.

While the general importance of contact chemoreception in final host selection is clearly recognized, few papers dealt specifically with this aspect of behaviour. In some the role of chemicals inhibiting feeding was stressed, but opposed to this view was the suggestion that specific stimulatory volatile chemicals may act at short range. These have been neglected in the past because of technical difficulties in their isolation, and a feature throughout the meeting was a growing appreciation of the need for a really sound understanding of the chemistry of host-plants.

The longer-term effects of host-plant relations have also been considered from a number of points of view. Several papers have emphasized the possibility of changes in host-plant selection, either for feeding or for oviposition, arising from sustained association with unusual host-plants. The concept of such induced changes is not new and the susceptibility of different insect species to such changes is variable, but the phenomenon may be more widespread than had previously been appreciated. This is important, not only for our understanding of insect/host-plant relationships, but in emphasizing the need for standard rearing techniques for experimental insects since in the absence of such standardization experimental differences could arise through changes induced before the experiment is started.

Variation in growth and reproductive capacity with the nutritive status of the host-plant were also discussed and some studies on aphids seemed to give an affirmative answer to the question: do insects eat what is good for them? The physiological processes underlying these nutritional effects were, however, wholly neglected and no consideration was given, for instance, to the possibility of the food or specific chemicals acting as signals governing the endocrine control of reproduction. The long-term adverse effects of secondary plant substances ("allelochemics"), as opposed to their short-term effects in reducing amounts of food ingested, have not generally been considered and this is another field requiring much more work.

The need to standardize and control the chemical quality of the food in all these studies has been emphasized as our understanding of the variability of natural foods increases, and this necessity has led to the production of artificial diets in some cases. Yet there is a frequent failure to define these diets adequately so that the competitiveness of the insects is reduced, a factor of vital importance in mass rearing for practical control, but also perhaps an indicator of other changes which could be important in behaviour studies. A need for the control of the quality of insects produced is clearly indicated and this entails the development of diets which really are suitable for a species in general and not just for a laboratory clone which is capable of surviving on it.

We have also had some discussion on the interpretation of results obtained in bioassay studies, but experimental design is still a relatively neglected field and our results may often be interpreted in various ways. Many experiments, for instance, do not enable us to differentiate between the effects of feeding and nutrition, or between repellent and antibiotic properties of a plant, and it is clear that more attention needs to be devoted to this aspect of our work.

Host-plant specificity has important considerations in the development of varietal resistance, which need not be complete to be of serious practical value. The empirical approach to the selection of resistant varieties is becoming less acceptable, and with an adequate knowledge of insect behaviour and physiology in relation to the host-plant, and of the genetics of the plant itself, it should be possible to "design" resistant plants.

In the case of cotton this has been shown to be feasible, given adequate resources and co-operation between scientists and industry. Not only is it feasible, but the economic yield can be increased in this way. This achievement is the practical outcome of basic studies on insect/host-plant interactions. It marks the coming-of-age of the field.

308

Finally, in several papers the evolutionary significance of the insect/host-plant interrelationship has been considered. First with regard to speciation within the insects, arising from genetic change perhaps following induced changes in host-plant preferences, and then, most important, a reconsideration of our attitude to the evolution of insects and their host-plants: co-evolution or subsequent (sequential) evolution. There was considerable support for the suggestion that allelochemics were not, in many cases, initially developed as defence mechanisms against insects since, it is argued, insects do not provide the major selective pressures on plants. Rather it is envisaged that the plants evolved their chemical complexes in response to other pressures, and then the insects evolved to fit the plants.

It was appreciated that this point of view did not exclude the possibility of co-evolution in some circumstances and this breadth of approach has been overall a most significant feature of the symposium. In all sections we have seen a movement away from hard and fast lines and a realization that, at least in our current state of knowledge, useful generalizations are not often to be drawn.

AUTHOR INDEX

Aasen, A. J. 45
Abdel-Rahman, A. H. 65
Adams, J. B. 177
Adkisson, P. L. 22
Adlung, K. G. 290
Afzal, M. 22
Agarwal, R. A. 13
Akeson, W. R. 53
Akey, D. H. 33, 177
Akimov, Yu. A. 119
Ali, M. 23
Alonso, C. 258
Applebaum, S. W. 83, 180
Arai, N. 151
Arbuthnot, K. D. 195
Arn, H. 178
Ascher, K. R. S. 53, 151
Ata, M. A. 281
Auclair, J. L. 29, 34, 178, 180
Augustine, M. G. 53
Avery, J. 200
Azmy, N. 151

Bailey, J. C. 21
Baker, J. E. 53, 200
Bakke, A. 235, 290
Banks, C. J. 178
Barr, B. A. 234
Barton Browne, L. 53
Bartkowski, J. 306
Bashford, M. A. 180
Bates, M. 113
Bielak, B. 60
Beck, S. D. 53, 99, 150, 177, 217, 221, 252
Bell, C. H. 65
Bennett, E. W. 214
Bennewitz, M. 179
Benschoter, C. A. 161
Benz ,G. 107
Berger, B. D. 83
Bernays, E. A. 35, 39, 41, 45, 53
Beroza, M. 136
Birks, P. R. 83
Blaney, W. M. 35, 39
Bongers, W. 28, 99, 300
Booth, C. D. 123
Borg, T. K. 200
Bornman, C. H. 178

Both, V. H. 306
Botha, C. E. J. 178
Bowmen, M. C. 136
Boyd, J. 151
Bradley, R. H. E. 178
Bragdon, F. G. 178
Bramstedt, F. 178
Breure, A. 92
Bridge, R. R. 161
Brierley, G. P. 201
Brown, E. S. 150
Brown, J. C. 161
Burton, R. L. 136
Bush, G. L. 94, 214
Butler, G. D. 161

Carlisle, D. B. 258
Cartier, J. J. 34, 178
Cavanagh, G. G. 172
Chambon, J. P. 184
Chapman, R. F. 35, 41, 45, 53, 307
Chararas, C. 290
Chatterji, S. M. 252
Chawla, S. S. 178
Chevin, H. 184
Chiang, H. C. 195
Chippendale, G. M. 252
Chu, H. M. 200
De Cicco, A. 132
Cirio, U. 214
Cleere, J. S. 178
Cloutier, M. 178
Coaker, T. H. 21, 85, 89
Constante, E. G. 214
Cook, A. G. 35, 47
Cooke, J. 178
Coutin, R. 207
Cowan, C. B. 161
Cowland, J. W. 21
Cress, D. C. 221
Csehi, É. 195
Culvenor, C. C. J. 45
Curtis, C. E. 21

Dabrowski, Z. T. 55, 60
Dadd, R. H. 53, 83, 178, 179, 306
Dahms, R. G. 45
Davidson, R. H. 53

311

Jepson, W. F. 253
Jermy, T. 28, 60, 77, 99, 109, 151, 245, 266, 273, 281, 300
Johnson, B. 83, 123
Johnson, C. G. 300
Johnson, R. A. 172
Jones, R. L. 136
Jones, W. J. 253
Josefsson, E. 208
Judge, F. D. 83
Juniper, B. E. 40

Karasev, V. S. 115
Katiyar, K. P. 253
Kellock, A. W. 45
Kendall, M. D. 39
Kennedy, J. S. 34, 89, 113, 121, 123, 179, 307
Kershaw, W. J. S. 123
Kerur, D. 132, 306
Kester, D. E. 45
Khalifa, A. 151
Kinzer, G. W. 234
Kjaer, A. 207
Klein, W. 39
Kleinjan, J. E. 83, 179
Kline, L. N. 234
Klingauf, F. 39, 179, 285
Kloft, W. 179
Klun, J. A. 46
Knight, V. A. 201
Kolattukudy, P. E. 39, 40
Kozár, F. 125, 127
Krieger, D. L. 178, 179
Krishnananda, N. 13
Krzymańska, J. 129
Kuijten, P. J. 91
Kunkel, H. 83, 179
Kurth, E. F. 118
Kuznetsov, N. N. 119
Kühn, A. 300

Labeyrie, V. 107, 108, 133, 136, 281
Lamb, K. P. 179
Langhlin, R. J. 132
La Pidus, J. B. 53
Laska, P. 155
Laster, M. L. 157, 161
Launois-Luong, H. 137
Laurema, S. 153, 155
Leal, M. P. 161
Leather, E. M. 45
LeBerre, J. R. 137, 184, 185, 300
Leckstein, P. M. 179
Lemonde, A. 132
van Lenteren, J. C. 281
Less, M. 132
Libbey, L. M. 234
Lipa, J. J. 306
Lipke, H. 306
Lishtvanova, L. N. 118, 119
Little, C. H. A. 259
Llewellyn, M. 83, 179

Long, W. H. 253
Loomis, R. S. 46
Loper, G. M. 161
Loschiavo, S. R. 53
Lubischew, A. A. 113
Ludlow, A. R. 123
Lukefahr, M. J. 161
Lum, P. T. M. 65
Lusis, O. 172

Ma, W. C. 39, 40, 53, 139, 151, 273
Maltais, J. B. 34, 178, 179
Manglitz, G. R. 53
Mann, J. D. 46
Margolis, H. 161
Markkula, M. 153, 155, 285
Martin, D. F. 161
Martin, J. T. 40
Mathez, F. C. 253
Matolcsy, G. 245
Matsumoto, Y. 221
Maxwell, F. G. 21, 157, 161, 221
McCaffery, A. R. 163
McCoy, C. E. 21
McFadden, M. W. 214
McGregor, S. E. 161
McIndoo, N. E. 300
McLean, D. L. 179
Mead, J. F. 132
Mehrotra, K. N. 53
Meier, R. 300
von Meier, W. 179
Meisner, J. 53, 151
Meredith, W. R., Jr. 157, 161
Merton, L. F. H. 172
Merz, E. 266
Meyer, J. R. 161
Meyer, V. G. 161
Michael, R. R. 234
Miles, P. W. 179
Minks, A. K. 172
Mittal, R. K. 22
Mittler, T. E. 83, 173, 178, 179, 180
Moericke, V. 179
Mohyuddin, A. J. 53
Monro, J. 136
Moon, H. D. 132
Moon, M. S. 180
Moore, C. A. 21
Moorhouse, J. E. 53, 300
Mordue, W. 172
Moreau, J. P. 181, 184, 189
Morgan, M. E. 234
Morton, R. A. 200
Mothes, K. 45, 46
Mound, L. A. 161
Mudd, S. H. 46
Müller, B. 281
Müller, F. P. 113, 187

Nagy, B. 191, 195
Nakamura, H. 281
Nalbandov, O. 39

313

Nation, J. I. 259
Nault, L. R. 180
Nayar, J. 39, 53
Neiswander, C. R. 195
Nixon, H. L. 178
Noble, L. W. 161
Noordink, J. P. W. 83, 178, 179
Norris, D. M. 53, 197, 200, 201, 221
Novák, V. 290
Nöcker-Wenzel, K. 39
Nunn, L. C. 132
Nye, I. W. B. 253

Osborne, D. J. 258

Pag, H. 94
Pankanin, M. 281
Pantanetti, P. 179
Panwar, V. P. S. 252
Parkhurst, R. M. 306
Parrott, P. J. 94
Parrott, W. L. 221
Parry, W. H. 180
Paskin, M. W. J. 300
Pass, B. C. 108
Pathak, M. D. 21
Pennel, J. T. 180
Porron, J. M. 178
Pessah, N. 200
Petavy, G. 172
Peterson, L. K. 21
Pettersson, J. 203, 208
Pfrimmer, T. R. 161
Pimentel, D. 136
Pitman, G. B. 234, 290
Pitts, C. W. 28
Plumb, M. A. 180
Pond, D. D. 21
Poos, F. W. 21
Pouzat, J. 65, 108, 136, 281
Prevett, P. F. 65
Prints, E. Ya. 127
Prints, Ya. I. 127
Prokopy, R. J. 136, 209, 214
Putnam, T. B. 234

Raccah, B. 83, 180
Rácz, V. 215
Rahn, R. 108, 136
Ramwell, P. W. 132
Ranney, C. D. 161
Rao, P. J. 53
Rathore, Y. S. 195
Raucourt, M. 300
Raven, P. H. 113
Reddy, A. S. 252
Reed, D. K. 22
Rees, C. J. C. 70
Reese, J. C. 217, 221
Reiser, R. 132, 306
Rejman, A. 60
Renwick, J. A. A. 235, 290
Retnakaran, A. 34

Rhyne, C. L. 161
Riad, A. A. 45
Richards, O. W. 65
Richardson, C. D. 132
Riddiford, L. M. 108, 136
Riggenbach, W. 94
Ritter, F. J. 53, 300
Robert, P. Ch. 136, 223, 227
Robinson, F. A. 259
Robinson, J. 46
Rodriquez, J. G. 60
Roe, J. M. 40
Roeder, K. D. 151
Rolley, F. 185
Roncero, A. V. 214
Rosen, H. 40
Rosental, J. M. 200, 201, 221
Rothstein, A. 200
Roubaud, E. 195
Rubtsov, I. A. 113
Rudinsky, J. A. 229, 234, 290
von Rudloff, E. 234
Rudnev, D. F. 118, 290
Russ, K. 237

Sacantanis, K. 214
Salama, H. S. 151, 281
Sandner, H. 281
Sang, J. H. 83, 178
Sartwell, Jr. C. 234
Sáringer, Gy. 65, 241, 245, 281
Schafer, R. 281
Schaefers, G. A. 83
Schanz, M. 300
Scheltes, P. 242
Schetters, C. 127
Scheurer, S. 255
Schmitz, R. F. 234
Schmutterer, H. 178, 253
Schoene, W. J. 94
Scholze, P. 259
Schoonhoven, L. M. 70, 77, 99, 151, 261,
      266, 273, 300
Schönherr, J. 235
Schuster, M. F. 157, 161
Schutte, H. R. 45, 46
Scott, K. M. 201
Scsegolev, V. N. 195
Sengonca, C. 179
Settlemire, C. T. 201
Sharma, G. C. 252
Shaw, Y. E. 132
Siddiqui, K. H. 252
Singer, G. 201
Singh, H. G. 22
Skinner, W. A. 306
Slafon, W. H. 132
Sloane, S. G. H. 132
Sloof, R. 99
Smelyanets, V. P. 118, 119
Smissman, E. E. 221
Smith, F. F. 21
Smith, L. W. 45

314

315

SUBJECT INDEX